CENTENNIAL OF GENERAL RELATIVITY

A Celebration

CENTENNIAL OF GENERAL RELATIVITY

A Celebration

Editor

César Augusto Zen Vasconcellos

Universidade Federal do Rio Grande do Sul, Brazil & ICRANet, Italy

World Scientific

NEW JERSEY • LONDON • SINGAPORE • BEIJING • SHANGHAI • HONG KONG • TAIPEI • CHENNAI

Published by

World Scientific Publishing Co. Pte. Ltd.
5 Toh Tuck Link, Singapore 596224
USA office: 27 Warren Street, Suite 401-402, Hackensack, NJ 07601
UK office: 57 Shelton Street, Covent Garden, London WC2H 9HE

Library of Congress Cataloging-in-Publication Data
Names: Vasconcellos, César A. Z., editor.
Title: Centennial of general relativity : a celebration / edited by: César Augusto Zen Vasconcellos (Universidade Federal do Rio Grande do Sul, Brazil & ICRANet, Italy).
Description: Singapore ; Hackensack, NJ : World Scientific Publishing Co. Pte. Ltd., [2016] | Includes bibliographical references.
Identifiers: LCCN 2016022165| ISBN 9789814699655 (hardcover ; alk. paper) | ISBN 9814699659 (hardcover ; alk. paper)
Subjects: LCSH: General relativity (Physics)--History. | Relativity (Physics)--History.
Classification: LCC QC173.6 .C46 2016 | DDC 530.11--dc23
LC record available at https://lccn.loc.gov/2016022165

British Library Cataloguing-in-Publication Data
A catalogue record for this book is available from the British Library.

Desk Editor: Ng Kah Fee

Typeset by Stallion Press
Email: enquiries@stallionpress.com

Printed in Singapore

To my wife Mônica, to my daughter Helena, to my son Marcio, to my stepdaughters Daniela, Marina and Barbara, and to Fiorella, with love.

Preface

The equations of the General Relativity Theory (GRT) formulated by Albert Einstein in 1915,

$$G_{\mu\nu} = -8\pi G \left(T_{\mu\nu} + \frac{\Lambda}{8\pi G} g_{\mu\nu} \right)$$

represent the foundation of our present understanding about the universe. The right-hand side of this equation describes the energy content of the universe, where the $(\Lambda/8\pi G)\, g_{\mu\nu}$ term was included by Einstein in a later formulation, for cosmological reasons, which modernly is interpreted as *dark energy*, the cause for the current cosmic acceleration, i.e., the observation that the universe appears to be expanding at an increasing rate. The left-hand side of this equation, on the other hand describes the geometry of spacetime. The equality between these two components of the equation means that in General Relativity mass and energy determine the geometry and concomitantly the curvature of spacetime, which represents in turn a manifestation of gravity, the warping in the fabric of spacetime.

Albert Einstein in 1917 stated:

> "since the introduction of the special principle of relativity has been justified, every intellect which strives after generalization must feel the temptation to venture the step towards the general principle of relativity".[1]

In a manuscript,[2] Einstein summarized his successful efforts to go beyond Special Relativity and mentioned the names of scientists that played an essential role in the development of General Relativity: Hermann Minkowski, Carl Friedrich Gauss, Bernhard Riemann, Elwin

[1]See *The Road to Relativity*, Gutfreund, H. & Renn, J. (Princeton University Press, Princeton and Oxford, USA, 2015).
[2]See reference in footnote 1.

Bruno Christoffel, Gregorio Ricci-Curbastro, Tulio Levi-Civita, and Marcel Grossmann.

Einstein was the only one however — despite highlighting the brilliant contribution of the others —, who had persistently followed his intuition since 1907. His brilliant intuition, combined with his persistence therefore represent the key to understand his remarkable trajectory in the development of General Relativity, a theory that has revolutionized our conceptions of the universe and our understanding of its evolution.

About the contents of the book

The first chapter of the book, by Norman K. Glendenning, is divided into two parts. The first part features with clarity and creativity, a review of the General Relativity Theory (GRT), to give insight into this remarkable theory to those readers who, from a general scientific and philosophical point of view, are interested in the most intriguing details of the mathematical apparatus and conceptions of GRT that opened a completely new perspective on the cosmos. In the second part, the author focus his attention on compact stars and into obtaining and interpreting the Oppenheimer–Tolman–Volkoff (TOV) equations.[3] TOV equations are essential in the understanding of the diversity of the very dense objects that comprise a substantial part of the universe — mainly white dwarfs, neutron stars, pulsars, quark stars, black holes — and which may provide a better understanding of the equation of state of nuclear matter and of the quark–gluon plasma, the Holy Grail of the primordial content of the universe.

In Chapter 2, following the topic of compact stars discussed in the previous chapter, Omair Zubairi and Fridolin Weber derive Tolman–Oppenheimer–Volkoff (TOV)-like stellar structure equations for deformed compact stellar objects, whose mathematical form is similar to the traditional TOV equation for spherical stars. The authors then solve these equations numerically for a given equation of state (EoS) and predict stellar properties such as masses and radii along with pressure and density profiles and investigate any changes from spherical models of compact stars. According to the authors, if rotating, deformed compact objects are among

[3]Editor's note: we adopt here, unlike Chapter 1, the denomination TOV equations, most often used in the literature. See: Oppenheimer, J. R. and Volkoff, G. M., *Phys. Rev.* **55**, 374 (1939); Tolman, R. C., *Phys. Rev.* **55**, 364 (1939).

the possible astrophysical sources emitting gravitational waves which could be detected.

Einstein in the years 1945 and 1946 developed an algebraic extension of General Relativity by introducing complex-valued fields on real spacetime in order to apply Hermitian symmetry to GRT.[4] Einstein introduced in his formalism a Hermitian metric whose real part is symmetric and describes the gravitational field while the imaginary part is antisymmetric and corresponds to the Maxwell field strengths. However, in this formulation the spacetime manifold remains real. Einstein soon recognized that this proposed unification does not satisfy the criteria that the metric tensor, $g_{\mu\nu}$, which characterizes the geometry of spacetime, should appear in GRT as a covariant entity with an underlying symmetry principle. In Chapter 3, Peter O. Hess and Walter Greiner discuss their new algebraic extension of the General Relativity Theory, named pc-GR, with pseudo-complex (pc) coordinates, which contains a minimal length and in addition requires the appearance of an energy–momentum tensor, related to vacuum fluctuations (dark energy), which are provoked by the presence of a central mass. An important result of their theory is that the dark energy density increases toward the central mass and avoids the appearance of an event horizon. Finally, in their contribution they present observational consequences related to *quasi-periodic oscillations* in accretion disks around the so-called *galactic black holes*, and discuss the structure of these disks. Additionally, in a note added in proof, the authors comment on the first direct measurements of gravitational waves announced in February[5] 2016 and discuss briefly some predictions of pc-GR on this matter.

Our understanding of the origin of the universe, of its evolution and the physical laws that govern its behavior — as well as the different states of matter that make up its evolutionary stage — reached in recent years levels never before imagined. This is due mainly to new and recent discoveries in astronomy and relativistic astrophysics as well as to experiments on particle and nuclear physics that overcame the traditional boundaries of knowledge on physics. As a result we have presently a new understanding of the universe in its two extreme domains, the very large and the very small: the recognition of the deep connections that exist between quarks and the

---

[4]Einstein, A. *Ann. Math.* **46**, 578 (1945); Einstein, A. and Strauss, E., *Ann. Math.* **47**, 731 (1946).

[5]Abbott, B. P. *et al.*, *Phys. Rev. Lett.* **116**, 061102 (2016) and Chapter 11.

cosmos. Based on our present understanding of this intimate relationship between quarks and cosmos, Renxin Xu and Yanjun Guo argue, in Chapter 4, that 3-flavour symmetry would be restored in macro/gigantic-nuclei compressed by gravity during a supernova event. The authors make conjectures in this chapter about the presence of strange matter in the universe and about its composition, more precisely, a condensed 3-flavour quark matter state. The authors also make predictions about the role of future advanced facilities (e.g., the Square Kilometre Array (SKA) radio telescope) which would provide clear evidence for strange stars. Additionally, considering that the applications of the TOV equations, as discussed in Chapter 1, are restricted to describe self-gravitational perfect fluids, Renxin Xu and Yanjun Guo discuss the extension of these equations to describe gravitational implications of solid strange stars with *anisotropic* local pressure in elastic matter, including the release of the elastic energy and the gravitational energy in those stars which is not negligible and may have significant astrophysical implications. Finally, the first direct measurements of gravitational waves[6] have also motivated Renxin Xu and Yanjun Guo to mention, in a note added in proof, that the proposed model of strange star with rigidity (i.e., strangeon star) is quite likely to be tested further by kilo-Hz gravitational wave observations.

The comprehension of the large-scale structure of the universe based on models of cold dark matter has been a major subject of study involving predictions of General Relativity and in particular inferences on dark matter from gravitational effects on visible matter. In Chapter 5, Roberto Sussman describes how a particular solution of the equations of General Relativity, the so-called *Szekeres Models*, can be used to construct assorted configurations of multiple non-spherical self-gravitating cold dark matter structures and to describe its evolution. According to the author, this approach is able to provide a fully relativistic non-perturbative coarse grained description of actually existing cosmic structure at various scales. As a consequence, this modeling allows an enormous range of potential applications to current astrophysical and cosmological problems.

The scientists' view of the universe, these days, has expanded in a way never before imagined by observing, in particular, electromagnetic waves in the infrared spectrum bands, ultraviolet, radio, optical and X-rays. In this context, modern cosmology serves as a guide that covers different aspects

[6]See footnote 5 and Chapter 11.

of a field of research in rapid and constant transformation. In Chapter 6, Marc Lachièze-Rey presents a qualitative and interesting discussion about the current view of modern cosmology and the contributions of Albert Einstein and Georges Lemaître on this topic. Marc Lachièze-Rey discusses in his contribution, topics as diverse as galaxies and the expanding universe, big-bang models, cosmic microwave background, modern cosmology, dark issues, cosmological constant and dark energy, the topology of spacetime, and the cosmic time.

There has been along the more recent history of physics questionings about the incompleteness of GRT and of Quantum Field Theory, as well as about deviations of the Standard Model. Similarly there has been questionings about the existence or not, in the beginning of the universe for distances of the order of Planck scale, of a singularity, the Big Bang, according to GRT, around 13.7 billion years ago, among others. Some of these questionings find in particle and astroparticle physics a safe haven for insights about the realm of quantum gravity and for a deeper knowledge about the content of the universe in its first moments. In this context, it is important to have a thorough knowledge about the latest results of experiments performed at the world's leading particle and astroparticle laboratory, the European Organization for Nuclear Research — CERN, to allow a better comprehension of the smaller constituents of the universe in its early stages and in this way learn more about the structure and content of the primordial plasma.[7] It is precisely in this context that the book presents a compilation of the latest experimental data in particle physics obtained by CERN, and in particular about the latest discovery of the Higgs particle. Chapters 7, 8, and 9, by Cristina Biino, Géraldine Conti, and Katharina Müller, from CERN, provide excellent reviews about the most impressive achievements of this outstanding laboratory and international collaboration. In Chapter 7, Cristina Biino, describes in detail the technical characteristics of the major experimental facilities at CERN, the latest data involving highlights of Standard

[7]As an example of CERN's contributions to a better understanding of the universe, we may ask if CERN's data might shed some light on any connection between the Higgs boson and gravity. CERN is also able to racetrack in the universe and create the quark–gluon plasma. In this context, CERN data might help to explain why protons and neutrons are 100 times more massive than quarks, what dark matter is, and how the universe came to its existence. In essence, CERN can recreate the physical conditions of the universe just fractions of a second after the Big Bang.

Model results from ATLAS and CMS — in particular the recent discovery of the Higgs boson — and the operational capability of this extraordinary laboratory in exploring the frontiers of physics in the description of the properties of the tiniest components in the first moments of the universe. In Chapter 8, Géraldine Conti describes an important number of searches for deviations from the Standard Model expectations performed with the LHC Run 1 data at ATLAS and CMS, referred to as "Beyond the Standard Model" (BSM) searches, which have been carried out in various areas of physics, including BSM Higgs, supersymmetry, exotic physics including searches for signatures of the graviton, dark matter and thermal black holes. In Chapter 9, Katharina Müller describes a wide range of selected physics results from the LHCb experiment, demonstrating its unique role, both as a heavy flavour experiment and as a general purpose detector in the forward region.

Numerous are the scientific objectives of Astrophysics, Astronomy and Cosmology: the search for a better understanding of the universe, its origin and its evolution; the discovery of the moment of the universe's creation and a more profound comprehension of the evolutionary history of stars and galaxies; the search for signatures of life on other worlds among many others. Research involving observation techniques in the spectral range of gamma rays, such as the stereoscopic imaging atmospheric Cherenkov technique developed in the 1980s and 1990s, presents extremely important perspectives for a better and deeper understanding of the cosmos. In Chapter 10, Ulisses Barres de Almeida focuses his contribution to this volume on the relativistic universe and on present and future results of *Teraelectronvolt Astronomy*. The main topics of his contribution are directed to the discussion of the importance of studies, through gamma-ray lenses, of galaxies, supernova remnants, starburst galaxies, pulsars and their environments, microquasars and black holes, active galaxies and supermassive black holes, blazars, and extragalactic cosmic rays. In perspective, the author discusses the role of the Cherenkov Telescope Array (CTA) in the process of evolution of our knowledge about the universe and its content and in particular about the future of astroparticle physics in South America.

In Chapter 11, the LIGO Scientific Collaboration and Virgo Collaboration teams report on two major scientific breakthroughs involving key predictions of Einstein's theory: the first direct detection of gravitational waves and the first observation of the collision and merger of a pair of black holes. This cataclysmic event, producing the gravitational-wave signal GW150914, took place in a distant galaxy more than one billion light years from the Earth. It was observed on September 14, 2015 by the two detectors of the Laser Interferometer Gravitational-Wave Observatory (LIGO), arguably the most sensitive scientific instruments ever constructed. LIGO estimated anew that the peak gravitational-wave power radiated during the final moments of the black hole merger was more than ten times greater than the combined light power from all the stars and galaxies in the observable universe. This remarkable discovery marks the beginning of an exciting new era of astronomy and open a gravitational-wave window on the universe.

General Relativity covers a series of fascinating notions and predictions about the geometry of spacetime; about the motion of massive objects; about the propagation and bending of light by gravity, which in particular originates the *gravitational lensing effect*: the frame-dragging of spacetime around rotating massive bodies, which can be observed as a multiple-image phenomenon of the same astronomical object in the sky; among many other conceptions and predictions. General Relativity introduces new paradigms in correspondence to classical gravitation: for instance, the comprehension of gravity as a manifestation of the spacetime curvature, the distortion of space and time by the presence of massive objects, the gravitational time dilation, the gravitational redshift and the gravitational time delay. The cosmological and astrophysical implications of General Relativity are inexhaustible. It predicts the existence of black holes and compact stars as end states of massive stars. The evidences for the existence of black holes and that they are responsible for the intense radiation emitted by micro-quasars and active galactic nuclei are significant. The predictions about the existence of dark matter and dark energy, among many other fascinating topics, give to this theory a conceptual beauty and a profound scientific wealth. The recent observation, for the first time, of gravitational waves, 100 years after the creation of the General Theory of Relativity, is a demonstration of the extraordinary vitality of this theory. In this book we focus on some of the most relevant predictions of General Relativity. We hope that the readers enjoy the reading.

Acknowledgments

We thank the authors of the contributions of the celebration book as well as Dimiter Hadjimichef (UFRGS, Porto Alegre, Brazil), Hugo Pérez Rojas

(ICIMAF, Havana, Cuba), Peter Hess (UNAM, Mexico City) for valuable comments and to Mônica Estrázulas for most of the creative suggestions on the book cover.

Porto Alegre, July 2016
César A. Zen Vasconcellos*

*Full Professor, Physics Institute, Universidade Federal do Rio Grande do Sul (UFRGS), Porto Alegre, Rio Grande do Sul, Brazil. E-mail: cesarzen@cesarzen.com.

Contents

Chapter 1

General Relativity and Compact Stars

Norman K. Glendenning
Nuclear Science Division, and
Institute for Nuclear and Particle Astrophysics
Lawrence Berkeley National Laboratory
University of California
1 Cyclotron Road
Berkeley, California 94720, EUA
nkglendenning@lbl.gov, nkg94708@earthlink.net

The chapter is devoted to General Relativity. The goal is to rigorously arrive at the equations that describe the structure of relativistic stars — the Oppenheimer–Volkoff equations —, the form that Einstein's equations take for spherical static stars. Two important facts emerge immediately. No form of matter whatsoever can support a relativistic star above a certain mass called the limiting mass. Its value depends on the nature of matter but the existence of the limit does not. The implied fate of stars more massive than the limit is that either mass is lost in great quantity during the evolution of the star or it collapses to form a black hole.

Contents

1. Introduction

"In the deathless boredom of the sidereal calm we cry with regret for a lost sun ..."

Jean de la Ville de Mirmont, L'Horizon Chimérique.

Compact stars — broadly grouped as neutron stars and white dwarfs — are the ashes of luminous stars. One or the other is the fate that awaits the cores of most stars after a lifetime of tens to thousands of millions of years. Whichever of these objects is formed at the end of the life of a particular luminous star, the compact object will live in many respects unchanged from the state in which it was formed. Neutron stars themselves can take several forms — hyperon, hybrid, or strange quark star. Likewise white dwarfs take different forms though only in the dominant nuclear species. A black hole is probably the fate of the most massive stars, an inaccessible region of spacetime into which the entire star, ashes and all, falls at the end of the luminous phase.

Neutron stars are the smallest, densest stars known. Like all stars, neutron stars rotate — some as many as a few hundred times a second. A star rotating at such a rate will experience an enormous centrifugal force that must be balanced by gravity else it will be ripped apart. The balance of the two forces informs us of the lower limit on the stellar density. Neutron stars are 10^{14} times denser than Earth. Some neutron stars are in binary orbit with a companion. Application of orbital mechanics allows an assessment of masses in some cases. The mass of a neutron star is typically 1.5 solar masses. We can therefore infer their radii: about ten kilometers. Into such a small object, the entire mass of our Sun and more, is compressed.

We infer the existence of neutron stars from the occurrence of supernova explosions (the release of the gravitational binding of the neutron star) and observe them in the periodic emission of pulsars. Just as neutron stars acquire high angular velocities through conservation of angular momentum, they acquire strong magnetic fields through conservation of magnetic flux during the collapse of normal stars. The two attributes, rotation and strong magnetic dipole field, are the principle means by which neutron stars can be detected — the beamed periodic signal of pulsars.

The extreme characteristics of neutron stars set them apart in the physical principles that are required for their understanding. All other stars can be described in Newtonian gravity with atomic and low-energy nuclear physics under conditions essentially known in the laboratory.[1] Neutron stars in their several forms push matter to such extremes of density that nuclear and particle physics — pushed to their extremes — are essential for their description. Further, the intense concentration of matter in neutron stars

[1]Luminous stars evolve through thermonuclear reactions. These are nuclear reactions induced by high temperatures but involving collision energies that are small on the nuclear scale. In some cases the reaction cross-sections can be measured with nuclear accelerators, and in others, measured cross-sections must be extrapolated to lower energy.

can be described only in General Relativity, Einstein's theory of gravity which alone describes the way the weakest force in nature arranges the distribution of the mass and constituents of the densest objects in the universe.

1.1. *Compact stars*

Of what are compact stars made? The name "neutron star" is suggestive and at the same time misleading. No doubt neutron stars are made of baryons like nucleons and hyperons but also likely contain cores of quark matter in some cases. We use "neutron star" in a generic sense to refer to stars as compact as described above. How does a star become so compact as neutron stars and why is there little doubt that they are made of baryons or quarks? The notion of a neutron star made from the ashes of a luminous star at the end point of its evolution goes back to 1934 and the study of supernova explosions by Baade and Zwicky [Baade & Zwicky (1934)].

During the luminous life of a star, part of the original hydrogen is converted in fusion reactions to heavier elements by the heat produced by gravitational compression. When sufficient iron — the end point of exothermic fusion — is made, the core containing this heaviest ingredient collapses and an enormous energy is released in the explosion of the star. Baade and Zwicky guessed that the source of such a magnitude as makes these stellar explosions visible in daylight and for weeks thereafter must be gravitational binding energy. This energy is released by the solar mass core as the star collapses to densities high enough to tear all nuclei apart into their constituents.

By a simple calculation one learns that the gravitational energy acquired by the collapsing core is more than enough to power such explosions as Baade and Zwicky were detecting. Their view as concerns the compactness of the residual star has since been supported by many detailed calculations, and most spectacularly by the supernova explosion of 1987 in the Large Magellanic Cloud, a nearby minor galaxy visible in the southern hemisphere. The pulse of neutrinos observed in several large detectors carried the evidence for an integrated energy release over 4π steradians of the expected magnitude.

The gravitational binding energy of a neutron star is about 10 percent of its mass. Compare this with the nuclear binding energy of 9 MeV per nucleon in iron which is one percent of the mass. We conclude that the release of gravitational binding energy at the death of a massive star is of the order ten times greater than the energy released by nuclear fusion reactions during the entire luminous life of the star. The evidence that the source of energy for a supernova is the binding energy of a compact star — a neutron star — is compelling. How else could a tenth of a solar mass of energy be generated and released in such a short time?

Neutron stars are more dense than was thought possible by physicists at the turn of the century. At that time astronomers were grappling with the thought of white dwarfs whose densities were inferred to be about a million times denser than the Earth. It was only following the discovery of the quantum theory and Fermi–Dirac statistics that very dense *cold* matter — denser than could be imagined on the basis of atomic sizes — was conceivable.

Prior to the discovery of Fermi–Dirac statistics, the high density inferred for the white dwarf Sirius seemed to present a dilemma. For while the high density was understood as arising from the ionization of the atoms in the hot star making possible their compaction by gravity, what would become of this dense object when ultimately it had consumed its nuclear fuel? Cold matter was known only in the atomic form it is on Earth with densities of a few grams per cubic centimeter. The great scientist Sir Arthur Eddington surmised for a time that the star had "got itself into an awkward fix" — that it must some how re-expand to matter of familiar densities as it cooled, but it had no remaining source of energy to do so.

The perplexing problem of how a hot dense body without a source of energy could cool persisted until R. H. Fowler "came to the rescue"[2] by showing that Fermi–Dirac degeneracy allowed the star to cool by remaining comfortably in a previously unknown state of cold matter, in this case a degenerate[3] electron state. A little later Baade and Zwicky conceived of a similar degenerate state as the final resting place for nucleons after the supernova explosion of a luminous star.

The constituents of neutron stars — leptons, baryons and quarks — are degenerate. They lie helplessly in the lowest energy states available to them. They must. Fusion reactions in the original star have reached the

[2]Eddington in an address in 1936 at Harvard University.

[3]Nucleons and electrons obey the Pauli exclusion principle, according to which each particle must occupy a different quantum state from the others. A degenerate state refers to the complete occupation of the lowest available energy states. In that event, no reaction and therefore no energy generation is possible — hence the name — degenerate state.

end point for energy release — the core has collapsed, and the immense gravitational energy converted to neutrinos has been carried away. The star has no remaining source of energy to excite the fermions. Only the Fermi pressure and the short-range repulsion of the nuclear force sustain the neutron star against further gravitational collapse — sometimes. At other times the mass is so concentrated that it falls into a black hole, a dynamical object whose existence and external properties can be understood in the Classical Theory of General Relativity.

1.2. *Compact stars and relativistic physics*

Classical General Relativity is completely adequate for the description of neutron stars, white dwarfs, and for the most part, the exterior region of black holes as well as some aspects of the interior.[4] Section 2 is devoted to General Relativity. The goal is to rigorously arrive at the equations that describe the structure of relativistic stars — the Oppenheimer–Volkoff equations — the form that Einstein's equations take for spherical static stars. Two important facts emerge immediately. No form of matter whatsoever can support a relativistic star above a certain mass called the limiting mass. Its value depends on the nature of matter but the existence of the limit does not. The implied fate of stars more massive than the limit is that either mass is lost in great quantity during the evolution of the star or it collapses to form a black hole.

Black holes — the most mysterious objects of the universe — are treated at the classical level and only briefly. The peculiar difference between time as measured at a distant point and on an object falling into the hole is discussed. And it is shown that in black holes there is no statics. Everything at all times must approach the central singularity. Unlike neutron stars and white dwarfs, the question of their internal constitution does not arise at the classical level. They are enclosed within a horizon from which no information can be received. The ultimate fate of black holes is unknown.

Luminous stars are known to rotate because of the Doppler broadening of spectral lines. Therefore their collapsed cores, spun up by conservation of angular momentum, may rotate very rapidly. Consequently, no account of compact stars would be complete without a discussion of rotation, its effects

[4]The density at which quantum gravity would be relevant is 10^{78} higher than found in neutron stars.

on the structure of the star and spacetime in the vicinity, the limits on rotation imposed by mass loss at the equator and by gravitational radiation, and the nature of compact stars that would be implied by very rapid rotation.

Rotating relativistic stars set local inertial frames into rotation with respect to the distant stars. An object falling from rest at great distance toward a rotating star would fall — not toward its center but would acquire an ever larger angular velocity as it approached. The effect of rotating stars on the fabric of spacetime acts back upon the structure of the stars and so is essential to our understanding.

1.3. *Compact stars and dense-matter physics*

The physics of dense matter is not as simple as the final resting place of stars imagined by Baade and Zwicky. The constitution of matter at the high densities attained in a neutron star — the particle types, abundances and their interactions — pose challenging problems in nuclear and particle physics. How should matter at supernuclear densities be described? In addition to nucleons, what exotic baryon species constitute it? Does a transition in phase from quarks confined in nucleons to the deconfined phase of quark matter occur in the density range of such stars? And how is the transition to be calculated? What new structure is introduced into the star? Do other phases like pion or kaon condensates play a role in their constitution?

In Fig. 1 we show a computation of the possible constitution and interior crystalline structure of a neutron star near the limiting mass of such stars. Only now are we beginning to appreciate the complex and marvelous structure of these objects. Surely the study of neutron stars and their astronomical realization in pulsars will serve as a guide in the search for a solution to some of the fundamental problems of dense many-body physics both at the level of nuclear physics — the physics of baryons and mesons — and ultimately at the level of their constituents — quarks and gluons. And neutron stars may be the only objects in which a Coulomb lattice structure (Fig. 1) formed from two phases of one and the same substance (hadronic matter) exists.

We do not know from experiment what the properties of superdense matter are. However we can be guided by certain general principles in our investigation of the possible forms that compact stars may take. Some of the possibilities lead to quite striking consequences that may in time be observable. The rate of discovery of new pulsars, X-ray neutron stars and

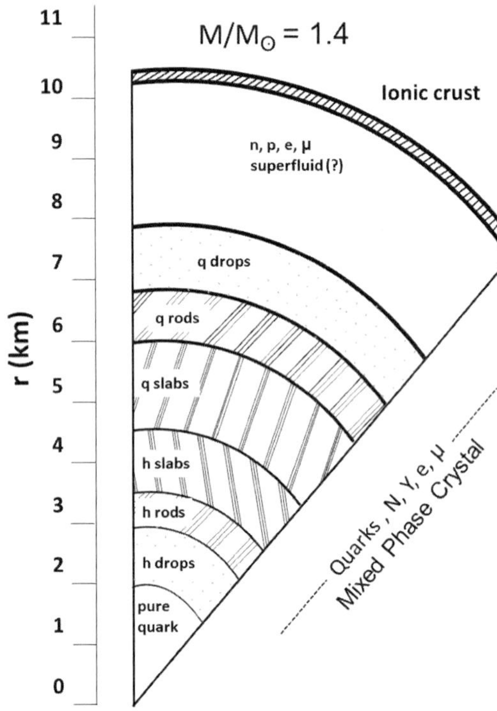

Fig. 1. A section through a neutron star model that contains an inner sphere of pure quark matter surrounded by a crystalline region of mixed hadronic and quark matter. The mixed phase region consists of various geometrical objects of the rare phase immersed in the dominant one hadronic drops, labeled by h, immersed in quark matter through to quark drops, labeled by h, immersed in hadronic matter. The particle composition of these regions is quarks, nucleons, hyperons, and leptons. A liquid of neutron star matter containing nucleons and leptons surrounds the mixed phase. A thin crust of heavy ions forms the stellar surface. [Glendenning (2001)]

other high-energy phenomena associated with neutron stars is astonishing, and was unforeseen a dozen years ago.

White dwarfs are the cores of stars whose demise is less spectacular than a supernova — a more quiescent thermal expansion of the envelope of a low mass star into a planetary nebula. White dwarf constituents are nuclei immersed in an electron gas and therefore arranged in a Coulomb lattice. White dwarfs are supported against collapse by Fermi pressure of degenerate electrons — while neutron stars — are supported by the Fermi pressure of degenerate nucleons. White dwarfs pose less severe and less fundamental problems than neutron stars. The nuclei will comprise varying proportions of helium, carbon, and oxygen, and in some cases heavier elements like magnesium, depending on how far in the chain of exothermic nuclear fusion reactions the precursor star burned before it was disrupted by instabilities leaving behind the dwarf. White dwarfs are barely relativistic.

Of a vastly different nature than neutron stars are *strange stars.* Like neutron stars they are, if they exist, very dense, of the same order as neutron stars. However their very existence hinges on a hypothesis that at first sight seems absurd. According to the hypothesis, sometimes referred to as the *strange-matter hypothesis*, quark matter — consisting of an approximately equal number of up, down and strange quarks — has an equilibrium energy per nucleon that is lower than the mass of the nucleon or the energy per nucleon of the most bound nucleus, iron. In other words, under the hypothesis, strange quark matter is the *absolute* ground state of the strong interaction.

We customarily find that systems, if not in their ground state, readily decay to it. Of course this is not always so. Even in well known objects like nuclei, there are certain excited states whose structure is such that the transition to the ground state is hindered. The first excited state of ^{180}Ta has a half-life of 10^{15} years, five orders of magnitude longer than the age of the universe! The strange-matter hypothesis is consistent with the present universe — a long-lived excited stat — if strange matter is the ground state. The structure of strange stars is fascinating as are some of their properties.

2. General Relativity

"Scarcely anyone who fully comprehends this theory
can escape its magic."

A. Einstein

"Beauty is truth, truth beauty — that is all
Ye know on Earth, and all ye need to know."

J. Keats

General Relativity — Einstein's theory of gravity — is the most beautiful and elegant of physical theories. Not only that; it is the foundation for our understanding of compact stars. Neutron stars and black holes owe their very existence to gravity as formulated by Einstein [Einstein (1916, 1951)]. Dense objects like neutron stars could also exist in Newton's theory, but they would be very different objects. Chandrasekhar found (in connection with white dwarfs) that all degenerate stars have a maximum possible mass. In Newton's theory such a maximum mass is attained asymptotically when all fermions whose pressure supports the star are ultra-relativistic. Under such conditions stars populated by heavy quarks would exist. Such unphysical stars do not occur in Einstein's theory.

Perhaps the beauty of Einstein's theory can be attributed to the essentially simple but amazing answer it provides to a fundamental question: what meaning is attached to the absolute equality of inertial and gravitational masses? If all bodies move in gravitational fields in precisely the same way, no matter what their constitution or binding forces, then this means that their motion has nothing to do with their nature, but rather with the *nature of spacetime.* And if spacetime determines the motion of bodies, then according to the notion of action and reaction, this implies that spacetime in turn is *shaped by bodies and their motion.*

Beautiful or not, the predictions of theory have to be tested. The first three tests of General Relativity were proposed by Einstein, the gravitational redshift, the deflection of light by massive bodies and the perihelion shift of Mercury. The latter had already been measured. Einstein computed the anomalous part of the precession to be 43 arc-seconds per century compared to the measurement of 42.98 ± 0.04. A fourth test was suggested by Shapiro in 1964 — the time delay in the radar echo of a signal sent to a planet whose orbit is carrying it toward superior conjunction[5] with the Sun. Eventually agreement to 0.1 percent with the prediction of Einstein's theory was achieved in these difficult and remarkable experiments. It should be remarked that all of the above tests involved weak gravitational fields.

The crowning achievement was the 20-year study by Taylor and his colleagues of the Hulse–Taylor pulsar binary discovered in 1974. Their work yielded a measurement of 4.22663 degrees per year for the periastron shift of the orbit of the neutron star binary and a measurement of the decay of the orbital period by $7.60 \pm 0.03 \times 10^{-7}$ seconds per year. This rate of decay agrees to less than 1% with careful calculations of the effect of energy loss through gravitational radiation as predicted by Einstein's

[5]Superior conjunction refers to the situation when the Earth and the planet are on opposite sides of the Sun.

theory [Taylor *et al.* (1992)]. A fuller discussion of these experiments and other intricacies involved in the tests of relativity can be found in the book by Will [Will (1995)]. Since these early experiments, more accurate tests are being made by Dick Manchester and collaborators at Parkes Obsevatory in Australia, who have discovered a closer binary pair of neutron stars — "We have verified GR to 0.1% already in two years — ten times better than the early experiment." (Private communication: R. N. Manchester, June 15, 2005).

The goal of this section is to provide a rigorous derivation of the Oppenheimer–Volkoff equations that describe the structure of relativistic stars. We start by briefly outlining the Special Theory of Relativity for it is an essential ingredient of General Relativity. Then we formulate the General Theory of Relativity and derive all parts of the theory that are necessary to our goal.

2.1. *Relativity*

> "The views of space and time which I wish to lay before you have sprung from the soil of experimental physics, and therein lies their strength. They are radical. Henceforth space by itself, and time by itself, are doomed to fade away into mere shadows, and only a kind of union of the two will preserve an independent reality."
>
> *H. Minkowski* [Minkowski (1915)]

The principle of relativity in physics goes back to Galileo who asserted that the laws of nature are the same in all *uniformly* moving laboratories. The relativity principle, stated in the narrow terms of reference frames in uniform motion, referred to as inertial frames, implies the existence of an absolute space. The notion of the absoluteness of time goes back to time immemorial. A *Galilean* transformation assumes the absoluteness of space and time:

$$x' = x - a - vt, \quad y' = y, \quad z' = z, \quad t' = t - b. \tag{1}$$

Newton's second law $F_x = m\,d^2x/dt^2$ is evidently invariant under this transformation if one assumes that force and mass are independent of the state of motion.

In contrast, Maxwell's equations do not take on the same form if subjected to a Galilean transformation whereas under a *Lorentz transformation* they do.[6] This fact led Einstein to the postulate that the speed of light is the same in all inertial systems and consequently that the principle of relativity should hold with respect to inertial frames connected by *Lorentz transformations*. That is the historical role that light speed played in the discovery of Special Relativity, and the reason for the undoubted influence that the Michelson–Morley experiment [Michelson & Morley (1887)] had on the early acceptance of the theory.

However, the underlying physics is quite different from how it appears in the historical development of the Special Theory. The speed of light need not have been postulated as an invariant. Minkowski realized soon after Einstein's epochal discovery in 1905 that the spacetime manifold of our world is not Euclidean space in which events unfold in an absolute foliated time.[7] Spacetime is a 'Minkowski' manifold having such a nature that $d\tau^2 \equiv k^2 dt^2 - dx^2 - dy^2 - dz^2$ is invariant in the absence of gravity. The constant k is a conversion factor between length and time. Voigt observed in 1887 that $\Box\phi = 0$ preserved its form under a transformation that differed from the Lorentz transformation by only a scale factor [Voigt (1887)]. In fact we will see shortly that the d'Alembertian $\Box$ is a Lorentz scalar. Consequently,

$$\left(\frac{1}{k^2} \frac{\partial^2}{\partial t^2} - \nabla^2 \right) \phi = 0 \tag{2}$$

informs us that a disturbance described by a wave equation for a massless particle in Minkowski spacetime propagates with velocity k in vacuum as viewed from this and any other reference frame connected to it by a Lorentz transformation. Hence, the constant k of the spacetime manifold is determined empirically by a measurement of the speed of light, c. In this way it is seen that the constancy of the speed of light is a *consequence* of the nature of the spacetime manifold in a gravity-free universe, or in a sufficiently small region of our gravity-filled universe. It is determined by the conversion factor between time and length of the manifold.

That the constancy of the speed of light is a consequence of the local spacetime manifold and not its determiner is most clearly illustrated by

---

[6]See, for example, Ref. [Bjorken & Drell (1965)] for the Lorentz invariant form of Maxwell's equations.

[7]Foliated time refers to the time of events as being arranged as pages in a book, one following the other, there never being a question of which preceded another.

a thought experiment proposed by Swiatecki [Swiatecki (1983)]. He shows that the invariance of the differential interval between spacetime events

$$d\tau^2 = k^2 dt^2 - dx^2 - dy^2 - dz^2 \tag{3}$$

can be verified (at least in principle) *without* resort to propagation of light signals, but with only measuring rods and clocks. And if it were technically feasible to perform the experiment with sufficient accuracy, k would be measured and its value would be found to equal c.

Minkowski's fundamental discovery of the nature of spacetime in the absence of gravity was inspired by Einstein's postulate of the constancy of the speed of light. However, the constancy of the speed of light is a consequence of the spacetime manifold of our universe and its value (as for any massless particle) is equal to the conversion factor between space and time, as we have seen. The Minkowski invariant describes the nature of our spacetime (in a suitably limited region); the speed of light and that of any other massless particle is equal to the conversion factor k between time and length, as emphasized by W. Swiatecki [Swiatecki (1983)]. In other words, Special Relativity is a consequence of the local spacetime manifold in which we live. The significance of the local restriction will become clear as we follow the development of the General Theory.

2.2. *Lorentz invariance*

The Special Theory of Relativity, which holds in the absence of gravity, plays a central role in physics. Even in the strongest gravitational fields the laws of physics must conform to it in a sufficiently small locality of any spacetime event. That was a fundamental insight of Einstein. Consequently, the Special Theory plays a central role in the development of the General Theory of Relativity and its applications.

2.2.1. *Lorentz transformations*

The Lorentz transformation leaves invariant the *proper time* or *differential interval* in Minkowski spacetime

$$d\tau^2 = dt^2 - dx^2 - dy^2 - dz^2 \quad (\text{units}\, c = 1) \tag{4}$$

as measured by observers in frames moving with constant relative velocity (called inertial frames because they move freely under the action of no forces). The Minkowski manifold also implies an absolute spacetime in which spacetime events that can be connected by a Lorentz transformation lie within the cone defined by $d\tau = 0$. Absolute means unaffected by any physical conditions. This was the same criticism that Einstein made of Newton's space and time, and the one that powered his search for a new theory in which the expression of physical laws does not depend on the frame of reference, but, nevertheless, in which Lorentz invariance would remain a *local* property of spacetime. We will develop the core of the General Theory which extends the relativity principle to arbitrary frames and therefore to a gravity-filled universe, not just unaccelerated frames in relative uniform motion; but here we review briefly the Special Theory.

A *pure* Lorentz transformation is one without spatial rotation, while a general Lorentz transformation is the product of a rotation in space and a pure Lorentz transformation. We recall the pure transformation, sometimes also referred to as a *boost*. For convenience, define

$$x^0 = t, \quad x^1 = x, \quad x^2 = y, \quad x^3 = z. \tag{5}$$

(In spacetime a point such as that above is sometimes referred to as an *event*.) The linear homogeneous transformation connecting two reference frames can be written

$$x'^{\mu} = \Lambda^{\mu}{}_{\nu} x^{\nu}. \tag{6}$$

(We shall use the convenient notation introduced by Einstein whereby repeated indices are summed — Greek over time and space, Roman over space.)

Any set of four quantities A^{μ} ($\mu = 0, 1, 2, 3$) that transforms under a change of reference frame in the same way as the coordinates is a *contravariant* Lorentz four-vector,

$$A'^{\mu} = \Lambda^{\mu}{}_{\nu} A^{\nu}. \tag{7}$$

The invariant interval (also variously called the proper time, the line element, or the separation formula) can be written

$$d\tau^2 = \eta_{\mu\nu} dx^{\mu} dx^{\nu}, \tag{8}$$

where $\eta_{\mu\nu}$ is the Minkowski metric which in rectilinear coordinates is

$$\eta_{\mu\nu} \equiv \begin{pmatrix} 1 & 0 & 0 & 0 \\ 0 & -1 & 0 & 0 \\ 0 & 0 & -1 & 0 \\ 0 & 0 & 0 & -1 \end{pmatrix}. \tag{9}$$

The condition of the invariance of $d\tau^2$ is

$$\eta_{\alpha\beta} dx^{\alpha} dx^{\beta} = \eta_{\mu\nu} dx'^{\mu} dx'^{\nu} = \eta_{\mu\nu} \Lambda^{\mu}{}_{\alpha} \Lambda^{\nu}{}_{\beta}\, dx^{\alpha} dx^{\beta}. \tag{10}$$

Since this holds for any dx^α, dx^β we conclude that the $\Lambda^\mu{}_\nu$ must satisfy the fundamental relationship assuring invariance of the proper time:

$$\eta_{\alpha\beta} = \eta_{\mu\nu}\Lambda^\mu{}_\alpha\Lambda^\nu{}_\beta. \tag{11}$$

Transformations that leave $d\tau^2$ invariant leave the speed of light the same in all inertial systems, because if $d\tau = 0$ in one system, it is true in all, and the content of $d\tau = 0$ is that $d\mathbf{x}/dt = 1$.

Let us find the transformation matrix $\Lambda^\mu{}_\alpha$, for the special case of a boost along the x-axis. In this case it is clear that

$$x'^2 = x^2, \quad x'^3 = x^3, \tag{12}$$

and, moreover, that x'^0 and x'^1 cannot involve x^2 and x^3. So,

$$\begin{aligned} x'^0 &= \Lambda^0{}_0 x^0 + \Lambda^0{}_1 x^1 \\ x'^1 &= \Lambda^1{}_0 x^0 + \Lambda^1{}_1 x^1 \end{aligned} \tag{13}$$

with the remaining Λ elements zero. So, the above quadratic form in Λ yields the three equations,

$$\begin{aligned} 1 &= (\Lambda^0{}_0)^2 - (\Lambda^1{}_0)^2 \\ -1 &= (\Lambda^0{}_1)^2 - (\Lambda^1{}_1)^2 \\ 0 &= \Lambda^0{}_0\Lambda^0{}_1 - \Lambda^1{}_0\Lambda^1{}_1. \end{aligned} \tag{14}$$

To get a fourth equation, suppose that the origins of the two frames in uniform motion coincide at $t = 0$ and the primed x-axis x'^1 is moving along x^1 with velocity v. That is, $x^1 = vt$ is the equation of the *primed* origin as it moves along the *unprimed* x-axis. The equation for the primed coordinate is

$$0 = x'^1 = \Lambda^1{}_0 x^0 + \Lambda^1{}_1 x^1 = (\Lambda^1{}_0 + \Lambda^1{}_1 v)t, \tag{15}$$

or

$$\Lambda^1{}_0 = -\Lambda^1{}_1 v. \tag{16}$$

The four equations can now be solved with the result,

$$\begin{aligned} \Lambda^0{}_0 &= \Lambda^1{}_1 = \gamma, \\ \Lambda^1{}_0 &= \Lambda^0{}_1 = -v\gamma, \\ \Lambda^2{}_2 &= \Lambda^3{}_3 = 1, \end{aligned} \tag{17}$$

where

$$\gamma \equiv (1 - v^2)^{-1/2} \equiv \cosh\theta, \quad v\gamma \equiv \sinh\theta, \quad v \equiv \tanh\theta. \tag{18}$$

So

$$\begin{aligned} x'^0 &= x^0 \cosh\theta - x^1 \sinh\theta, \\ x'^1 &= -x^0 \sinh\theta + x^1 \cosh\theta, \\ x'^2 &= x^2, \quad x'^3 = x^3. \end{aligned} \tag{19}$$

The combination of two boosts in the same direction, say v_1 and v_2, corresponds to $\theta = \theta_1 + \theta_2$. A boost in an arbitrary direction with the *primed* axis having velocity $\mathbf{v} = (v^1, v^2, v^3)$ relative to the *unprimed* is

$$\begin{aligned} \Lambda^0{}_0 &= \gamma, \\ \Lambda^0{}_j &= \Lambda^j{}_0 = -v^j\gamma, \\ \Lambda^j{}_k &= \Lambda^k{}_j = \delta^j_k + (\gamma - 1)v^j v^k/\mathbf{v}^2. \end{aligned} \tag{20}$$

For a spatial rotation, say in the x–y plane, the transformation for a positive rotation about the common z-axis is

$$\begin{aligned} x'^1 &= x^1 \cos\omega + x^2 \sin\omega, \\ x'^2 &= -x^1 \sin\omega + x^2 \cos\omega, \\ x'^0 &= x^0, \quad x'^3 = x^3. \end{aligned} \tag{21}$$

Transformation of vectors according to either of the above, or a product of them, preserves the invariance of the interval $d\tau^2$. For convenience they can be written in matrix form as

$$\Lambda \equiv \begin{pmatrix} \cosh\theta & -\sinh\theta & 0 & 0 \\ -\sinh\theta & \cosh\theta & 0 & 0 \\ 0 & 0 & 1 & 0 \\ 0 & 0 & 0 & 1 \end{pmatrix}, \tag{22}$$

$$R \equiv \begin{pmatrix} 1 & 0 & 0 & 0 \\ 0 & \cos\omega & \sin\omega & 0 \\ 0 & -\sin\omega & \cos\omega & 0 \\ 0 & 0 & 0 & 1 \end{pmatrix}. \tag{23}$$

2.2.2. *Covariant vectors*

Two contravariant Lorentz vectors such as

$$A^\mu = (A^0, A^1, A^2, A^3) \tag{24}$$

and B^μ may be used to create a *scalar* product (Lorentz scalar)

$$A' \cdot B' \equiv \eta_{\mu\nu} A'^{\mu} B'^{\nu} = \eta_{\mu\nu} \Lambda^{\mu}{}_{\alpha} \Lambda^{\nu}{}_{\beta} A^{\alpha} B^{\beta} = \eta_{\alpha\beta} A^{\alpha} B^{\beta} \equiv A \cdot B. \tag{25}$$

Because of the minus signs in the Minkowski metric we have

$$A \cdot B = A^0 B^0 - \mathbf{A} \cdot \mathbf{B}, \tag{26}$$

and the *covariant* Lorentz vector is defined by

$$A_\mu \equiv (A^0, -A^1, -A^2, -A^3). \tag{27}$$

A covariant Lorentz vector is obtained from its contravariant dual by the process of lowering indices with the metric tensor,

$$A_\mu = \eta_{\mu\nu} A^\nu. \tag{28}$$

Conversely, raising of indices is achieved by

$$A^\mu = \eta^{\mu\nu} A_\nu. \tag{29}$$

It is straightforward to show that

$$\eta^{\mu\alpha} \eta_{\alpha\nu} \equiv \eta^{\mu}_{\nu} = \delta^{\mu}_{\nu}, \tag{30}$$

where

$$\delta^{\mu}_{\nu} = \begin{cases} 1 & \text{if } \mu = \nu \\ 0 & \text{otherwise} \end{cases} \tag{31}$$

is the Kronecker delta. It follows that

$$\eta_{\mu\nu} = \eta^{\mu\nu}. \tag{32}$$

The Lorentz transformation for a covariant vector is written in analogy with that of a contravariant vector:

$$A'_\mu = \Lambda_{\mu}{}^{\nu} A_\nu. \tag{33}$$

To obtain the elements $\Lambda_{\mu}{}^{\nu}$ we write the above in two different ways,

$$\eta_{\mu\beta} \Lambda^{\beta}{}_{\alpha} A^{\alpha} = \eta_{\mu\beta} A'^{\beta} = A'_\mu = \Lambda_{\mu}{}^{\nu} A_\nu = \Lambda_{\mu}{}^{\nu} \eta_{\nu\alpha} A^{\alpha}. \tag{34}$$

This holds for arbitrary A^μ so

$$\Lambda_{\mu}{}^{\nu} = \eta_{\mu\alpha} \Lambda^{\alpha}{}_{\beta} \eta^{\beta\nu}. \tag{35}$$

Using (30) in the above we get the inverse relationship

$$\Lambda^{\mu}{}_{\nu} = \eta^{\mu\alpha} \Lambda_{\alpha}{}^{\beta} \eta_{\beta\nu}. \tag{36}$$

Multiplying (35) by $A^{\mu}{}_{\sigma}$, summing on μ, and employing the fundamental condition of invariance of the proper time (11) we find

$$\Lambda^{\mu}{}_{\sigma}\Lambda_{\mu}{}^{\tau} = \delta^{\tau}_{\sigma}. \tag{37}$$

We can now invert (6) and find that $\Lambda_{\mu}{}^{\nu}$ is the inverse Lorentz transformation,

$$x^{\mu} = \Lambda_{\nu}{}^{\mu} x'^{\nu}. \tag{38}$$

The elements of the inverse transformation are given in terms of (17) or (20) by (35). We have

$$\begin{aligned} \Lambda_0{}^0 &= \Lambda_1{}^1 = \gamma, \\ \Lambda_1{}^0 &= \Lambda_0{}^1 = v\gamma, \\ \Lambda_2{}^2 &= \Lambda_3{}^3 = 1. \end{aligned} \tag{39}$$

A boost in an arbitrary direction with the primed axis having velocity $\mathbf{v} = (\mathbf{v}^1, \mathbf{v}^2, \mathbf{v}^3)$ relative to the unprimed is

$$\begin{aligned} \Lambda_0{}^0 &= \gamma, \\ \Lambda_0{}^j &= \Lambda_j{}^0 = v^j\gamma, \\ \Lambda_j{}^k &= \Lambda_k{}^j = \delta^j_k + (\gamma - 1)v^j v^k/\mathbf{v}^2. \end{aligned} \tag{40}$$

The *four-velocity* is a vector of particular interest and defined as

$$u^{\mu} = \frac{dx^{\mu}}{d\tau}. \tag{41}$$

Because $d\tau$ is an invariant scalar and dx^{μ} is a vector, u^{μ} is obviously a contravariant vector. From the expression for the invariant interval we have

$$d\tau = \sqrt{1 - \mathbf{v}^2}\, dt, \quad \mathbf{v} = \frac{d\mathbf{r}}{dt}, \tag{42}$$

with $\mathbf{r} = (x^1, x^2, x^3)$; it therefore follows that

$$u^0 \equiv \frac{dt}{d\tau} = \gamma, \qquad u^i \equiv \frac{dx^i}{d\tau} = \frac{dx^i}{dt}\frac{dt}{d\tau} = v^i\gamma, \tag{43}$$

or

$$u^{\mu} = \gamma(1, v^1, v^2, v^3), \quad u_{\mu} = \gamma(1, -v^1, -v^2, -v^3), \quad u^{\mu}u_{\mu} = 1. \tag{44}$$

The transformation of a tensor under a Lorentz transformation follows from (7) and (33) according to the position of the indices; for example,

$$T'^{\mu\nu} = \Lambda^{\mu}{}_{\alpha}\Lambda^{\nu}{}_{\beta}T^{\alpha\beta}. \tag{45}$$

We note that according to (11), the Minkowski metric $\eta^{\mu\nu}$ is a tensor; moreover, it has the same constant form in every Lorentz frame.

2.2.3. *Energy and momentum*

The relativistic analogue of Newton's law $F = ma$ is

$$F^{\mu} = m\frac{d^2x^{\mu}}{d\tau^2}, \tag{46}$$

and the four-momentum is

$$p^{\mu} = m\frac{dx^{\mu}}{d\tau}. \tag{47}$$

Hence, from (41) and (42)

$$\begin{aligned} p^0 &\equiv E = m\gamma, \\ \mathbf{p} &= m\gamma\mathbf{v}. \end{aligned} \tag{48}$$

2.2.4. *Energy–momentum tensor of a perfect fluid*

A perfect fluid is a medium in which the pressure is isotropic in the rest frame of each fluid element, and shear stresses and heat transport are absent. If at a certain point the velocity of the fluid is $\mathbf{v}$, an observer with this velocity will observe the fluid in the neighborhood as isotropic with an energy density ε and pressure p. In this local frame the energy–momentum tensor is

$$T'^{\mu\nu} \equiv \begin{pmatrix} \varepsilon & 0 & 0 & 0 \\ 0 & p & 0 & 0 \\ 0 & 0 & p & 0 \\ 0 & 0 & 0 & p \end{pmatrix}. \tag{49}$$

As viewed from an arbitrary frame, say the laboratory system, let this fluid element be observed to have velocity $\mathbf{v}$. According to (38) we obtain the transformation

$$T^{\mu\nu} = \Lambda_{\alpha}^{\ \mu}\Lambda_{\beta}^{\ \nu}T'^{\alpha\beta}. \tag{50}$$

The elements of the transformation are given by (39) in the case that the fluid element is moving with velocity v along the laboratory x-axis, or by (40) if it has the general velocity $\mathbf{v}$. It is easy to check that in the arbitrary frame

$$T^{\mu\nu} = -p\eta^{\mu\nu} + (p+\varepsilon)u^{\mu}u^{\nu} \tag{51}$$

and reduces to the diagonal form above when $\mathbf{v} = 0$. We have used the four-velocity defined above by (43). Relative to the laboratory frame it is the four-velocity of the fluid element.

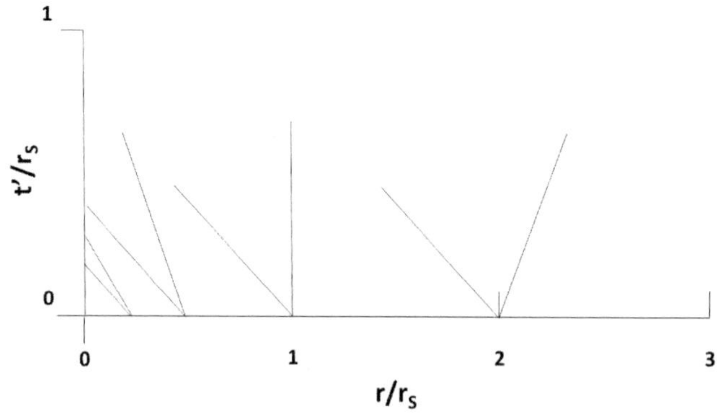

Fig. 2. The possible futures of any event at the vertex of each cone lies within the cone. Light propagates along the cone itself. On the scale of distance relative to the Schwarzschild radius of the black hole, the cones narrow and are tipped toward the black hole. At the critical radius, the outer edge of the cone is vertical; not even light can escape. Within the black hole, light can propagate only inward, as with anything else.

2.2.5. *Light cone*

For vanishing proper time intervals, $d\tau = 0$ given by (4) defines a cone (Fig. 2) in the four-dimensional space x^μ with the time axis as the axis of the cone. Events separated from the vertex event for which the proper time, (or invariant interval) vanishes ($d\tau = 0$), are said to have null separation. They can be connected to the event at the vertex by a light signal. Events separated from the vertex by a real interval $d\tau^2 > 0$ can be connected by a subluminal signal — a material particle can travel from one event to the other. An event for which $d\tau^2 < 0$ refers to an event outside the two cones; a light signal cannot join the vertex event to such an event. Therefore, events in the cone with t greater than that of the vertex of the cone lie in the future of the event at the vertex, while events in the other cone lie in its past. Events lying outside the cone are not causally connected to the vertex event.

2.3. *Scalars, vectors, and tensors in curvilinear coordinates*

In the last section we dealt with inertial frames of reference in flat spacetime. We now wish to allow for curvilinear coordinates. Our scalars, vectors, and tensors now refer to a point in spacetime. Their components refer to the reference frame at that point.

A scalar field $S(x)$ is a function of position, but its value does not depend on the coordinate system. An example is the temperature as registered on thermometers located in various rooms in a house. Each registered temperature may be different, and therefore is a function of position, but independent of the coordinates used to specify the locations:

$$S'(x') = S(x). \tag{52}$$

A vector is a quantity whose components change under a coordinate transformation. One important vector is the displacement vector between adjacent points. Near the point x^μ we consider another, $x^\mu + dx^\mu$. The four displacements dx^μ are the components of a vector. Choose units so that time and distance are measured in the same units ($c = 1$). In Cartesian coordinates we can write the *invariant interval* $d\tau$ of the Special Theory of Relativity, sometimes called the *proper time*, as

$$d\tau^2 = (dx^0)^2 - (dx^1)^2 - (dx^2)^2 - (dx^3)^2. \tag{53}$$

Under a coordinate transformation from these rectilinear coordinates to arbitrary coordinates, $x^\mu \rightarrow x'^{\,\mu}$, we have (from the rules of partial differentiation)

$$dx'^{\,\mu} = \frac{\partial x'^{\,\mu}}{\partial x^\nu} dx^\nu. \tag{54}$$

As before, repeated indices are summed. We can also write the inverse of the above equation and substitute for the spacetime differentials in the invariant interval to obtain an equation of the form

$$d\tau^2 = g_{\mu\nu} dx^\mu dx^\nu, \tag{55}$$

where the $g_{\mu\nu}$ are defined in terms of products of the partial derivatives of the coordinate transformation.

Depending on the nature of the coordinate system, say rectilinear, oblique, or curvilinear, or on the presence of a gravitational field, the invariant interval may involve bilinear products of different dx^μ and the $g_{\mu\nu}$ will be functions of position and time. The $g_{\mu\nu}$ are field quantities — the components of a tensor called the metric tensor. Because the $g_{\mu\nu}$ appear in a quadratic form (55), we may take them to be symmetric:

$$g_{\mu\nu} = g_{\nu\mu}. \tag{56}$$

In regions of spacetime for which the rectilinear system of the Special Theory of Relativity holds, the metric tensor $g_{\mu\nu}$ is equal to the Minkowski tensor (9). In fact, as we shall see, Special Relativity holds *locally* anywhere at any time. We shall refer to reference frames in which the metric is given by the Minkowski tensor as *Lorentz frames.*

The invariant interval or proper time $d\tau$ is real for a timelike interval and imaginary for a spacelike.[8] The notation *proper time* is seen to be appropriate because, when two events occur at the same space point, what remains of the invariant interval is dt.

Any four quantities A^μ that transform as dx^μ comprise a *contravariant* vector

$$A'^{\mu} = \frac{\partial x'^{\mu}}{\partial x^\nu} A^\nu, \tag{57}$$

and

$$g_{\mu\nu} A^\mu A^\nu \equiv A^2 \tag{58}$$

is its invariant length. It is obviously invariant under the same transformations that leave (53) invariant because the four quantities A^μ form a four-vector like dx^μ.

A *covariant* vector can be obtained through the process of *lowering* indices with the metric tensor:

$$A_\mu = g_{\mu\nu} A^\nu. \tag{59}$$

In terms of this vector, the magnitude equation (58) can be written as

$$A_\mu A^\mu = A^2. \tag{60}$$

Let A^μ and B^μ be distinct contravariant vectors. Then so is $A^\mu + \lambda B^\mu$ for all finite λ. The quantity

$$g_{\mu\nu}(A^\mu + \lambda B^\mu)(A^\nu + \lambda B^\nu) \tag{61}$$

is the invariant squared length. Because this is true for all λ, the coefficient of each power of λ is also an invariant; for the linear term we find

$$g_{\mu\nu}(A^\mu B^\nu + B^\mu A^\nu) = 2 g_{\mu\nu} A^\mu B^\nu, \tag{62}$$

where we have used the symmetry of $g_{\mu\nu}$. Thus, we obtain the invariant scalar product of two vectors:

$$g_{\mu\nu} A^\mu B^\nu = A_\nu B^\nu = A \cdot B. \tag{63}$$

To derive the transformation law for a covariant vector use the fact, just proven, that $A_\mu B^\mu$ is a scalar. Then using the law of transformation

[8]The opposite convention $ds^2 = -d\tau^2$ could be employed. The interval ds is often referred as the line element.

of a contravariant vector (57), we have

$$A'_\mu B'^\mu = A_\alpha B^\alpha = A_\alpha \frac{\partial x^\alpha}{\partial x'^\mu} B'^\mu, \tag{64}$$

where A'_μ is the same vector as A_μ, but referred to the primed reference frame. From the above equation it follows that

$$\left(A'_\mu - \frac{\partial x^\alpha}{\partial x'^\mu} A_\alpha \right) B'^\mu = 0. \tag{65}$$

Because B^μ is any vector, the quantity in brackets must vanish; thus we have the law of transformation of a covariant vector,

$$A'_\mu = \frac{\partial x^\nu}{\partial x'^\mu} A_\nu. \tag{66}$$

Compare this transformation law with that of (57).

Let the determinant of $g_{\mu\nu}$ be g,

$$g = \det |g_{\mu\nu}|. \tag{67}$$

As long as g does not vanish, the equations (59) can be inverted. Let the coefficients of the inverse be called $g^{\mu\nu}$. Then find

$$A^\nu = g^{\nu\mu} A_\mu. \tag{68}$$

Multiply (59) by $g^{\alpha\mu}$ and sum on μ with the result

$$A^\alpha = g^{\alpha\mu} A_\mu = g^{\alpha\mu} g_{\mu\nu} A^\nu, \tag{69}$$

or

$$(g^{\alpha\mu} g_{\mu\nu} - \delta^\alpha_\nu) A^\nu = 0, \tag{70}$$

where δ^α_ν is the Kronecker delta. Because this equation holds for any vector, we have

$$g^\alpha_\beta \equiv g^{\alpha\mu} g_{\mu\beta} = \delta^\alpha_\beta. \tag{71}$$

The two g's, one with subscripts, the other with superscripts, are inverses. In the same way as $g_{\mu\nu}$ can be used to lower an index, $g^{\mu\nu}$ can be used to raise one. Both are symmetric:

$$g_{\mu\nu} = g_{\nu\mu}, \quad g^{\mu\nu} = g^{\nu\mu}. \tag{72}$$

The derivative of a scalar field $S(x) = S'(x')$ with respect to the components of a contravariant position vector yields a covariant vector field and, vice versa:

$$\frac{\partial S}{\partial x'^{\mu}} = \frac{\partial x^{\nu}}{\partial x'^{\mu}} \frac{\partial S}{\partial x^{\nu}}. \tag{73}$$

Accordingly, we shall sometimes use the abbreviations

$$\partial_{\mu} \equiv \frac{\partial}{\partial x^{\mu}} \quad \text{and} \quad \partial^{\mu} \equiv \frac{\partial}{\partial x_{\mu}}, \tag{74}$$

especially in writing Lagrangians of fields. In relativity it is also useful to have an even more compact notation for the coordinate derivative — the "comma subscript":

$$S_{,\mu} \equiv \frac{\partial S}{\partial x^{\mu}}. \tag{75}$$

The d'Alembertian,

$$\Box = \partial_{\mu} \partial^{\mu}, \tag{76}$$

is manifestly a scalar.

Tensors are similar to vectors, but with more than one index. A simple tensor is one formed from the product of the components of two vectors, $A^{\mu} B^{\nu}$. But this is special because of the relationships between its components. A general tensor of the second rank can be formed by a sum of such products:

$$T^{\mu\nu} = A^{\mu} B^{\nu} + C^{\mu} D^{\nu} + \cdots \tag{77}$$

The superscripts can be lowered as with a vector, either one index, or both,

$$T_{\mu}{}^{\nu} = g_{\mu\alpha} T^{\alpha\nu}, \quad T^{\nu}{}_{\mu} = T^{\nu\alpha} g_{\alpha\mu}, \quad T_{\mu\nu} = g_{\mu\alpha} g_{\nu\beta} T^{\alpha\beta}. \tag{78}$$

Similarly, we may have tensors of higher rank, either contravariant with respect to all indices, or covariant, or mixed. The position of the indices on the mixed tensor (the lower to the left or right of the upper) refers to the position of the index that was lowered. If $T^{\mu\nu}$ is symmetric, then $T^{\mu}{}_{\nu} = T_{\nu}{}^{\mu}$ and it is unimportant to keep track of the position of the index that has been lowered (or raised). But if $T^{\mu\nu}$ is antisymmetric, then the two orderings differ by a sign.

If two of the indices on a tensor, one a superscript the other a subscript, are set equal and summed, the rank is reduced by two. This process is called *contraction*. If it is done on a second-rank mixed tensor, the result is a scalar,

$$S = T^{\mu}{}_{\mu} = T_{\mu}{}^{\mu}. \tag{79}$$

When $T^{\mu\nu}$ is antisymmetric, the contractions $T_{\mu}{}^{\mu}$ and $T^{\mu}{}_{\mu}$ are identically zero.

The test of tensor character is whether the object in question transforms under a coordinate transformation in the obvious generalization of a vector. For example,

$$T'^{\mu}{}_{\nu} = \frac{\partial x'^{\mu}}{\partial x^{\alpha}} \frac{\partial x^{\beta}}{\partial x'^{\nu}} T^{\alpha}{}_{\beta} \tag{80}$$

is a tensor.

In general, we deal with curved spacetime in General Relativity. We must therefore deal with curvilinear coordinates. Vectors and tensors at a point in such a spacetime have components referring to the axis at that point. The components will change according to the above laws, depending on the way the axes change at that point. Therefore, the metric tensors $g_{\mu\nu}$, $g^{\mu\nu}$ cannot be constants. They are field quantities which vary from point to point. As we shall see, they can be referred to collectively as the gravitational field. Because the formalism of this section is expressed by local equations, it holds in curved spacetime, for curved spacetime is flat in a sufficiently small locality.

Because the derivative of a scalar field is a vector (73), one might have thought that the derivative of a vector field is a tensor. However, by checking the transformation properties one finds that this supposition is not true.

We have referred invariably to the $g_{\mu\nu}$ as tensors. Now we show that this is so. Let A^{μ}, B^{ν} be arbitrary vector fields, and consider two coordinate systems such that the same point P has the coordinates x^{μ} and x'^{μ} when referred to the two systems, respectively. Then we have

$$g'_{\alpha\beta} A'^{\alpha} A'^{\beta} = g_{\mu\nu} A^{\mu} A^{\nu} = g_{\mu\nu} \frac{\partial x^{\mu}}{\partial x'^{\alpha}} \frac{\partial x^{\nu}}{\partial x'^{\beta}} A'^{\alpha} A'^{\beta}. \tag{81}$$

Because this holds for arbitrary vectors, we find

$$g'_{\alpha\beta} = g_{\mu\nu} \frac{\partial x^{\mu}}{\partial x'^{\alpha}} \frac{\partial x^{\nu}}{\partial x'^{\beta}}, \tag{82}$$

which, by comparison with (66), shows that $g_{\mu\nu}$ is a covariant tensor. Similarly $g^{\mu\nu}$ is a contravariant tensor:

$$g'^{\alpha\beta} = g^{\mu\nu} \frac{\partial x'^{\alpha}}{\partial x^{\mu}} \frac{\partial x'^{\beta}}{\partial x^{\nu}}. \tag{83}$$

These are called the *fundamental tensors*. Of course, the above tensor character of the metric is precisely what is required to make the square of the interval $d\tau^2$ of (55) an invariant, as is trivially verified.

Mixed tensors of arbitrary rank transform, for each index, according to the transformation laws (57, 66) depending on whether the index is a superscript or a subscript, as can be derived in obvious analogy to the above manipulations.

Tensors and tensor algebra are very powerful techniques for carrying the consequences discovered in one frame to another. That the linear combination of tensors of the same rank and arrangement of upper and lower indices is also a tensor; that the direct product of two tensors of the same or different rank and arrangement of indices, $A^{\mu\cdots}_{\alpha\cdots}B^{\mu\cdots}_{\cdots} = T^{\mu\cdots\nu\cdots}_{\alpha\cdots}$ is also a tensor; and that contraction (defined above) of a pair of indices, one upper, one lower produces a tensor of rank reduced by two — are all easy theorems that we do not need to prove, but only note in passing. Of particular note, if the difference of two tensors of the same transformation rule vanishes in one frame, then it vanishes in all (i.e., the two tensors are equal in all frames).

2.4. *Principle of equivalence of inertia and gravitation*

> "The possibility of explaining the numerical equality of inertia and gravitation by the unity of their nature gives to the general theory of relativity, according to my conviction, such a superiority over the conceptions of classical mechanics, that all the difficulties encountered in development must be considered as small in comparison."
>
> A. Einstein [Einstein (1951)]

Eötvös established that all bodies have the same ratio of inertial to gravitational mass with high precision [Eötvös (1890)]. With an appropriate choice of units, the two masses are equal for all bodies to the accuracy established for the ratio. One might have expected such conceptually different properties, one having to do with inertia to motion (m_I), the other with "charge" (m_G), in an expression of mutual attraction between bodies, to be entirely different. The relation between the force exerted by the gravitational attraction of a body of mass M at a distance R upon the object, and the acceleration imparted to it are expressed by Newton's equation, valid for weak fields and small material velocities:

$$m_I a = G\frac{m_G M}{R^2}. \tag{84}$$

Einstein reasoned that the near equality of two such different properties must be more than mere coincidence and that inertial and gravitational masses must be exactly equal: $m_I = m_G = m$. The mass drops out! In that case all bodies experience precisely the same acceleration in a gravitational field, as was presaged by Galileo's experiments centuries earlier. For all other forces that we know, the acceleration is inverse to the mass.

The equivalence of inertial and gravitational mass is established to high accuracy for atomic and nuclear binding energies.[9] Moreover, as a result of very careful lunar laser-ranging experiments, the Earth and Moon are found to fall with equal acceleration toward the Sun to a precision of almost 1 part in 10^{13}, better than the most accurate Eötvös-type experiments on laboratory bodies. This exceedingly important test involving bodies of different gravitational binding was conceived by Nordvedt [Nordvedt (1968)]. The essentially null result establishes the so-called strong statement of equivalence of inertial and gravitational mass: Free bodies — no matter their nature or constituents, nor how much or little those constituents are bound, nor by what force — all move in the spacetime of an arbitrary gravitational field as if they were identical test particles! *Because their motion has nothing to do with their nature, it evidently has to do with the nature of spacetime.*

Einstein felt certain that a deep meaning was attached to the equivalence; "The experimentally known matter independence of the acceleration of fall is $\cdots$ a powerful argument for the fact that the relativity postulate has to be extended to coordinate systems which, relative to each other, are in non-uniform motion" [Einstein (1920)]. This conviction led him to the formulation of the *equivalence principle.* The equivalence principle provides the link between the physical laws as we discern them in our laboratories and their form under any circumstance in the universe — more precisely, in arbitrarily strong and varying gravitational fields. It also provides a tool for the development of the theory of gravitation itself, as we shall see throughout the sequel.

The universe is populated by massive objects moving relative to one another. The gravitational field may be arbitrarily changing in time and space. However, the presence of gravity cannot be detected in a sufficiently small reference frame falling freely with a particle under no influence other

[9]Eötvös experiments on such diverse media as wood, platinum, copper, glass, and other materials involve different molecular, atomic, and nuclear binding energies and different ratios of neutrons and protons.

than gravity. The particle will remain at rest in such a frame. It is a *local inertial frame.* A local inertial frame and a *local Lorentz frame* are synonymous. The laws of Special Relativity hold in inertial frames and therefore in the neighborhood of a *freely falling frame.* In this way the relativity principle is extended to arbitrary gravitational fields.

Associated with a given spacetime event there are an infinity of locally inertial frames related by Lorentz transformations. All are equivalent for the description of physical phenomena in a sufficiently small region of spacetime. So we arrive at a statement of the equivalence principle: *At every spacetime point in an arbitrary gravitational field (meaning anytime and anywhere in the universe), a local inertial (Lorentz) frame can be chosen so that the laws of physics take on the form they have in Special Relativity.* This is the meaning of the equality of inertial and gravitational masses that Einstein sought. The restricted validity of inertial frames to small localities of any event suggested the very fruitful analogy with local flatness on a curved surface.

Einstein went further than the above statement of the equivalence principle. He spoke of the laws of nature rather than just the laws of physics. It seems entirely plausible that the extension is true, but we deal here only with physics.

The equivalence principle has great power. It is the instrument by which all the special relativistic laws of physics — valid in a gravity-free universe — can be generalized to a gravity-filled universe. We shall see how Einstein was able to give dynamic meaning to the spacetime continuum as an integral part of the physical world quite unlike the conception of an absolute spacetime in which the rest of physical processes take place.

2.4.1. *Photon in a gravitational field*

Employing the conservation of energy and Newtonian physics, Einstein reasoned that the gravitational field acts on photons. Let a photon be emitted from z_1 vertically to z_2, and only for simplicity, let the field be uniform. A device located at z_2 converts its energy on arrival to a particle of mass m with perfect efficiency. The particle drops to z_1 where its energy is now $m + mgh$, where g is the acceleration due to the uniform field. A device at z_1 converts it into a photon of the same energy as possessed by the particle. The photon again is directed to z_2. If the original (and each succeeding photon) does not lose energy $(h\nu)gh$ in climbing the gravitational field equal to the energy gained by the particle in dropping in the field, we would have a device that creates energy. By the law of conservation of energy Einstein discovered the gravitational redshift, commonly designated by the factor z and equal in this case to gh. The shift in energy of a photon by falling (in this case blue-shifted) in the Earth's gravitational field has been directly confirmed in an experiment performed by Pound and Rebka [Pound & Rebka (1960)].

In the above discussion the equivalence principle entered when the photon's inertial mass ($h\nu$) was used also as its gravitational mass in computing the gravitational work. One can also see the role of the equivalence principle by considering a pulse of light emitted over a distance h along the axis of a spaceship in uniform acceleration g in outer space. The time taken for the light to reach the detector is $t = h$ (we use units $G = c = 1$). The difference in velocity of the detector acquired during the light travel time is $v = gt = gh$, the Doppler shift z in the detected light. This experiment, carried out in the gravity-free environment of a spaceship whose rockets produce an acceleration g, must yield the same result for the energy shift of the photon in a uniform gravitational field g according to the equivalence principle. The Pound-Rebka experiment can therefore be regarded as an experimental proof of the equivalence principle.

We may regard a radiating atom as a clock, with each wave crest regarded as a tick of the clock. Imagine two identical atoms situated one at some height above the other in the gravitational field of the Earth. Since, by dropping in the gravitational field, the light is blue-shifted when compared to the radiation of an identical atom (clock) at the bottom, the clock at the top is seen to be running faster than the one at the bottom. Therefore, identical clocks, stationary with respect to the Earth, run at different rates according to their different heights above the Earth. Time flows at different rates in different gravitational fields.

The trajectory of photons is also bent by the gravitational field. Imagine a freely falling elevator in a constant gravitational field. Its walls constitute an inertial frame as guaranteed by the equivalence principle. Therefore, a photon (as for a free particle) directed from one wall to the opposite along a path parallel to the floor will arrive at the other wall at the same height from which it started. But relative to the Earth, the elevator has fallen during the traversal time. Therefore the photon has been detected toward the Earth and follows a curved path as observed from a frame fixed on the Earth.

2.4.2. *Tidal gravity*

Einstein predicted that a clock near a massive body would run more slowly than an identical distant clock. In doing so he arrived at a hint of the deep connection of the structure of spacetime and gravity. Two parallel straight lines never meet in the gravity-free, flat spacetime of Minkowski. A single inertial frame would suffice to describe all of spacetime. In formulating the equivalence principle (knowing that gravitational fields are not uniform and constant but depend on the motion of gravitating bodies and the position where gravitational effects are experienced), Einstein understood that only in a suitably small locality of spacetime do the laws of Special Relativity hold. Gravitational effects will be observed on a larger scale. *Tidal gravity* refers to the deviation from uniformity of the gravitational field at nearby points.

These considerations led Einstein to the notion of spacetime curvature. Whatever the motion of a free body in an arbitrary gravitational field, it will follow a straight-line trajectory over any small locality as guaranteed by the equivalence principle. And in a gravity-endowed universe, free particles whose trajectories are parallel in a local inertial frame, will not remain parallel over a large region of spacetime. This has a striking analogy with the surface of a sphere on which two straight lines that are parallel over a small region do meet and cross. What if in fact the particles are freely falling in curved spacetime? In this way of thinking, the law that free particles move in straight lines remains true in an arbitrary gravitational field, thus obeying the principle of relativity in a larger sense. Any sufficiently small region of curved spacetime is locally flat. The paths in curved spacetime that have the property of being locally straight are called geodesics.

2.4.3. *Curvature of spacetime*

Let us now consider a thought experiment. Two nearby bodies released from rest above the Earth follow parallel trajectories over a small region of their trajectories, as we know from the equivalence principle. But if holes were drilled in the Earth through which the bodies could fall, the bodies would meet and cross at the Earth's center. So there is clearly no single Minkowski spacetime that covers a large region or the whole region containing a massive body.

Einstein's view was that spacetime curvature caused the bodies to cross, bodies that in this curved spacetime were following straight line paths in every small locality, just as they would have done in the whole of Minkowski (flat) spacetime in the absence of gravitational bodies. The presence of gravitating bodies denies the existence of a global inertial frame. Spacetime can be flat everywhere only if there exists such a global frame. Hence, spacetime is curved by massive bodies. In their presence a test particle follows a geodesic path, one that is always locally straight. The concept of a "gravitational force" has been replaced by the curvature of spacetime, and the natural free motions of particles in it are defined by geodesics.

2.4.4. *Energy conservation and curvature*

Interestingly, the conservation of energy can also be used to inform us that spacetime is curved. Consider a static gravitational field. Let us conjecture that spacetime is flat so that the Minkowski metric holds; we will arrive at a contradiction.

Imagine the following experiment performed by observers and their apparatus at rest with respect to the gravitational field and their chosen Lorentz frame in the supposed flat spacetime of Minkowski. At a height z_1 in the field, let a monochromatic light signal be emitted upward a height h to $z_2 = z_1 + h$. Let the pulse be emitted for a specific time dt_1 during which N wavelengths (or photons) are emitted. Let the time during which they are received at z_2 be measured as dt_2. (Because the spacetime is assumed to be described by the Minkowski metric and the source and receiver are at rest in the chosen frame, the proper times and coordinate times are equal.)

Because the field in the above experiment is static, the path in the $z{-}t$ plane will have the same shape for both the beginning and ending of the pulse (as for each photon) as they trace their path in the Minkowski space we postulate to hold. The trajectories will not be lines at 45 degrees because of the field, but the curved paths will be congruent; a translation in time will make the paths lie one upon the other. Therefore $d\tau_2 = dt_2 = dt_1 = d\tau_1$ will be measured at the stationary detector if spacetime is Minkowskian. In this case, the frequency (and hence the energy received at z_2) is the same as that sent from z_1. But this cannot be. The photons comprising the signal must lose energy in climbing the gravitational field (see Section 2.4.1). The conjecture that spacetime in the presence of a gravitational field is Minkowskian must therefore be false. We conclude that the presence of the gravitational field has caused spacetime to be curved. Such a line of reasoning was first conceived by Schild [Schild (1960, 1962)].

2.5. *Gravity*

> "I was sitting in a chair at the patent office at Bern when all of a sudden a thought occurred to me: 'If a person falls freely he will not feel his own weight.' I was startled. This simple thought had a deep impression on me. It impelled me toward a theory of gravitation."
>
> A. Einstein [Ishiwara (1916)]

Massive bodies generate curvature. Galaxies, stars, and other bodies are in motion; therefore the curvature of spacetime is everywhere changing. For this reason there is no "prior geometry". There are no immutable reference frames to which events in spacetime can be referred. Indeed, the changing geometry of spacetime and of the motion and arrangement of mass-energy in spacetime are inseparable parts of the description of physical processes. This is a very different idea of space and time from that of Newton and even of the Special Theory of Relativity. We now take up the unified discussion of gravitating matter and motion.

The power of the equivalence principle in informing us so simply that spacetime must be curved by the presence of massive bodies in the universe suggests a fruitful way of beginning. Following Weinberg [Weinberg (1972)], or indeed, following the notion expressed by Einstein in the quotation above, we seek the connection between an arbitrary reference frame and a reference frame that is freely falling with a particle that is moving only under the influence of an arbitrary gravitational field. In this freely falling and therefore locally inertial frame, the particle moves in a straight line. Denote the coordinates by ξ^α. The equations of motion are

$$\frac{d^2\xi^\alpha}{d\tau^2} = 0, \tag{85}$$

and the invariant interval (or proper time) between two neighboring spacetime events expressed in this frame, from (8), is

$$d\tau^2 = \eta_{\alpha\beta} d\xi^\alpha d\xi^\beta. \tag{86}$$

The freely falling coordinates may be regarded as functions of the coordinates x^μ of any arbitrary reference frame — curvilinear, accelerated, or rotating. We seek the connection between the equations of motion in the freely falling frame and the arbitrary one which, for example, might be the laboratory frame. From the chain rule for differentiation we can rewrite

(85) as

$$0 = \frac{d}{d\tau}\left(\frac{\partial \xi^\alpha}{\partial x^\mu}\frac{dx^\mu}{d\tau}\right)$$
$$= \frac{\partial \xi^\alpha}{\partial x^\mu}\frac{d^2 x^\mu}{d\tau^2} + \frac{\partial^2 \xi^\alpha}{\partial x^\mu \partial x^\nu}\frac{dx^\mu}{d\tau}\frac{dx^\nu}{d\tau}.$$

Multiply by $\partial x^\lambda/\partial \xi^\alpha$, and use the chain rule again to obtain

$$\frac{dx^\lambda}{dx^\mu} = \frac{\partial x^\lambda}{\partial \xi^\alpha}\frac{\partial \xi^\alpha}{\partial x^\mu} = \delta^\lambda_\mu. \tag{87}$$

The equation of motion of the particle in an arbitrary frame when the particle is moving in an arbitrary gravitational field therefore is

$$\frac{d^2 x^\lambda}{d\tau^2} + \Gamma^\lambda_{\mu\nu}\frac{dx^\mu}{d\tau}\frac{dx^\nu}{d\tau} = 0. \tag{88}$$

Here $\Gamma^\lambda_{\mu\nu}$, defined by

$$\Gamma^\lambda_{\mu\nu} \equiv \frac{\partial x^\lambda}{\partial \xi^\alpha}\frac{\partial^2 \xi^\alpha}{\partial x^\mu \partial x^\nu}, \tag{89}$$

is called the *affine connection*. The affine connection is symmetric in its lower indices.

The path defined by equation (88) is called a geodesic, the extremal path in the spacetime of an arbitrary gravitational field. We do not see here that it is an extremal, but this is hinted at inasmuch as it defines the same path of (85), the straight-line path of a free particle as observed from its freely falling frame. In the next section we will see that a geodesic path is locally a straight line.

The invariant interval (86) can also be expressed in the arbitrary frame by writing $d\xi^\alpha = (\partial \xi^\alpha/\partial x^\mu)dx^\mu$ so that

$$d\tau^2 = g_{\mu\nu}dx^\mu dx^\nu \tag{90}$$

with

$$g_{\mu\nu} = \frac{\partial \xi^\alpha}{\partial x^\mu}\frac{\partial \xi^\beta}{\partial x^\nu}\eta_{\alpha\beta}. \tag{91}$$

In the new and arbitrary reference frame, the second term of (88) causes a deviation from a straight-line motion of the particle in this frame. Therefore, the second term represents the effect of the gravitational field. (To be sure, the connection coefficients also represent any other non-inertial effects that may have been introduced by the choice of reference frame, such as rotation.)

The affine connection (89) appearing in the geodesic equation clearly plays an important role in gravity, and we study it further. We first show that the affine connection is a non-tensor, and then show how it can be expressed in terms of the metric tensor and its derivatives. In this sense the metric behaves as the gravitational potential and the affine connection as the force. Write $\Gamma^{\lambda}_{\mu\nu}$ expressed in (89) in another coordinate system x'^{μ} and use the chain rule several times to rewrite it:

$$\begin{aligned}\Gamma'^{\lambda}_{\mu\nu} &= \frac{\partial x'^{\lambda}}{\partial x^{\rho}}\frac{\partial x^{\rho}}{\partial \xi^{\alpha}}\frac{\partial}{\partial x'^{\mu}}\left(\frac{\partial x^{\sigma}}{\partial x'^{\nu}}\frac{\partial \xi^{\alpha}}{\partial x^{\sigma}}\right)\\ &= \frac{\partial x'^{\lambda}}{\partial x^{\rho}}\frac{\partial x^{\rho}}{\partial \xi^{\alpha}}\left[\frac{\partial x^{\sigma}}{\partial x'^{\nu}}\frac{\partial x^{\tau}}{\partial x'^{\mu}}\frac{\partial^2 \xi^{\alpha}}{\partial x^{\tau}\partial x^{\sigma}} + \frac{\partial^2 x^{\sigma}}{\partial x'^{\mu}\partial x'^{\nu}}\frac{\partial \xi^{\alpha}}{\partial x^{\sigma}}\right]\\ &= \frac{\partial x'^{\lambda}}{\partial x^{\rho}}\frac{\partial x^{\tau}}{\partial x'^{\mu}}\frac{\partial x^{\sigma}}{\partial x'^{\nu}}\Gamma^{\rho}_{\tau\sigma} + \frac{\partial x'^{\lambda}}{\partial x^{\rho}}\frac{\partial^2 x^{\rho}}{\partial x'^{\mu} x'^{\nu}}. \end{aligned} \tag{92}$$

According to the transformation laws of tensors developed in Section 1.2.3, the second term on the right spoils the transformation law of the affine connection. It is therefore a nontensor.

Let us now obtain the expression of the affine connection in terms of the derivatives of the metric tensor. Form the derivative of (82):

$$\frac{\partial}{\partial x'^{\kappa}} g'_{\mu\nu} = \frac{\partial}{\partial x'^{\kappa}}\left(g_{\rho\sigma}\frac{\partial x^{\rho}}{\partial x'^{\mu}}\frac{\partial x^{\sigma}}{\partial x'^{\nu}}\right). \tag{93}$$

Take the derivatives and form the following combination and find that it is equal to the above derivative:

$$\begin{aligned}\frac{\partial g'_{\kappa\nu}}{\partial x'^{\mu}} + \frac{\partial g'_{\kappa\mu}}{\partial x'^{\nu}} - \frac{\partial g'_{\mu\nu}}{\partial x'^{\kappa}} &= \frac{\partial x^{\tau}}{\partial x'^{\kappa}}\frac{\partial x^{\rho}}{\partial x'^{\mu}}\frac{\partial x^{\sigma}}{\partial x'^{\nu}}\left(\frac{\partial g_{\sigma\tau}}{\partial x^{\rho}} + \frac{\partial g_{\rho\tau}}{\partial x^{\sigma}} - \frac{\partial g_{\rho\sigma}}{\partial x^{\tau}}\right)\\ &\quad + 2g_{\rho\sigma}\frac{\partial x^{\sigma}}{\partial x'^{\kappa}}\frac{\partial^2 x^{\rho}}{\partial x'^{\mu}\partial x'^{\nu}}. \end{aligned} \tag{94}$$

Multiply this equation by $\frac{1}{2}$ and then multiply the left and right sides by the left and right sides, respectively, of the law of transformation (83), namely,

$$g'^{\lambda\kappa} = g^{\alpha\beta}\frac{\partial x'^{\lambda}}{\partial x^{\alpha}}\frac{\partial x'^{\kappa}}{\partial x^{\beta}}. \tag{95}$$

Use the chain rule and rename several dummy indices to obtain

$$\left\{\begin{matrix}\lambda\\ \mu\nu\end{matrix}\right\}' = \frac{\partial x'^{\lambda}}{\partial x^{\rho}}\frac{\partial x^{\tau}}{\partial x'^{\mu}}\frac{\partial x^{\sigma}}{\partial x'^{\nu}}\left\{\begin{matrix}\rho\\ \tau\rho\end{matrix}\right\} + \frac{\partial x'^{\lambda}}{\partial x^{\rho}}\frac{\partial^2 x^{\rho}}{\partial x'^{\mu}\partial x'^{\nu}} \tag{96}$$

where the prime on $\{\}$ means that it is evaluated in the x'^{μ} frame and the symbol stands for

$$\left\{ \begin{matrix} \lambda \\ \mu\nu \end{matrix} \right\} = \frac{1}{2} g^{\lambda\kappa} \left[\frac{\partial g_{\kappa\nu}}{\partial x^{\mu}} + \frac{\partial g_{\kappa\mu}}{\partial x^{\nu}} - \frac{\partial g_{\mu\nu}}{\partial x^{\kappa}} \right]. \tag{97}$$

This is called a *Christoffel symbol of the second kind.* It is seen to transform in exactly the same way as the affine connection (92). Subtract the two to obtain

$$\left[\Gamma^{\lambda}_{\mu\nu} - \left\{ \begin{matrix} \lambda \\ \mu\nu \end{matrix} \right\} \right]' = \frac{\partial x'^{\lambda}}{\partial x^{\rho}} \frac{\partial x^{\tau}}{\partial x'^{\mu}} \frac{\partial x^{\sigma}}{\partial x'^{\nu}} \left[\Gamma^{\rho}_{\tau\sigma} - \left\{ \begin{matrix} \rho \\ \tau\sigma \end{matrix} \right\} \right]. \tag{98}$$

This shows that the difference is a tensor. According to the equivalence principle, at anyplace and anytime there is a local inertial frame ξ^{α} in which the effects of gravitation are absent, the metric is given by (9), and $\Gamma^{\lambda}_{\mu\nu}$ vanishes (compare (85) and (88)). Because the first derivatives of the metric tensor vanish in such a local inertial system, the Christoffel symbol also vanishes. Because the difference of the affine connection and the Christoffel symbol is a tensor which vanishes in this frame, the difference vanishes in all reference frames. So everywhere we find

$$\Gamma^{\lambda}_{\mu\nu} = \left\{ \begin{matrix} \lambda \\ \mu\nu \end{matrix} \right\} = \frac{1}{2} g^{\lambda\kappa} \Big(g_{\kappa\nu,\mu} + g_{\kappa\mu,\nu} - g_{\mu\nu,\kappa} \Big). \tag{99}$$

We use the "comma subscript" notation introduced earlier to denote differentiation (75).

Sometimes it is useful to have the superscript lowered on the affine connection

$$\Gamma_{\kappa\mu\nu} = g_{\kappa\lambda} \Gamma^{\lambda}_{\mu\nu}. \tag{100}$$

It is equal to the Christoffel symbol of the first kind

$$\Gamma_{\kappa\mu\nu} = \left[\begin{matrix} \kappa \\ \mu\nu \end{matrix} \right] = \frac{1}{2} \Big(g_{\kappa\nu,\mu} + g_{\kappa\mu,\nu} - g_{\mu\nu,\kappa} \Big). \tag{101}$$

The above formulas provide a means of computing the affine connection from the derivatives of the metric tensor and will prove very useful. It is trivial from the above to prove that

$$\Gamma_{\kappa\mu\nu} + \Gamma_{\mu\kappa\nu} = g_{\mu\kappa,\nu}. \tag{102}$$

2.5.1. *Mathematical definition of local Lorentz frames*

Spacetime is curved globally by the massive bodies in the universe. Therefore, we need to define mathematically the meaning of "local Lorentz frame". In a rectilinear Lorentz frame the metric tensor is $\eta_{\mu\nu}$ (9). Therefore, in the local region around an event P (a point in the four-dimensional spacetime continuum), the metric tensor, its coordinate derivatives, and the affine connection have the following values:

$$g_{\mu\nu}(P) = \eta_{\mu\nu}\,, \quad g_{\mu\nu,\alpha}(P) = 0, \quad \Gamma^{\lambda}_{\mu\nu}(P) = 0. \tag{103}$$

The third of these equations follows from the second and from (99). All local effects of gravitation disappear in such a frame. The geodesic equation (88) defining the path followed by a free particle in an arbitrary gravitational field becomes locally the equation of a uniform straight line, in accord with the equivalence principle. Of course, physical measurements are always subject to the precision of the measuring devices. The extent of the local region around P, in which the above equations will hold and in which spacetime is said to be flat, will depend on the accuracy of the devices and therefore their ability to detect deviations from the above conditions as one measures further from P.

2.5.2. *Geodesics*

In the Special Theory of Relativity a free particle remains at rest or moves with constant velocity in a straight line. A straight line is the shortest distance between two points in Euclidean three-dimensional space. In Minkowski spacetime a straight line is the longest interval between two events, as we shall shortly see. Both situations are covered by saying that a straight line is an extremal path between two points. We shall show that in an arbitrary gravitational field, a particle moving under the influence of only gravity, follows a path that is, in the sense that we shall define, the straightest line possible in curved spacetime.

We first show that a straight-line path between two events in Minkowski spacetime maximizes the proper time. This is easily proved. Orient the axis so that the two events marking the ends of the path, A and B, lie on the t-axis with coordinates $(0,0,0,0)$ and $(T,0,0,0)$, and consider an alternate path in the $t-x$ plane that consists of two straight-line segments that pass from A to B through $(T/2, R/2, 0, 0)$. The proper time as measured on the second path is

$$\tau = 2\sqrt{(T/2)^2 - (R/2)^2} = \sqrt{T^2 - R^2}. \tag{104}$$

For any finite R, τ is smaller than the proper time along the straight-line path from A to B, namely, T. Therefore, a straight-line path is a maximum in proper time. We have referred to the equation of motion of a particle moving freely in an arbitrary gravitational field (88) as a geodesic equation. In general, a geodesic that is not null (a null geodesic, as is the case for a light particle, has $d\tau = 0$), is the extremal path of

$$\int_A^B d\tau \tag{105}$$

where A and B refer to spacetime events on the geodesic. To prove this result, let $x^\mu(\tau)$ denote the coordinates along the geodesic path, parameterized by the proper time, and let $x^\mu(\tau) + \delta x^\mu(\tau)$ denote a neighboring path with the same end points, A to B. From

$$d\tau^2 = g_{\mu\nu} dx^\mu dx^\nu , \tag{106}$$

we have to first order in the variation,

$$\begin{aligned} 2d\tau\delta(d\tau) &= \delta g_{\mu\nu} dx^\mu dx^\nu + 2g_{\mu\nu} dx^\mu \delta(dx^\nu) \\ &= dx^\mu dx^\nu g_{\mu\nu,\lambda} \delta x^\lambda + 2g_{\mu\nu} dx^\mu d(\delta x^\nu). \end{aligned} \tag{107}$$

Recalling the four-velocity, $u^\mu = dx^\mu/d\tau$, we have

$$\delta(d\tau) = \left(\frac{1}{2} u^\mu u^\nu g_{\mu\nu,\lambda} \delta x^\lambda + g_{\mu\lambda} u^\mu \frac{d}{d\tau} \delta x^\lambda \right) d\tau. \tag{108}$$

Thus

$$\delta \int_A^B d\tau = \int_A^B \left[\frac{1}{2} u^\mu u^\nu g_{\mu\nu,\lambda} - \frac{d}{d\tau} (g_{\mu\lambda} u^\mu) \right] \delta x^\lambda d\tau \tag{109}$$

where an integration by parts in the second term was performed. Because the variation of the path δx^λ is arbitrary save for its end points being zero, we obtain as the extremal condition,

$$\frac{d}{d\tau}(g_{\mu\lambda} u^\mu) - \frac{1}{2} u^\mu u^\nu g_{\mu\nu,\lambda} = 0. \tag{110}$$

The first and second terms can be rewritten:

$$\begin{aligned} \frac{d}{d\tau}(g_{\mu\lambda} u^\mu) &= g_{\mu\lambda} \frac{du^\mu}{d\tau} + g_{\mu\lambda,\nu} u^\mu u^\nu , \\ g_{\mu\lambda,\nu} u^\mu u^\nu &= \frac{1}{2}(g_{\mu\lambda,\nu} + g_{\lambda\nu,\mu}) u^\mu u^\nu. \end{aligned} \tag{111}$$

Now using the relationship (101), we find

$$g_{\mu\lambda} \frac{du^\mu}{d\tau} + \Gamma_{\lambda\mu\nu} u^\mu u^\nu = 0. \tag{112}$$

Multiplying by $g^{\sigma\lambda}$ and summing on λ, we obtain the geodesic equation (88):

$$\frac{du^\sigma}{d\tau} + \Gamma^\sigma_{\mu\nu} u^\mu u^\nu = 0. \tag{113}$$

This completes the proof that the path defined by the geodesic equation, the equation of motion of a particle in a purely gravitational field, extremizes the proper time between any two events on the path.

The straight-line path between two events in Minkowski spacetime maximizes the interval between the events. We proved that a geodesic path, in the general case that a gravitational field is present, will be an extremum, but if the spacetime separation of the ends of the path is large, there may be two geodesic paths, one of minimum and one of maximum length. The geodesic path of a particle in spacetime is frequently referred to as its *world line*. A world line is a continuous sequence of points in spacetime; it represent the history of a particle or photon.

In a region of spacetime sufficiently small that the Minkowski metric holds (the existence of which locality is guaranteed by the equivalence principle), we see that the geodesic equation reduces to that for uniform straight-line motion,

$$\frac{du^\mu}{d\tau} = 0\,. \tag{114}$$

Therefore, the path of a particle moving under the influence of a general gravitational field will be locally straight. But we know that no global Lorentz frame exists in the presence of gravitating bodies; therefore, geodesic paths will in general be curved. However, in the above sense they will be as straight as possible in curved spacetime.

2.5.3. *Comparison with Newton's gravity*

We confirm the assertion made earlier that the metric tensor $g_{\mu\nu}$ takes the place in General Relativity that the Newtonian potential occupies in Newton's theory. Of course this must be done in a weak field situation for it is only there that Newton's theory applies. For this reason, of the ten independent $g_{\mu\nu}$'s, only one can be involved in the correspondence.

We consider a particle moving slowly in a weak static gravitational field. From the Special Theory of Relativity we have

$$d\tau = (dt^2 - d\mathbf{r}^2)^{1/2} = (1 - d\mathbf{v}^2)^{1/2} dt, \quad \mathbf{v} = \frac{d\mathbf{r}}{dt}, \tag{115}$$

where boldface symbols denote three-vectors. The slowly moving assumption is

$$\frac{d\mathbf{r}}{dt} \ll \frac{dt}{d\tau} \approx 1. \tag{116}$$

So the geodesic equation (88) can be written with the neglect of the velocity terms as

$$\frac{d^2x^\mu}{d\tau^2} + \Gamma^\mu_{00}\left(\frac{dt}{d\tau}\right)^2 = 0. \tag{117}$$

Because the field is static, the time derivatives of $g_{\mu\nu}$ vanish. Consequently,

$$\Gamma^\mu_{00} = \frac{1}{2}g^{\mu\nu}(2g_{\nu 0,0} - g_{00,\nu}) = -\frac{1}{2}g^{\mu\nu}g_{00,\nu}(1 - \delta^\nu_0). \tag{118}$$

Because the field is weak we may take

$$g_{00} = (1 + \delta)\eta_{00}, \tag{119}$$

where $\delta \ll 1$ and similarly for the other $g_{\mu\nu}$. To first order in the small quantities, we have

$$\Gamma^\mu_{00} = -\frac{1}{2}\eta^{\mu\nu}\eta_{00}\frac{d\delta}{dx^\nu}(1 - \delta^\nu_0). \tag{120}$$

Thus the geodesic equations become

$$\frac{d^2\mathbf{r}}{d\tau^2} = -\frac{1}{2}\left(\frac{dt}{d\tau}\right)^2 \nabla\delta, \qquad \frac{d^2t}{d\tau^2} = 0. \tag{121}$$

The second of these tells us that $\tau = at + b$. So we may write the first as

$$\frac{d^2\mathbf{r}}{dt^2} = -\frac{1}{2}\nabla\delta. \tag{122}$$

Newton's equation is

$$\frac{d^2\mathbf{r}}{dt^2} = -\nabla V, \tag{123}$$

where V is the gravitational potential. Comparing, we have

$$g_{00} = 1 + 2V. \tag{124}$$

In particular, if the gravitational field is produced by a body of mass M,

$$V = -\frac{GM}{r} \quad \rightarrow \quad g_{00} = 1 - \frac{2GM}{r}, \tag{125}$$

where G is Newton's constant. Thus we see for weak fields how the metric is related to the Newtonian potential.

2.6. *Covariance*

2.6.1. *Principle of general covariance*

Physical laws in their form ought to be independent of the frame in which they are expressed and of the location in the universe, that is, independent of the gravitational field. The principle of general covariance states that a law of physics holds in a general gravitational field if it holds in the absence of gravity and its form is invariant to any coordinate transformation. Physical laws frequently involve spacetime derivatives of scalars, vectors, or tensors. We have seen that the derivative of a scalar is a vector but that the ordinary derivative of a vector or a tensor is not a tensor (Section 1.2.3). Therefore, we need a type of derivative — a covariant derivative — that reduces to ordinary differentiation in the absence of gravity and which retains its form under any coordinate transformation, that is, in any gravitational field.

2.6.2. *Covariant differentiation*

Take the derivative of the expression of the covariant vector transformation law (66),

$$\frac{dA'_\mu}{dx'^\rho} = \frac{\partial x^\nu}{\partial x'^\mu}\frac{\partial x^\sigma}{\partial x'^\rho}\frac{\partial A_\nu}{\partial x^\sigma} + \frac{\partial^2 x^\nu}{\partial x'^\rho \partial x'^\mu}A_\nu. \tag{126}$$

If only the first term were present we would have the correct transformation law for a covariant tensor. Now multiply the left and right sides of (92) by the left and right sides of (66), respectively, and rearrange to find

$$\Gamma'^\lambda_{\mu\nu}A'_\lambda = \frac{\partial x^\alpha}{\partial x'^\mu}\frac{\partial x^\beta}{\partial x'^\nu}\Gamma^\kappa_{\alpha\beta}A_\kappa + \frac{\partial^2 x^\kappa}{\partial x'^\mu \partial x'^\nu}A_\kappa. \tag{127}$$

Subtracting the above two equations after renaming dummy indices of summation, we get

$$\left(\frac{dA'_\mu}{dx'^\nu} - \Gamma'^\lambda_{\mu\nu}A'_\lambda\right) = \frac{\partial x^\alpha}{\partial x'^\mu}\frac{\partial x^\beta}{\partial x'^\nu}\left(\frac{dA_\alpha}{dx^\beta} - \Gamma^\lambda_{\alpha\beta}A_\lambda\right), \tag{128}$$

which proves the tensor character of the quantity in brackets. This we call the *covariant derivative* of a covariant vector. We denote it by

$$A_{\mu;\nu} \equiv \frac{dA_\mu}{dx^\nu} - \Gamma^\lambda_{\mu\nu}A_\lambda, \tag{129}$$

and the "semicolon subscript" shall denote the covariant derivative, and imply the operations shown on the right. The covariant derivative of a covariant vector is a second-rank covariant tensor which reduces to ordinary differentiation in inertial frames — and therefore locally in any gravitational field.

Through similar manipulations we find the covariant derivative of a contravariant vector,

$$A^{\mu}{}_{;\nu} \equiv \frac{dA^{\mu}}{dx^{\nu}} + \Gamma^{\mu}_{\sigma\nu} A^{\sigma}. \tag{130}$$

This is a second-rank mixed tensor because its transformation law is

$$\left(\frac{dA'^{\mu}}{dx'^{\nu}} + \Gamma'^{\mu}_{\lambda\nu} A'^{\lambda}\right) = \frac{\partial x'^{\mu}}{\partial x^{\alpha}} \frac{\partial x^{\beta}}{\partial x'^{\nu}} \left(\frac{dA^{\alpha}}{dx^{\beta}} + \Gamma^{\alpha}_{\kappa\beta} A^{\kappa}\right). \tag{131}$$

The covariant derivative of a mixed tensor of arbitrary order can be obtained by successive application of the above two rules to each index; there is one ordinary derivative of the tensor and an affine connection for each index with sign as indicated by the above.

In particular, the covariant derivative of the metric tensor is

$$g_{\mu\nu;\lambda} = \frac{\partial g_{\mu\nu}}{\partial x^{\lambda}} - \Gamma^{\rho}_{\mu\lambda} g_{\rho\nu} - \Gamma^{\rho}_{\nu\lambda} g_{\rho\mu}. \tag{132}$$

In a local inertial frame, where the affine connection and the derivative of the metric tensor vanish, we see that the covariant derivative of the metric tensor vanishes in that frame. But because this itself is a tensor, it must vanish in all frames. Similarly, for the covariant derivative of $g^{\mu\nu}$,

$$g_{\mu\nu;\lambda} = 0 = g^{\mu\nu}{}_{;\lambda}. \tag{133}$$

2.6.3. *Geodesic equation from the covariance principle*

As an important example of the application of the covariant derivative, consider the four-velocity of a free particle in a Lorentz frame in the absence of gravity. We denote the four-velocity by $u^{\mu} = dx^{\mu}/d\tau$ and its equation of motion is $du^{\mu}/d\tau = 0$, or equivalently in differential form,

$$du^{\mu} = 0. \tag{134}$$

The covariant derivative (131) was introduced to preserve the vector or tensor character so that a law expressed in such form is preserved in form for all coordinate transformations in accord with the principle of relativity. The equation expressing the law is said to be covariant if its form is preserved. Therefore the law of free motion (134) in a Lorentz frame in the absence of gravity is generalized to frames in arbitrary gravitational fields by requiring that the covariant differential of the four-velocity vanish:

$$0 = u^{\mu}{}_{;\nu} dx^{\nu} = \frac{du^{\nu}}{dx^{\nu}} dx^{\nu} + \Gamma^{\mu}_{\sigma\nu} u^{\sigma} dx^{\nu}$$
$$= du^{\nu} + \Gamma^{\mu}_{\sigma\nu} u^{\sigma} dx^{\nu}. \tag{135}$$

Dividing the above equation by $d\tau$ yields the expected result — the geodesic equation (88) — the equation of motion derived previously for a free particle in an arbitrary gravitational field:

$$\frac{d^2 x^{\lambda}}{d\tau^2} + \Gamma^{\lambda}_{\mu\nu} \frac{dx^{\mu}}{d\tau} \frac{dx^{\nu}}{d\tau} = 0. \tag{136}$$

This is an example of the application of the principle of general covariance and it is seen to rest on the equivalence principle, which assures us that a Lorentz frame can be erected locally.

To restate the principle briefly, *any law that holds in the special theory of relativity and in the absence of gravity can be generalized by replacing the metric* $\eta_{\mu\nu}$ *by* $g_{\mu\nu}$ *and replacing ordinary derivatives by covariant derivatives.*

We obtain an additional result that we need later, namely, the equations of motion for the covariant components of the four-velocity. The law of motion of a free particle in the special theory, expressed in differential form as in (134), implies at once that $du_{\mu} = g_{\mu\nu} du^{\nu} = 0$. The covariant translation of this fact is

$$0 = u_{\mu;\nu} dx^{\nu} = \frac{du_{\mu}}{dx^{\nu}} dx^{\nu} - \Gamma^{\lambda}_{\mu\nu} u_{\lambda} dx^{\nu} \tag{137}$$

or

$$\frac{d^2 x_{\mu}}{d\tau^2} - \Gamma^{\lambda}_{\mu\nu} \frac{dx_{\lambda}}{d\tau} \frac{dx_{\nu}}{d\tau} = 0. \tag{138}$$

This is the equation corresponding to (136) for the covariant acceleration. We carry the analysis a step further. Examine the second term on the left

$$\Gamma^{\lambda}_{\mu\nu} u_{\lambda} u^{\nu} = \frac{1}{2} g^{\lambda\kappa} (g_{\kappa\nu,\mu} + g_{\kappa\mu,\nu} - g_{\mu\nu,\kappa}) u_{\lambda} u^{\nu}$$
$$= \frac{1}{2} (g_{\kappa\nu,\mu} + g_{\kappa\mu,\nu} - g_{\mu\nu,\kappa}) u^{\kappa} u^{\nu}. \tag{139}$$

Because of the symmetry of the product $u^{\kappa} u^{\nu}$, the last two terms in the bracket cancel. We are left with

$$\frac{du_{\mu}}{d\tau} = \frac{1}{2} g_{\kappa\nu,\mu} u^{\kappa} u^{\nu}. \tag{140}$$

This proves that if all the $g_{\alpha\beta}$ are independent of some coordinate component, say x^{μ}, then the covariant velocity u_{μ} is a constant along the particle's trajectory. We will use this result in a much later section during a discussion of the phenomenon of dragging of local inertial frames by a rotating star (according to which a body dropped freely from a great distance falls, not toward the star's center, but is dragged ever more strongly in the sense of the star's rotation).

2.6.4. *Covariant divergent and conserved quantities*

The element of four-volume transforms under coordinate change as

$$dx'^0 dx'^1 dx'^2 dx'^3 = J dx^0 dx^1 dx^2 dx^3, \tag{141}$$

where J is the Jacobian of the transformation,

$$J = \det\left|\frac{\partial x'^\rho}{\partial x^\mu}\right|. \tag{142}$$

For brevity the four-volume element is often written dx^4.

The transformation law for the metric tensor is

$$g_{\mu\nu} = \frac{\partial x'^\alpha}{\partial x^\mu} g'_{\alpha\beta} \frac{\partial x'^\beta}{\partial x^\nu}. \tag{143}$$

We may regard this as an element in the product of three matrices. The corresponding determinant equation is

$$g = J g' J = J^2 g' \tag{144}$$

where $g = \det|g_{\mu\nu}|$ and is a negative quantity as can be verified by looking at the Minkowski metric. Thus, we may write

$$\sqrt{-g} = J\sqrt{-g'}. \tag{145}$$

If $S = S'$ is a scalar field, then

$$\int_{V_4} S\sqrt{-g}\, d^4x = \int_{V_4} S\sqrt{-g'}\, J d^4x = \int_{V_4} S'\sqrt{-g'}\, d^4x'. \tag{146}$$

is an invariant where V_4 is a prescribed four-volume. The quantity

$$\mathcal{S} \equiv S\sqrt{-g} \tag{147}$$

is called a *scalar density*, and its integral over a region of spacetime is invariant to a coordinate transformation. Also, and very important to us, $\sqrt{-g}\, d^4x$ is the invariant volume element.

The covariant derivative of a vector A^μ is given by (130). If we contract indices, according to (79) we have a scalar. This is the covariant divergence of A^μ:

$$A^\mu{}_{;\mu} \equiv A^\mu{}_{,\mu} + \Gamma^\mu_{\nu\mu} A^\nu . \tag{148}$$

From (99) we find

$$\Gamma^\nu_{\mu\nu} = \frac{1}{2} g^{\nu\kappa}(g_{\kappa\nu,\mu} + g_{\kappa\mu,\nu} - g_{\mu\nu,\kappa}). \tag{149}$$

Interchange the names of the dummy summation indices in the second term on the right to see that it cancels the third. Thus

$$\Gamma^\nu_{\mu\nu} = \frac{1}{2} g^{\nu\kappa} g_{\kappa\nu,\mu}. \tag{150}$$

We need still another result. Denote the cofactor of the element $g_{\alpha\beta}$ by $C^{\alpha\beta}$. The determinant $g = \det|g_{\alpha\beta}|$ can be expanded in any of the set of minors (i.e., any $\alpha = 0, 1, 2,$ or 3) in the equation

$$g = g_{(\alpha)\beta} C^{(\alpha)\beta} \quad (\text{no sum on}\, \alpha). \tag{151}$$

Because the cofactor contains no elements $g_{(\alpha)\beta}$, we find

$$\frac{\partial g}{\partial g_{\alpha\nu}} = \frac{\partial (g_{(\alpha)\beta} C^{(\alpha)\beta})}{\partial g_{\alpha\nu}} = \frac{\partial g_{\alpha\mu}}{\partial g_{\alpha\nu}} C^{\alpha\mu} = \delta^\nu_\mu C^{\alpha\mu} = C^{\alpha\nu}. \tag{152}$$

Therefore,

$$g_{,\alpha} = \frac{\partial g}{\partial g_{\mu\nu}} g_{\mu\nu,\alpha} = C^{\mu\nu} g_{\mu\nu,\alpha}. \tag{153}$$

We need the expression

$$C^{\mu\nu} = g g^{\mu\nu}, \tag{154}$$

which can be proved by multiplying by $g_{\mu\nu}$ and summing only over ν,

$$g_{(\mu)\nu} C^{(\mu)\nu} = g_{(\mu)\nu} g^{(\mu)\nu} g = g. \tag{155}$$

This is the determinant expansion in minors (151). Thus, we have derived the result

$$g_{,\alpha} = g g^{\mu\nu} g_{\mu\nu,\alpha}. \tag{156}$$

Hence,

$$\Gamma^\nu_{\mu\nu} = \frac{1}{2} g^{-1} g_{,\mu} = \frac{1}{2}\Big(ln(-g)\Big)_{,\mu} = \frac{1}{\sqrt{-g}}\Big(\sqrt{-g}\Big)_{,\mu}. \tag{157}$$

We can use this to rewrite the covariant divergence of A^μ as

$$A^\mu_{;\mu} = \frac{1}{\sqrt{-g}}\Big(\sqrt{-g} A^\mu\Big)_{,\mu}. \tag{158}$$

With (157) in (148), we obtain the important result for the covariant divergence,

$$\sqrt{-g}A^{\mu}{}_{;\mu} = \left(\sqrt{-g}A^{\mu}\right)_{,\mu}. \tag{159}$$

The left side is a scalar density. From the invariance of the integral of a scalar density over a prescribed four-volume, we have the invariant

$$\int_{V_4} \sqrt{-g}A^{\mu}{}_{;\mu}d^4x = \int_{V_4} \left(\sqrt{-g}A^{\mu}\right)_{,\mu} d^4x. \tag{160}$$

The right side can be converted to a surface integral over a three-volume at a definite time x^0 by Gauss' theorem.

If the covariant divergence vanishes, we get a conservation law as follows:

$$A^{\mu}{}_{;\mu} = 0 \longrightarrow \left(\sqrt{-g}A^{\mu}\right)_{,\mu} = 0. \tag{161}$$

As a result, we obtain

$$\left(\sqrt{-g}A^{0}\right)_{,0} = -\left(\sqrt{-g}A^{m}\right)_{,m} \quad \text{(summed over } m = 1-3\text{)}. \tag{162}$$

Integrate the above expression over a three-volume at definite time x^0 to find

$$\begin{aligned}\frac{\partial}{\partial x^0}\int_V \sqrt{-g}A^0 d^3x &= -\int_V \left(\sqrt{-g}A^m\right)_{,m} d^3x \\ &= -\int_S \sqrt{-g}\mathbf{A}\cdot d\mathbf{S}.\end{aligned} \tag{163}$$

If there is no three-current $\sqrt{-g}\mathbf{A}$ crossing the surface, then the quantity of density $\sqrt{-g}A^0$ contained within V is constant,

$$\int_V \sqrt{-g}A^0 d^3x = \text{constant}. \tag{164}$$

This quantity is frequently referred to as the total charge of whatever A^{μ} represents.

We can apply precisely the same reasoning to the covariant divergence of an antisymmetric tensor:

$$\text{If } A^{\mu\nu} = -A^{\nu\mu}, \quad \text{then } \sqrt{-g}A^{\mu\nu}{}_{;\nu} = \left(\sqrt{-g}A^{\mu\nu}\right)_{,\nu}, \tag{165}$$

where the quantity on the left is a *vector density* according to the previous section. Similarly we can derive conservation laws for the three-volume integral of the four densities $\sqrt{-g}A^{\mu 0}$ if the covariant divergence vanishes and there is no three-flux through the surface of the volume. However, if the tensor is not antisymmetric, the above theorem does not generally apply in curved spacetime to a tensor of more than one index.

2.7. *Riemann curvature tensor*

The order of ordinary differentiation in flat spacetime does not matter. The order of covariant differentiation does matter in curved spacetime. From an investigation of this fact we arrive at a measure of curvature.

2.7.1. *Second covariant derivative of scalars and vectors*

If we take the covariant derivative of a scalar twice and then invert the order, the answer is easily verified to be the same:

$$S_{;\mu;\nu} = S_{;\mu,\nu} - \Gamma^{\alpha}_{\mu\nu} S_{;\alpha} = S_{,\mu,\nu} - \Gamma^{\alpha}_{\mu\nu} S_{;\alpha}, \tag{166}$$

where we use the fact in the second equality that the covariant derivative of a scalar is the ordinary derivative $S_{;\mu} = S_{,\mu}$. The above result is symmetrical in μ, ν.

However for vectors and tensors, a changed order of differentiation in general produces a different result. The operations involved, all defined above, are many but straightforward. The result for the vector A_σ is

$$A_{\sigma;\mu;\nu} - A_{\sigma;\nu;\mu} = A_\rho R^{\rho}_{\sigma\mu\nu}, \tag{167}$$

where

$$R^{\rho}_{\sigma\mu\nu} \equiv \Gamma^{\rho}_{\sigma\nu,\mu} - \Gamma^{\rho}_{\sigma\mu,\nu} + \Gamma^{\alpha}_{\sigma\nu}\Gamma^{\rho}_{\alpha\mu} - \Gamma^{\alpha}_{\sigma\mu}\Gamma^{\rho}_{\alpha\nu} \tag{168}$$

is the *Riemann–Christoffel curvature tensor*. We know that it is a tensor because the left side of (167) is a tensor and A_ν is any vector. Riemann is the only tensor that can be constructed from the metric tensor and its first and second derivatives (cf. Ref. [Weinberg (1972)], p. 133).

2.7.2. *Symmetries of the Riemann tensor*

The Riemann tensor has a number of symmetry properties that can be easily derived from the above expression:

$$\begin{aligned} R^{\mu}_{\nu\rho\sigma} &= -R^{\mu}_{\nu\sigma\rho}, \\ R^{\alpha}_{\nu\rho\sigma} + R^{\alpha}_{\sigma\nu\rho} + R^{\alpha}_{\rho\sigma\nu} &= 0. \end{aligned} \tag{169}$$

Lowering the index on the Riemann tensor, we get

$$R_{\rho\sigma\,\mu\nu} = g_{\rho\alpha} R^{\alpha}_{\sigma\,\mu\nu}. \tag{170}$$

The additional symmetries follow:

$$\begin{aligned} R_{\mu\nu\,\rho\sigma} &= -R_{\nu\mu\,\rho\sigma} = -R_{\mu\nu\,\sigma\rho} \\ R_{\mu\nu\,\rho\sigma} &= R_{\rho\sigma\,\mu\nu} = R_{\sigma\rho\,\nu\mu}. \end{aligned} \tag{171}$$

As a consequence of the symmetries only 20 of the $4^4 = 256$ components of the Riemann tensor are independent. In two dimensions there are 15 such symmetry relationships. Consequently, there are $2^4 - 15 = 1$ independent components of the Riemann tensor, namely, the Gaussian curvature. (See Ref. [Berry (1976)] p. 60.)

We shall encounter two additional objects that are obtained from the Riemann tensor, the Ricci tensor,

$$R_{\mu\nu} = R^{\rho}_{\mu\nu\rho}, \tag{172}$$

and the *scalar curvature*

$$R = g^{\mu\nu} R_{\mu\nu}. \tag{173}$$

Multiply the left and right side of (171) by $g^{\mu\sigma}$ and then rename indices to find

$$R_{\mu\nu} = R_{\nu\mu}. \tag{174}$$

Because of this symmetry, when we raise an index on the Ricci tensor, it is unnecessary to preserve the location,

$$R^{\mu}{}_{\nu} = R_{\nu}{}^{\mu} = R^{\mu}_{\nu}. \tag{175}$$

From the definition of the Ricci tensor in terms of the Riemann tensor, we have the following explicit expression:

$$R_{\mu\nu} = \Gamma^{\alpha}_{\mu\alpha,\nu} - \Gamma^{\alpha}_{\mu\nu,\alpha} - \Gamma^{\alpha}_{\mu\nu}\Gamma^{\beta}_{\alpha\beta} + \Gamma^{\alpha}_{\mu\beta}\Gamma^{\beta}_{\nu\alpha}. \tag{176}$$

The first term might appear to contradict the assertion that $R_{\mu\nu}$ is symmetric in μ, ν. However the result (157) proves that the Ricci tensor is symmetric.

2.7.3. *Test for flatness*

If spacetime is flat, then we may choose a rectilinear coordinate system in which case the metric tensor is a constant throughout spacetime. Then according to (99) the nontensor $\Gamma^{\lambda}_{\mu\nu}$ vanishes in this frame in all spacetime. So also do the derivatives of $\Gamma^{\lambda}_{\mu\nu}$. Therefore the Riemann tensor (168) vanishes everywhere at all times in flat spacetime. Because this is a statement about a tensor, it is true in any coordinate system, rectilinear or not. The converse is true but more difficult to prove: If the Riemann tensor vanishes, spacetime is flat. We prove this later in Section 1.2.9.3.

2.7.4. *Second covariant derivative of tensors*

An arbitrary second-rank tensor can be expressed as the sum of products $A_\mu B_\nu$. It is simpler to start by examining the second covariant derivative of such a product:

$$\begin{aligned}(A_\mu B_\nu)_{;\rho;\sigma} &= (A_{\mu;\rho}B_\nu + A_\mu B_{\nu;\rho})_{;\sigma} \\ &= A_{\mu;\rho;\sigma}B_\nu + A_\mu B_{\nu;\rho;\sigma} + A_{\mu;\rho}B_{\nu;\sigma} + A_{\mu;\sigma}B_{\nu;\rho}.\end{aligned} \tag{177}$$

Interchange ρ, σ, and subtract to find

$$\begin{aligned}(A_\mu B_\nu)_{;\rho;\sigma} &= (A_\mu B_\nu)_{;\sigma;\rho} \\ &= A_\mu(B_{\nu;\rho;\sigma} - B_{\nu;\sigma;\rho}) + (A_{\mu;\rho;\sigma} - A_{\nu;\sigma;\rho})B_\nu \\ &= A_\mu B_\alpha R^\alpha_{\nu\rho\sigma} + A_\alpha R^\alpha_{\mu\rho\sigma} B_\nu.\end{aligned} \tag{178}$$

We can form an arbitrary linear combination of such products of first-rank tensors to obtain the result for a general tensor,

$$T_{\mu\nu;\rho;\sigma} - T_{\mu\nu;\sigma;\rho} = T_{\mu\alpha}R^\alpha_{\nu\rho\sigma} + T_{\alpha\nu}R^\alpha_{\mu\rho\sigma}. \tag{179}$$

2.7.5. *Bianchi identities*

The Bianchi identities are extremely important for the further development of the theory of gravity, allowing us to prove that the Einstein tensor, which we come to next, has vanishing divergence.

Apply the above result to the particular case that the second-rank tensor is the covariant derivative of a vector $T_{\mu\nu} = A_{\mu;\nu}$

$$A_{\mu;\nu;\rho;\sigma} - A_{\mu;\nu;\sigma;\rho} = A_{\mu;\alpha}R^\alpha_{\nu\rho\sigma} + A_{\alpha;\nu}R^\alpha_{\mu\rho\sigma}. \tag{180}$$

Now write down the additional two equations obtained from this by cyclic permutation of the indices $(\nu\rho\sigma)$, and add the three equations. First study the left side of the sum. Use (167) to get

$$\begin{aligned}LHS &= (A_{\mu;\nu;\rho} - A_{\mu;\rho;\nu})_{;\sigma} + (A_{\mu;\sigma;\nu} - A_{\mu;\nu;\sigma})_{;\rho} + (A_{\mu;\rho;\sigma} - A_{\mu;\sigma;\rho})_{;\nu} \\ &= (A_\alpha R^\alpha_{\mu\nu\rho})_{;\sigma} + (A_\alpha R^\alpha_{\mu\sigma\nu})_{;\rho} + (A_\alpha R^\alpha_{\mu\rho\sigma})_{;\nu}.\end{aligned} \tag{181}$$

Using (169) in the sum of the right-hand sides of the cyclic permutation, we are left with

$$RHS = A_{\alpha;\nu}R^\alpha_{\mu\rho\sigma} + A_{\alpha;\sigma}R^\alpha_{\mu\nu\rho} + A_{\alpha;\rho}R^\alpha_{\mu\sigma\nu}. \tag{182}$$

Equating left and right sides and cancelling common terms, we find

$$A_\alpha (R^\alpha_{\mu\nu\rho;\sigma} + R^\alpha_{\mu\sigma\nu;\rho} + R^\alpha_{\mu\rho\sigma;\nu}) = 0. \tag{183}$$

Because A_α is any vector,

$$R^\alpha_{\mu\nu\rho;\sigma} + R^\alpha_{\mu\sigma\nu;\rho} + R^\alpha_{\mu\rho\sigma;\nu} = 0. \tag{184}$$

In addition to the symmetry relationships derived earlier, the Riemann tensor satisfies the differential equations above known as the *Bianchi identities.*

2.7.6. *Einstein tensor*

Let us multiply the differential equations for the Bianchi identities (184) by $g^{\mu\nu}$, contract σ with α, and use the fact, already established, that the covariant derivatives of the metric tensor vanish:

$$\begin{aligned} 0 &= g^{\mu\nu}(R^\alpha_{\mu\nu\rho;\sigma} + R^\alpha_{\mu\sigma\nu;\rho} + R^\alpha_{\mu\rho\sigma;\nu}) \\ &= (g^{\mu\nu} R^\alpha_{\mu\nu\rho})_{;\alpha} + (g^{\mu\nu} R^\alpha_{\mu\alpha\nu})_{;\rho} + (g^{\mu\nu} R^\alpha_{\mu\rho\alpha})_{;\nu}. \end{aligned} \tag{185}$$

Examine each term in brackets using the Riemann tensor symmetries. The first term is

$$g^{\mu\nu} R^\alpha_{\mu\nu\rho} = g^{\mu\nu} g^{\alpha\beta} R_{\beta\mu\nu\rho} = g^{\mu\nu} g^{\alpha\beta} R_{\mu\beta\rho\nu} = g^{\alpha\beta} R^\nu_{\beta\rho\nu} = g^{\alpha\beta} R_{\beta\rho} = R^\alpha_\rho. \tag{186}$$

The second term is

$$g^{\mu\nu} R^\alpha_{\mu\alpha\nu} = -g^{\mu\nu} R^\alpha_{\mu\nu\alpha} = -g^{\mu\nu} R_{\mu\nu} = -R. \tag{187}$$

The third term is

$$g^{\mu\nu} R^\alpha_{\mu\rho\alpha} = g^{\mu\nu} R_{\mu\rho} = R^\nu_\rho. \tag{188}$$

Now put these results back into their brackets with the covariant derivatives as indicated in (185) to obtain

$$0 = R^\alpha_{\rho;\alpha} - R_{;\rho} + R^\nu_{\rho;\nu} = 2R^\alpha_{\rho;\alpha} - R_{;\rho}. \tag{189}$$

Multiply by $g^{\mu\rho}$, and note that

$$\begin{aligned} g^{\mu\rho} R^\alpha_{\rho;\alpha} &= \left(g^{\mu\rho} R^\alpha_\rho\right)_{;\alpha} = R^{\mu\alpha}{}_{;\alpha} = R^{\mu\nu}{}_{;\nu} \\ g^{\mu\rho} R_{;\rho} &= g^{\mu\nu} R_{;\nu} \end{aligned} \tag{190}$$

to arrive immediately at the vanishing divergence

$$\left(R^{\mu\nu} - \frac{1}{2} g^{\mu\nu} R\right)_{;\nu} = 0. \tag{191}$$

The object in the brackets is called the *Einstein curvature tensor*,

$$G^{\mu\nu} \equiv R^{\mu\nu} - \frac{1}{2} g^{\mu\nu} R. \tag{192}$$

The Einstein tensor is constructed from the Riemann curvature tensor and has an identically vanishing covariant divergence. It is symmetric and of second rank. Einstein was motivated to seek a tensor that contained no differentials of the $g^{\mu\nu}$ higher than the second — a tensor which was a linear homogeneous combination of terms linear in the second derivative or quadratic in the first (in analogy with Poisson's equation for the gravitational potential in Newton's theory:

$$\nabla^2 V = 4\pi\rho, \tag{193}$$

where ρ is the mass density generating the field). For the expression of energy and momentum conservation, it is important that the divergence vanish. The energy-momentum tensor of matter accomplishes this and is of second rank.

2.8. *Einstein's field equations*

"The geometry of spacetime is not given; it is determined by matter and its motion"[10]

W. Pauli, 1919

We know that other bodies will experience gravity in the vicinity of massive bodies. So mass is a source of gravity, and from the Special Theory of Relativity we must say in general that mass and energy are sources. We have just seen that a construction from the Riemann curvature tensor, namely, Einstein's tensor, has vanishing covariant divergence. We have three possibilities,

$$G^{\mu\nu} = 0, \tag{194}$$

or

$$G^{\mu\nu} = kT^{\mu\nu}, \tag{195}$$

where $T^{\mu\nu}$ is a symmetric divergenceless tensor constructed from the mass-energy properties of the material medium, or

$$G^{\mu\nu} = kT^{\mu\nu} + \Lambda g^{\mu\nu}. \tag{196}$$

[10]Very importantly, the converse is also true.

The constant Λ is the so-called *cosmological constant.* It was not present in the original theory and was added to obtain a static cosmology before it was known that the universe is expanding. Einstein regarded its numerical value as a matter to be settled by experiment — "The curvature constant [Λ] is, however, essentially determinable, and an increase in the precision of data derived from observations will enable us in the future to fix its sign and determine its value" [Einstein (1932)].

It is apparent that the cosmological constant corresponds to a constant energy density $\Lambda/(8\pi)$ and a constant pressure of the same numerical value but of opposite sign. The cosmological constant is sometimes referred to as the vacuum energy density. In any case it is small; its value has been recently measures by Perlmutter. Its effect is indeed cosmological; stellar structure is unaffected by it. We need not consider the cosmological term further.

The first set of differential equations (194) are those that must be satisfied by the metric in empty space outside material bodies and energy concentrations. An example is the gravitational fields outside a star.

The second set of differential equations (195) determine the gravitational fields $g^{\mu\nu}$ inside a spacetime region of mass-energy and in addition determine how the mass-energy is arranged by gravity. With appropriate $T^{\mu\nu}$ it would provide the equations of stellar structure. We have yet to fix the constant k. This can be done by looking to the weak field limit where the General Theory of Relativity should agree with Newton's well-tested, weak-field theory.

There are several remarkable notes we can make at this point. Einstein's field equations tell spacetime how to curve and mass-energy how to configure itself and how to move. Spacetime acts upon matter and in turn is acted upon by matter. This was Einstein's intuition and motivation in seeking a theory that placed spacetime and matter as co-determiners in nature. He was displeased with the Special Theory of Relativity as anything but a local theory, for it gave spacetime an absolute status.

Second, the Einstein field equations are nonlinear in the fields $g^{\mu\nu}$. (This can be verified by tracing back through the objects from which the Einstein tensor is constructed.) Nonlinearity means that the gravitational field interacts with itself. This is because the field carries energy, and mass-energy in any form is a source of gravity. The nonlinearity of the Einstein equations accounts for some of the extraordinary phenomena encountered in strong gravity, including black holes [Oppenheimer & Snyder (1939)] and the reversal of the centrifugal force in their vicinity [Abramowicz & Prasanna (1990)].

We have seen in (191) that the Einstein tensor has identically vanishing covariant divergence. Hence (195) requires of the matter tensor that

$$T^{\mu\nu}{}_{;\nu} = 0. \tag{197}$$

The corresponding equation in flat space is

$$T^{\mu\nu}{}_{,\nu} = 0. \tag{198}$$

Vanishing of the ordinary divergence of the energy–momentum tensor in the Special Theory of Relativity corresponds to the conservation of energy and momentum. However, (197) does not assure us of the constancy of any quantity in time. In fact (195) ensures that matter and the gravitational fields exchange energy, or in other words do work on each other, for it is the divergence of $G^{\mu\nu} - kT^{\mu\nu}$ that vanishes. So neither matter nor the gravitational field can by itself conserve energy in any sense. No contradiction exists with laboratory experiments performed on Earth. Over the dimensions of a typical laboratory, spacetime is essentially flat, and nothing that could be done in a laboratory could possibly disturb this flatness in any perceptible way.

This brings us back to the comparison of the weak-field limit between Newton's theory and Einstein's. The inverse-square law of the force between massive objects is not required by the inner structure of Newton's theory. He could have postulated an inverse α law, that force $F \sim Mm/r^{\alpha}$, and then attempted to fit α to the astronomical data of the solar system. Depending on what weight was given to the precession of planets, one would have found a value of α close to two.

Einstein's theory does not possess the flexibility of Newton's in this regard. We saw in (125) that Einstein predicts precisely the inverse square law. In this sense, he could claim as his own all the successes of the Newtonian theory in explaining the motion of planets in the solar system. They were computed with the inverse-square law, there being no flexibility in the choice of the power in his theory.

Concerning the precession of planets, an isolated planet in orbit about the Sun under an inverse square law is an ellipse whose orientation is fixed in space. However the total precession of the orbit of Mercury is observed to be about 5600 seconds/century. Most of this is caused by the fact that an Earth-bound observer is not in an inertial frame far from the Sun. For example, suppose that Mercury did not orbit about the Sun, but instead held a fixed position. Nevertheless, from the Earth it would be appear to move, sometimes to the left of the Sun, sometimes to the right, and alternately passing in front of and in back of the Sun. Taking account of this correction to the apparent motion of Mercury due to the Earth's own motion, the precession of Mercury is about 574 seconds/century. This value is about 43 seconds/century larger than the precession computed by Newtonian physics as due to the perturbation of the orbit by other planets, a small but disturbing discrepancy. An early triumph of Einstein was that he calculated, within the observational errors, the precise value of the excess precession. In Newton's theory only mass contributes to gravity, whereas in Einstein's theory the kinetic energy of the motion of the planets contributes as well.

2.9. *Relativistic stars*

Einstein's field equations are completely general and simple in appearance. However, they are exceedingly complicated because of their nonlinear character and because spacetime and matter act upon each other. As already remarked, there is no prior geometry of spacetime. There are a few cases in which solutions can be found in closed form. One of the most important closed-form solutions is the Schwarzschild metric outside a static spherical star. Another is the Kerr metric outside a rotating black hole. Einstein's equations can also be solved numerically as the coupled differential equations for the interior structure of a spherical static star, which are called the Oppenheimer–Volkoff equations for stellar structure.

In this section we take up the important problem of deriving the equations that govern spacetime and the arrangement of matter in the case of relativistic spherical static stars. They are the basic equations that underlie the development of neutron star models. They also demonstrate the mathematical existence of Schwarzschild black holes. They can also be used to develop white dwarf models, though Newtonian gravity is a good approximation for these stars.

2.9.1. *Metric in static isotropic spacetime*

We seek solutions to Einstein's field equations in static isotropic regions of spacetime such as would be encountered in the interior and exterior regions of static stars. Under these conditions the $g_{\mu\nu}$ are independent of time

($x^0 \equiv t$) and $g^{0m} = 0$. We choose spatial coordinates $x^1 = r$, $x^2 = \theta$, and $x^3 = \phi$. The most general form of the line element is then

$$d\tau^2 = U(r)\,dt^2 - V(r)\,dr^2 - W(r)\,r^2\Big(d\theta^2 + \sin^2\theta\,d\phi^2\Big). \tag{199}$$

We may replace r by any function of r without disturbing the spherical symmetry. We do so in such a way that $W(r) \equiv 1$. Then we may write

$$d\tau^2 = e^{2\nu(r)}dt^2 - e^{2\lambda(r)}dr^2 - r^2 d\theta^2 - r^2\sin^2\theta\,d\phi^2, \tag{200}$$

where λ, ν are functions only of r. Comparing with

$$d\tau^2 = g_{\mu\nu}dx^\mu dx^\nu, \tag{201}$$

we read off[11]

$$g_{00} = e^{2\nu(r)},\quad g_{11} = e^{2\lambda(r)},\quad g_{22} = -r^2,\quad g_{33} = -r^2\sin^2\theta,\quad g_{\mu\nu} = g^{\mu\nu} = 0 \tag{202}$$

with ($\mu \neq \nu$). Hence, from $g_{\mu\nu}g^{\nu\rho} = \delta^\rho_\mu$, we have in this special case

$$g_{\mu\mu} = 1/g^{\mu\mu} \quad \text{(not summed)}. \tag{203}$$

According to its definition as a contraction of the Riemann tensor, the Ricci tensor can be written

$$R_{\mu\nu} = \Gamma^\alpha_{\mu\alpha,\nu} - \Gamma^\alpha_{\mu\nu,\alpha} - \Gamma^\alpha_{\mu\nu}\Gamma^\beta_{\alpha\beta} + \Gamma^\alpha_{\mu\beta}\Gamma^\beta_{\nu\alpha}. \tag{204}$$

We can derive the nonvanishing affine connections (99), which are symmetric in their lower indices, from the metric tensor whose general form for static isotropic regions was derived above:

$$\begin{aligned}
&\Gamma^1_{00} = \nu' e^{2(\nu-\lambda)}, && \Gamma^0_{10} = \nu',\\
&\Gamma^1_{11} = \lambda', && \Gamma^2_{12} = \Gamma^3_{13} = 1/r,\\
&\Gamma^1_{22} = -re^{-2\lambda}, && \Gamma^3_{23} = \cot\,\theta,\\
&\Gamma^1_{33} = -r\sin^2\theta\,e^{-2\lambda}, && \Gamma^2_{33} = -\sin\,\theta\,\cos\,\theta.
\end{aligned} \tag{205}$$

[11]Note that in (200) some authors use the opposite signs for time and space components, and some use the functions ν, λ but without the factor 2, or use different notation altogether for the metric. Great care has to be exercised in using results from different sources.

The primes denote differentiation with respect to r. Hence, for static isotropic spacetime

$$
\begin{aligned}
R_{00} &= \left(-\nu'' + \lambda'\nu' - \nu'^2 - \frac{2\nu'}{r}\right) e^{2(\nu-\lambda)}, \\
R_{11} &= \nu'' - \lambda'\nu' + \nu'^2 + \frac{2\lambda'}{r}, \\
R_{22} &= (1 + r\nu' - r\lambda')\, e^{-2\lambda} - 1, \\
R_{33} &= R_{22} \sin^2\theta.
\end{aligned} \tag{206}
$$

2.9.2. *The Schwarzschild solution*

In the empty space outside a static star Einstein's equation is $G_{\mu\nu} = 0$, or equivalently

$$R_{\mu\nu} = \frac{1}{2} g_{\mu\nu} R. \tag{207}$$

Multiply by $g^{\alpha\mu}$, and the sum on the dummy index to find

$$R^{\alpha}_{\nu} = \frac{1}{2} \delta^{\alpha}_{\nu} R. \tag{208}$$

Contract by setting $\alpha = \nu$, and sum to get

$$R = 2R \quad \longrightarrow \quad R = 0. \tag{209}$$

Hence, the vanishing of Einstein's tensor implies

$$G_{\mu\nu} = 0 \quad \longrightarrow \quad R = 0, \quad R_{\mu\nu} = 0. \tag{210}$$

In empty space, Einstein's equation is equivalent to the vanishing of the Ricci tensor or, equivalently, the scalar curvature. Its form for static isotropic spacetime was worked out in the previous section.

From the vanishing of R_{00}, R_{11} we find that

$$\lambda' + \nu' = 0. \tag{211}$$

(Do not confuse ν and λ when used to denote indices and when used to denote the metric functions as in the above equation.) For large r, space must be unaffected by the star and therefore flat so that λ and ν tend to zero; therefore

$$\lambda + \nu = 0. \tag{212}$$

Using these results in $R_{22} = 0$, we find that

$$(1 + 2r\nu')e^{2\nu} = 1. \tag{213}$$

This condition integrates to

$$g_{00} \equiv e^{2\nu} = 1 - \frac{2GM}{r} \quad (r > R), \tag{214}$$

where M is the constant of integration, and we introduced Newton's constant. Having studied the Newtonian approximation, one identifies M with the mass of the star. From the foregoing results,

$$g_{11} = -e^{2\lambda} = -e^{-2\nu} = -\left(1 - \frac{2GM}{r}\right)^{-1} \quad (r > R). \tag{215}$$

This completes the derivation of the *Schwarzschild solution* of 1916 of Einstein's equations outside a spherical static star. It was the first exact solution found for Einstein's equations. The proper time is

$$d\tau^2 = \left(1 - \frac{2GM}{r}\right) dt^2 - \left(1 - \frac{2GM}{r}\right)^{-1} dr^2 - r^2 d\theta^2 - r^2 \sin^2\theta\, d\phi^2 \quad (r > R) \tag{216}$$

where R, in this context, denotes the radius of the star.

Let us summarize the Schwarzschild solution found above:

$$\begin{aligned} g_{00}(r) &= e^{2\nu(r)} = \left(1 - \frac{2GM}{r}\right), \quad r > R, \\ g_{11}(r) &= -e^{2\lambda(r)} = -\left(1 - \frac{2GM}{r}\right)^{-1}, \quad r > R, \\ g_{22}(r) &= -r^2, \\ g_{33}(r) &= -r^2 \sin^2\theta. \end{aligned} \tag{217}$$

Notice that the Schwarzschild metric is singular at the radius $r = r_S \equiv 2GM$. This does not mean that spacetime itself is singular at that radius, but only that this particular metric is. Other nonsingular metrics have been found, in particular, the Kruskal-Szerkeres metric [Kruskal (1960); Szerkeres (1960)]. However, further analysis shows that if r_S lies outside the star where the Schwarzschild solution holds, then it is a black hole — no particle or even light can leave the region $r < r_S$. This radius r_S is called the *Schwarzschild radius* or *singularity* or *horizon*. But because the above metric holds only outside the star, r_S has no special significance if it is smaller than the radius of the star. For then a different metric holds inside the star which does not possess a singularity. We come to this solution shortly.

2.9.3. *Riemann tensor outside a Schwarzschild star*

If spacetime is flat, then the Riemann curvature tensor vanishes (Section 1.2.7.3). We are now prepared to address the converse (albeit not rigorously): if spacetime is curved, some components of the Riemann tensor are finite (which components, of course, will depend upon how convoluted spacetime is).

The metric tensor and, indeed, the affine connection for the empty space outside a massive body were computed in the preceding section. We have seen in Section 1.2.4.3 that massive bodies curve spacetime. So we know that the Schwarzschild metric tensor refers to curved spacetime. Referring to the definition of the Riemann tensor (168) and the specific form that the affine connection takes for a static spherical star (205), we can compute

$$R^{1}_{010} = \left(\nu'' + 2\nu'^{2} - \nu'\lambda'\right) e^{2(\nu-\lambda)}. \tag{218}$$

Thus we exhibit at least one nonvanishing component of the Riemann tensor in the curved spacetime outside a Schwarzschild star. This suggests that the Riemann tensor is not identically zero in curved spacetime. An actual proof that if the Riemann tensor is finite then spacetime is curved requires the formulation of parallel transport, which we do not take up here. We declare, without rigorous proof, that the Riemann tensor vanishes if and only if spacetime is flat. Notice that, far from an isolated star where spacetime approaches flatness, the Riemann tensor approaches zero as it should.

2.9.4. *Energy–momentum tensor of matter*

From the success of Newtonian physics in describing celestial mechanics and other weak gravitational field phenomena, we know that mass is a source of gravity. From the experimental verifications of the Special Theory of Relativity, we know that all forms of energy are equivalent and must contribute equally as sources of gravity. Normally, of course, it is mass that dominates, and the average mass density in the solar system and in the universe is very small; that is why Newtonian physics is so accurate under the typical conditions mentioned above.

An essential aspect of Einstein's curvature tensor is that it automatically has vanishing covariant divergence (191). It is also a symmetric second-rank tensor. Accordingly, mass-energy — the source of the gravitational field — must be incorporated into a divergenceless, symmetric, second-rank tensor in flat space. As a tensor, it can be transcribed immediately to its form in an arbitrary spacetime frame by the general covariance principle. Such a tensor is the energy–momentum tensor.

In other parts of this book we shall be interested in specific theories of dense matter from which we will be able to explicitly construct the energy-momentum tensor of the theory. Here we are interested in the general form such a tensor takes. Frequently, matter may be regarded as a perfect fluid. The fluid velocity is assumed to vary continuously from point to point. The perfect fluid energy–momentum tensor in the Special Theory of Relativity can be expressed in terms of the local values of the pressure p and energy density ε as in (51). The General Relativistic energy–momentum tensor can be written immediately using the Principle of General Covariance:

$$T^{\mu\nu} = -pg^{\mu\nu} + (p+\varepsilon)u^{\mu}u^{\nu}, \qquad g_{\mu\nu}u^{\mu}u^{\nu} = 1. \tag{219}$$

In the above equations, u^{μ} is the local fluid four-velocity

$$u^{\mu} = \frac{dx^{\mu}}{d\tau} \tag{220}$$

and satisfies (220) because of (55).

The pressure and total energy density (including mass) are related by the equation of state of matter, frequently written in either form

$$p = p(\varepsilon) \quad \text{or} \quad \varepsilon = \varepsilon(p) \tag{221}$$

where p and ϵ are the pressure and energy density (including mass) in the local rest-frame of the fluid. In the next section we shall see how the equations for stellar structure involve these quantities and this relationship.

2.9.5. *The Oppenheimer–Volkoff equations*

We are now prepared to derive the differential equations for the structure of a static, spherically symmetric, relativistic star. For the region outside a star, we found that the vanishing of the Einstein tensor was equivalent to the vanishing of the Ricci tensor or the scalar curvature. This is not the case for the interior of the star. We need both the Ricci tensor and scalar curvature to construct the Einstein tensor. The general form of the metric for a static isotropic spacetime was obtained in (202). From Section 1.2.9.1 we find the scalar curvature,

$$\begin{aligned} R = g^{\mu\nu}R_{\mu\nu} &= e^{-2\nu}R_{00} - e^{2\lambda}R_{11} - \frac{2}{r^2}R_{22} \\ &= ge^{2\lambda}\left\{-2\nu'' + 2\lambda'\nu' - 2\nu'^2 - \frac{2}{r^2} + 4\frac{\lambda'}{r} - 4\frac{\nu'}{r}\right\} + \frac{2}{r^2}. \end{aligned} \tag{222}$$

It is more convenient to work with mixed tensors. For example,

$$G_0{}^0 = R_0{}^0 - \frac{1}{2}R \tag{223}$$

is obtained with the results of Section 1.2.9.1 for a static isotropic field, namely,

$$g_0^0 = g_{0\nu}g^{0\nu} = g_{00}g^{00} = 1. \tag{224}$$

So using results obtained earlier in this section we can find that the components of the Einstein tensor are

$$\begin{aligned}
r^2 G_0{}^0 &\equiv e^{-2\lambda}(1 - 2r\lambda') - 1 = -\frac{d}{dr}[r(1 - e^{-2\lambda})], \\
r^2 G_1{}^1 &\equiv e^{-2\lambda}(1 + 2r\nu') - 1, \\
G_2{}^2 &\equiv e^{-2\lambda}\left(\nu'' + \nu'^2 - \lambda'\nu' + \frac{\nu' - \lambda'}{r}\right), \\
G_3{}^3 &= G_2{}^2.
\end{aligned} \tag{225}$$

Because of the assumption that the star is static, the three-velocity of every fluid element is zero, so

$$u^\mu = 0 \ \ (\mu \neq 0), \quad u^0 = 1/\sqrt{g_{00}} \tag{226}$$

according to (220). The energy–momentum tensor expressed as a mixed tensor, we have the nonzero components in the present metric,

$$T_0{}^0 = \epsilon, \qquad T_\mu{}^\mu = -p \qquad (\mu \neq 0). \tag{227}$$

So the (00) component of the Einstein equations gives

$$r^2 G_0{}^0 = -\frac{d}{dr}\left\{r\left(1 - e^{-2\lambda(r)}\right)\right\} = kr^2 T_0{}^0 = kr^2\varepsilon(r). \tag{228}$$

This can be integrated immediately to yield

$$e^{-2\lambda(r)} = 1 + \frac{k}{r}\int_0^r \varepsilon(r)r^2 dr. \tag{229}$$

Let us define

$$M(r) \equiv 4\pi \int_0^r \varepsilon(r)r^2 dr, \tag{230}$$

and let R denote the radius of the star, the radial coordinate exterior to which the pressure vanishes. Zero pressure defines the edge of the star because zero pressure can support no material against the gravitational attraction from within. Denote the corresponding value of $M(R)$ by

$$M \equiv M(R). \tag{231}$$

Now comparing (125, 214, 215) we see that, to obtain agreement with the Newtonian limit, we must choose

$$k = -8\pi G \tag{232}$$

and interpret M as the *gravitational mass* of the star. Therefore, $M(r)$ is referred to as the *included mass* within the coordinate r. So Einstein's field equations can now be written

$$G^{\mu\nu} = -8\pi G T^{\mu\nu}. \tag{233}$$

From the above, we have found so far that

$$g_{11}(r) = -e^{2\lambda(r)} = -\left(1 - \frac{2GM(r)}{r}\right)^{-1}, \tag{234}$$

which agrees with (215), but now we see that $g_{11}(r)$ has the same form inside and outside the star although it is the included mass $M(r)$, not the total mass, that appears in the interior solution.

Having learned the constant of proportionality in Einstein's equations (233), let us now write out the field equations for a spherically symmetric static star, including the one we have already solved. In passing we note that our solution gives a relationship between the included mass $M(r)$ at any radial coordinate and the metric function $g_{11}(r)$ or $\lambda(r)$, but we have yet to learn how to compute one or the other. The differential equations from (225) are

$$G_0{}^0 \equiv e^{-2\lambda}\left(\frac{1}{r^2} - \frac{2\lambda'}{r}\right) - \frac{1}{r^2} = -8\pi G\,\varepsilon(r), \tag{235}$$

$$G_1{}^1 \equiv e^{-2\lambda}\left(\frac{1}{r^2} + \frac{2\nu'}{r}\right) - \frac{1}{r^2} = 8\pi G\,p(r), \tag{236}$$

$$G_2{}^2 \equiv e^{-2\lambda}\left(\nu'' + \nu'^2 - \lambda'\nu' + \frac{\nu' - \lambda'}{r}\right) = 8\pi\,Gp(r), \tag{237}$$

$$G_3{}^3 = G_2{}^2 = 8\pi G\,p(r). \tag{238}$$

The last equation contains no information additional to that provided by those preceding it.

To simplify notation, choose units so that $G = c = 1$. Solve (235) to find

$$-2r\lambda' = \left(1 - 8\pi r^2 \varepsilon\right) e^{2\lambda} - 1 \tag{239}$$

and (236) to find

$$2r\nu' = \left(1 + 8\pi r^2 p\right) e^{2\lambda} - 1. \tag{240}$$

Take the derivative of (240) and then multiply by r:

$$2r\nu' + 2r^2\nu'' = \left[2r\lambda'\left(1 + 8\pi r^2 p\right) + \left(16\pi r^2 p + 8\pi r^2 p'\right)\right] e^{2\lambda}\Big]. \tag{241}$$

Solve for ν'' using (239, 240):

$$2r^2\nu'' = 1 + \left(16\pi r^2 p + 8\pi r^2 p'\right) e^{2\lambda} - \left(1 + 8\pi r^2 p\right)\left(1 - 8\pi r^2 \varepsilon\right) e^{4\lambda}. \tag{242}$$

Square (240) to obtain the result

$$2r^2\nu'^{\,2} = \frac{1}{2}\left(1 + 8\pi r^2 p\right)^2 e^{4\lambda} - \left(1 + 8\pi r^2 p\right) e^{2\lambda} + \frac{1}{2}. \tag{243}$$

The last four numbered equations provide expressions for λ', ν', ν'', and $\nu'^{\,2}$ in terms of p, p', ε, and $e^{2\lambda}$ the latter of which, according to (234), can be expressed in terms of the included mass. Therefore the metric can be eliminated altogether by substitution of the above results into the remaining field equation (237). After a number of cancellations, we emerge with the result

$$\frac{dp}{dr} = -\frac{[p(r) + \varepsilon(r)][M(r) + 4\pi r^3 p(r)]}{r[r - 2M(r)]}. \tag{244}$$

This and (230) represent the reduction of Einstein's equations for the interior of a spherical, static, relativistic star. These equations are frequently referred to as the Oppenheimer–Volkoff equations. The stars they describe — static and spherically symmetric — are sometimes referred to as Schwarzschild stars.

Given an equation of state (221), the stellar structure equations (230) and (244) can be solved simultaneously for the radial distribution of pressure, $p(r)$, and hence for the distribution of mass-energy density $\varepsilon(r)$.

Moreover, in any detailed theory of dense matter, the baryon and lepton populations are obtained as a function of density; hence the distribution of particle populations in a star can be found coincident with a solution of the Oppenheimer–Volkoff equations.

It may seem curious that the expression (230) for mass has precisely the same form as one would write in nonrelativistic physics for the mass whose distribution is given by $\varepsilon(r)$. How can this be, inasmuch as we know that spacetime is curved by mass and mass in turn is moved and arranged by spacetime in accord with Einstein's equations? The answer is that (230) is not a prescription for computing the total mass of an arbitrary distribution $\varepsilon(r)$. There are no arbitrary distributions in gravity; rather $\varepsilon(r)$ is precisely prescribed by another of Einstein's equations (244). As such, M comprises the mass of the star and its gravitational field. Because of the mutual interaction of mass-energy and spacetime, there is no meaning to the question "What is the mass of the star?" in isolation from the field energy. That is why we refer to M as the gravitational mass or the mass-energy of the star. It is the only type of mass that enters Einstein's theory and it is the only stellar mass to which we will refer in this book. Therefore, we shall generally refer to a star's mass as simply the mass without the adjective "gravitational". Sometimes a so-called proper mass is defined. It appears nowhere in Einstein's equations and is an artifact.

It does make sense to inquire about the mass of the totality of nucleons in a star if they were dispersed to infinity. This mass is referred to as the baryon mass. The difference between gravitational mass and baryon mass, if negative, is the gravitational binding of the star. As we shall find, the gravitational binding is of the order of 100 MeV per nucleon in stars near the mass limit as compared to 10 MeV binding by the strong force in nuclei.

Notice that, according to (244), the pressure is a monotonic decreasing function from the inside of the star to its edge because all the factors in (244) are positive, leaving the explicit negative sign. This makes sense. Any region is weighted down by all that lies above. We have assumed that the denominator in (244) is positive. Overall this is true of the Earth, the Sun, and a neutron star. In fact, $2M/R < 8/9$ for any static stable star. It can also be shown that $2M(r)/r < 1$ for all regions of a stable star [Hartle (1973)]; so indeed we are justified in taking the last factor in (244) as positive.

In (216) we saw a singularity in the Schwarzschild solution if a star lies within $r = 2M$. Such stars are highly relativistic objects called black holes.

No light or particle can escape from within their Schwarzschild radius. A luminous star is highly non-relativistic. A neutron star is relativistic. Newtonian gravity would not produce the same results as General Relativity. This fact is clear, given that $2M$ can be as large as $\frac{8}{9}R$ for a neutron star, which makes the denominator of (244) a large correction (as much as 9 instead of 1).

We already have an expression for the radial metric function both inside and outside a star. It is sometimes useful to have the time metric function g_{00}. No general expression for the solution can be obtained, as for g_{11}, (234). However using the latter in (240) we obtain a differential equation

$$\frac{d\nu}{dr} = \frac{M(r) + 4\pi r^3 p(r)}{r[r - 2M(r)]}. \tag{245}$$

The solution must match the exterior solution (215). This is easily accomplished. If $\nu(r)$ is a solution, we can add any constant to it and still have a solution. We obtain the correct condition at R if we make the change

$$\nu(r) \longrightarrow \nu(r) - \nu(R) + \frac{1}{2}\ln\left(1 - \frac{2M}{R}\right), \quad r \leq R. \tag{246}$$

We can start the integration at $r = 0$ with any convenient value of $\nu(0)$, say zero.

Alternately, once the Oppenheimer–Volkoff equations have been solved so that $p(r)$ and hence $\epsilon(r)$ are known, one can find $\nu(r)$ by integration of

$$\frac{d\nu}{dr} = -\frac{1}{p+\epsilon}\frac{dp}{dr}, \tag{247}$$

namely,

$$\nu(r) = -\int_0^r \frac{1}{p+\epsilon}\frac{dp}{dr} + \text{constant}, \quad \nu(\infty) = 0. \tag{248}$$

The Oppenheimer–Volkoff equations can be integrated from the origin with the initial conditions $M(0) = 0$ and an arbitrary value for the central energy density $\varepsilon(0)$, until the pressure $p(r)$ becomes zero at, say R. Because zero pressure can support no overlying matter against the gravitational attraction, R defines the gravitational radius of the star and $M(R)$ its gravitational mass. For the given equation of state, there is a unique relationship between the mass and central density $\varepsilon(0)$. So for each possible equation of state, there is a unique family of stars, parameterized by, say, the central density or the central pressure. Such a family is often referred to as the single parameter sequence of stars corresponding to the given equation of state.

2.9.6. *Gravitational collapse and limiting mass*

In Newtonian physics mass alone generates gravity. In the Special Theory of Relativity mass is equivalent to energy, so in the general theory all forms of energy contribute to gravity. It is surprising that pressure also plays a most consequential role in the structure of relativistic stars beyond the role it plays in Newtonian gravity. Pressure supports stars against gravity, but surprisingly, it ultimately assures the gravitational collapse of relativistic stars whose mass lies above a certain limit.

Pressure appears together with energy density in determining the monotonic decrease of pressure (244) in a relativistic star. Gravity acts to compress the material of the star. As it does so, the pressure of the material is increased toward the center. But inasmuch as pressure appears on the right side of the equation, this increase serves to further enhance the grasp of gravity on the material. Therefore, for stars of increasing mass, for which the supporting pressure must correspondingly increase, the pressure gradient (which is negative) is increased in magnitude, making the radius of the star smaller because its edge necessarily occurs at $p = 0$. As a consequence, if the mass of a relativistic star exceeds a critical value, there is no escape from gravitational collapse to a black hole [Harrison *et al.* (1965)]. Whatever the equation of state, the one-parameter sequence of stable configurations belonging to that equation of state is terminated by a maximum-mass compact star. The mass of this star is referred to as the mass limit or limiting mass of the sequence.

2.10. *Action principle in gravity*

We arrived at Einstein's equations by noting the vanishing divergence of the Einstein curvature tensor and equating it to the energy-momentum tensor of matter as the source of the gravitational field. We did not comment on how the energy-momentum tensor might be obtained. In general, this tensor is not given but must be calculated from a theory of matter. In what frame should the theory be solved? Evidently in the general frame of the gravitational field. But this is an entirely different problem than is normally solved in many-body theory.

We are accustomed to solving problems in nuclear and particle theory in flat spacetime (or even flat space) in which the constant Minkowski metric $\nu_{\mu\nu}(r)$, appears, not a general and as yet unspecified field $g_{\mu\nu}(x)$, A tacit assumption is made in passing from the energy–momentum tensor in a Lorentz frame (51) to its form (220) in a general frame by means of general covariance, as was done in deriving the Oppenheimer–Volkoff equations of stellar structure. The local region over which Lorentz frames extend is assumed to be sufficiently large that the equations of motion of the matter fields can be solved in a Lorentz frame, that is, in the absence of gravity, and the corresponding energy–momentum tensor constructed from the solution for such a region.

As we shall see in the next chapter, the local inertial frames in the gravitational field of neutron stars (and therefore for the less dense white dwarfs and all other stars) are actually sufficiently extensive that the matter from which they are constituted can be described by theories in flat Minkowski spacetime. We shall refer to such a situation as a partial decoupling of matter from gravity. In other words, the equations of motion for the matter and radiation fields can be solved in Minkowski spacetime. The solutions will provide the means of calculating the energy density and pressure of matter ε and p throughout the star. But the general metric functions of gravity $g_{\mu\nu}$, reappear on the right side of Einstein's field equations in the energy–momentum tensor, (219), when referred to a general frame in accord with the principle of general covariance. Therefore the gravitational fields $g_{\mu\nu}(x)$, still appear on both sides of Einstein's field equations, and matter in bulk shapes spacetime just as spacetime shapes and moves matter in bulk. However, the local structure of matter is determined only by the equations of motion in Minkowski spacetime.

There are conceivable situations where the partial decoupling just described may not hold. In that case the equations of motion themselves contain, not the Minkowski metric tensor (a diagonal tensor with constant elements), but the general, spacetime-dependent, metric tensor. This is the fully coupled problem and obviously would be enormously dificult to solve. While we do not encounter this situation in this chapter (see as an example where strong coupling is used, Refs. [Lee *et al.* (1987); Glendenning *et al.* (1987)], nonetheless it is worth seeing in symbolic form what the fully coupled problem looks like. The expectation that the stress–energy tensor should be obtained in general from a theory of matter by solving the field equations of the theory in a general gravitational field will be verified. It is also interesting to see Einstein's equations emerge from a variational principle.

We employ the gravitational action principle. As in all cases, the Lagrangian of gravity ought to be a scalar. We have encountered the Ricci scalar curvature $R = g^{\mu\nu}R$, and from it, as Hilbert did, the Lagrangian density can be formed (with a prefactor that can be known only in hindsight):

$$\mathcal{L}_g = -\frac{1}{16\pi G}R\sqrt{-g}. \tag{249}$$

Here G is Newton's constant and g is the determinant of the metric $g_{\mu\nu}$, which is negative for our choice of signature for the metric. (Recall, for example, the Minkowski metric, or the Schwarzschild metric.) We also define the Lagrangian density

$$\mathcal{L}_m = L_m\sqrt{-g} \tag{250}$$

from the Lagrangian L_m of the matter and radiation fields ϕ. The total action is

$$I = \int \Big(\mathcal{L}_g + \mathcal{L}_m\Big) d^4x. \tag{251}$$

The coupled field equations for the matter and metric functions emerge as the conditions that yield vanishing variation of the action with respect to all the fields — the gravitational fields described by $g_{\mu\nu}$, and matter fields described by ϕ. The manipulations are quite tedious and are relegated to the next section. The field equations obtained are

$$\frac{\partial L_m}{\partial\phi} - \partial_\mu \frac{\partial L_m}{\partial(\partial\phi)} = 0, \tag{252}$$

$$G^{\mu\nu} = -8\pi G T^{\mu\nu}, \tag{253}$$

where $G^{\mu\nu} \equiv R^{\mu\nu} - \frac{1}{2}g^{\mu\nu}R$ is the Einstein tensor. The first of the field equations reduces to the familiar Euler–Lagrange equations in the limit of weak gravitational fields ($g_{\mu\nu} \to \eta_{\mu\nu}$). We shall encounter the Euler–Lagrange equations in studying theories of dense nuclear matter. The second are Einstein's field equations (233).

The matter–radiation energy–momentum tensor that emerges from the variational principle is given by

$$T^{\mu\nu} \equiv -g^{\mu\nu}L_m + 2\frac{\partial L_m}{\partial g_{\mu\nu}}. \tag{254}$$

The second term is

$$2\frac{\partial L_m}{\partial g_{\mu\nu}} = 2\frac{\partial L_m}{\partial(\partial_\alpha\phi)}\frac{1}{2}\frac{\partial(g_{\alpha\beta}\partial^\beta\phi)}{\partial g_{\mu\nu}} = \frac{\partial L_m}{\partial(\partial_\mu\phi)}\partial^\nu\phi. \tag{255}$$

Combining these results yields the canonical form of the energy–momentum tensor in field theory (for example, see Ref. [Bjorken & Drell (1965)]) except that the Minkowski tensor is replaced by the general metric of gravity. Thus we have

$$T^{\mu\nu} = -g^{\mu\nu} L_m + \sum_{\phi} \frac{\partial L_m}{\partial(\partial_\mu \phi)} g^{\nu\alpha} \partial_\alpha \phi, \tag{256}$$

where the sum is over the various fields ϕ in L_m. In this way we see how the equations couple all matter and gravitational fields, $\phi(x)$, ..., $g^{\mu\nu}$ in the general case.

2.10.1. *Derivations*

We write down most of the steps in deriving the Einstein field and matter-radiation equations from the variation of the action. The gravitational and matter fields will be subjected to arbitrary variations except the values and first derivatives will be kept constant on the boundaries. We concentrate on the gravitational part because that is the hardest. First, from (71) and (102) we readily obtain by differentiation,

$$g^{\alpha\mu}{}_{,\sigma} g_{\mu\nu} + g^{\alpha\mu} \Big(\Gamma_{\nu\mu\sigma} + \Gamma_{\mu\nu\sigma} \Big) = 0. \tag{257}$$

Multiply by $g^{\beta\nu}$ and sum on ν to find

$$g^{\alpha\beta}{}_{,\sigma} = -g^{\alpha\mu} \Gamma^{\beta}_{\mu\sigma} - g^{\beta\nu} \Gamma^{\alpha}_{\nu\sigma}. \tag{258}$$

Next, evaluate

$$\begin{aligned} \Big(g^{\mu\nu} \sqrt{-g} \Big)_{,\sigma} &= g^{\mu\nu}{}_{,\sigma} \sqrt{-g} + g^{\mu\nu} \Big(\sqrt{-g} \Big)_{,\sigma} \\ &= -g^{\mu\rho} \Gamma^{\nu}_{\rho\sigma} \Big(\sqrt{-g} \Big) - g^{\nu\rho} \Gamma^{\mu}_{\rho\sigma} \Big(\sqrt{-g} \Big) + g^{\mu\nu} \Big(\sqrt{-g} \Big)_{,\sigma}. \end{aligned} \tag{259}$$

Use (157) to find

$$\Big(g^{\mu\nu} \sqrt{-g} \Big)_{,\sigma} = \Big(-g^{\mu\rho} \Gamma^{\nu}_{\rho\sigma} - g^{\nu\rho} \Gamma^{\mu}_{\rho\sigma} + g^{\mu\nu} \Gamma^{\rho}_{\sigma\rho} \Big) \sqrt{-g}. \tag{260}$$

Now set $\sigma = \nu$ and contract. After a cancelation of two terms, find,

$$\Big(g^{\mu\nu} \sqrt{-g} \Big)_{,\nu} = -g^{\nu\alpha} \Gamma^{\mu}_{\alpha\nu} \sqrt{-g}. \tag{261}$$

The above results can now be employed to rewrite the gravitational action (where, as usual, we set $G = c = 1$ whenever convenient):

$$16\pi I_g = \int R \sqrt{-g} \, d^4x. \tag{262}$$

From (176) define

$$L \equiv g^{\mu\nu}\Big(\Gamma^{\sigma}_{\mu\nu}\Gamma^{\rho}_{\sigma\rho} - \Gamma^{\rho}_{\mu\sigma}\Gamma^{\sigma}_{\nu\rho}\Big);$$

$$M \equiv g^{\mu\nu}\Big(\Gamma^{\sigma}_{\mu\sigma,\nu} - \Gamma^{\sigma}_{\mu\nu,\sigma}\Big). \tag{263}$$

With these definitions the scalar curvature becomes

$$R = M - L. \tag{264}$$

Use (260) and (261) to find

$$M\sqrt{-g} = \Big(g^{\mu\nu}\Gamma^{\sigma}_{\mu\sigma}\sqrt{-g}\Big)_{,\nu} - \Big(g^{\mu\nu}\Gamma^{\sigma}_{\mu\nu}\sqrt{-g}\Big)_{,\sigma}$$

$$-\Big(g^{\mu\nu}\sqrt{-g}\Big)_{,\nu}\Gamma^{\sigma}_{\mu\sigma} + \Big(g^{\mu\nu}\sqrt{-g}\Big)_{,\sigma}\Gamma^{\sigma}_{\mu\nu}. \tag{265}$$

The first two terms are perfect differentials and so contribute nothing under the integral because of the vanishing of the fields and their derivatives on the boundaries. After some manipulation, the remaining two terms are found to be $2L\sqrt{-g}$. Consequently,

$$16\pi I_g = \int L\sqrt{-g}\, d^4x$$

$$= \int g^{\mu\nu}\Big(\Gamma^{\sigma}_{\mu\nu}\Gamma^{\rho}_{\sigma\rho} - \Gamma^{\rho}_{\mu\sigma}\Gamma^{\sigma}_{\nu\rho}\Big)\sqrt{-g}\, d^4x. \tag{266}$$

Now evaluate the variation, examining separately the two terms in the integrand. Use (157) and (261) to rewrite the variation of the first term;

$$\delta\Big(g^{\mu\nu}\Gamma^{\sigma}_{\mu\nu}\Gamma^{\rho}_{\sigma\rho}\sqrt{-g}\Big) = \Gamma^{\alpha}_{\mu\nu}\delta\Big(g^{\mu\nu}\Big(\sqrt{-g}\Big)_{,\alpha}\Big)$$

$$-\Gamma^{\beta}_{\alpha\beta}\delta\Big(g^{\alpha\nu}\sqrt{-g}\Big)_{,\nu} - \Gamma^{\beta}_{\alpha\beta}\Gamma^{\alpha}_{\mu\nu}\delta\Big(g^{\mu\nu}\sqrt{-g}\Big). \tag{267}$$

Use (258) to develop the variation of the second term in (153) and find

$$\delta\Big(g^{\mu\nu}\Gamma^{\rho}_{\mu\sigma}\Gamma^{\sigma}_{\nu\rho}\sqrt{-g}\Big) = -\Gamma^{\alpha}_{\nu\beta}\delta\Big(g^{\nu\beta}{}_{,\sigma}\sqrt{-g}\Big) - \Gamma^{\beta}_{\mu\alpha}\Gamma^{\alpha}_{\nu\beta}\delta\Big(g^{\mu\nu}\sqrt{-g}\Big). \tag{268}$$

Assemble the above two results and introduce the perfect differentials with compensating terms to form the variation of the integrand of (153):

$$\delta\mathcal{L}_g = \Big[\Gamma^{\alpha}_{\mu\nu}\delta\Big(g^{\mu\nu}\sqrt{-g}\Big)\Big]_{,\alpha} - \Big[\Gamma^{\beta}_{\alpha\beta}\delta\Big(g^{\alpha\nu}\sqrt{-g}\Big)\Big]_{,\nu}$$

$$-\Big(\Gamma^{\alpha}_{\mu\nu,\alpha} - \Gamma^{\beta}_{\mu\beta,\nu} - \Gamma^{\beta}_{\mu\alpha}\Gamma^{\alpha}_{\nu\beta} + \Gamma^{\beta}_{\alpha\beta}\Gamma^{\alpha}_{\mu\nu}\Big)\,\delta\Big(g^{\mu\nu}\sqrt{-g}\Big). \tag{269}$$

The first two terms are perfect differentials and yield zero because the variations vanish on the boundary. The remaining bracket is the Ricci tensor (176). Therefore, we have for the variation of the gravitational action

$$\delta I_g = \frac{1}{16\pi} \int R_{\mu\nu} \delta\left(g^{\mu\nu}\sqrt{-g}\right) d^4x. \tag{270}$$

Next, again from (71) deduce that

$$\delta g^{\alpha\nu} = -g^{\mu\alpha} g^{\nu\beta} \delta g_{\alpha\beta}. \tag{271}$$

Also, from (156) find

$$2\sqrt{-g}\left(\sqrt{-g}\right)_{,\alpha} = \left(\sqrt{-g}\sqrt{-g}\right)_{,\alpha} = g_{,\alpha} = g g^{\mu\nu} g_{\mu\nu,\alpha}. \tag{272}$$

Consequently,

$$\left(\sqrt{-g}\right)_{,\alpha} = \frac{1}{2}\sqrt{-g}\, g^{\mu\nu} g_{\mu\nu,\alpha}. \tag{273}$$

With this result we now have

$$\delta\left(g^{\mu\nu}\sqrt{-g}\right) = -\left(g^{\mu\alpha}g^{\nu\beta} - \frac{1}{2}g^{\mu\nu}g^{\alpha\beta}\right)\sqrt{-g}\,\delta g_{\alpha\beta}. \tag{274}$$

Recalling the definition of Einstein's curvature tensor, we have obtained

$$\delta I_g = -\frac{1}{16\pi} \int G^{\alpha\beta}\sqrt{-g}\,\delta g_{\alpha\beta}\, d^4x. \tag{275}$$

Thus we derive Einstein's field equation in empty space from the vanishing of the variation of the gravitational action.

If we add the action of matter and radiation fields to the gravitational action, we get the total action. Under arbitrary variations of the gravitational and other fields we insist that the total action vanish. This leads to the Euler–Lagrange equations (252) for the matter–radiation fields and to the Einstein equations (253) for the gravitational fields. Note, however, that the equations of motion for the matter–radiation fields contain the gravitational fields $g^{\mu\nu}$ and they reduce to the usual form in Minkowski spacetime only when the $g^{\mu\nu}$ can be replaced by the Minkowski tensor.

Return now to the total action (251) and vary all fields. Also use the result (275) for the variation of $\mathcal{L}_g$

$$\delta I = \int \Bigg[\left\{ -\frac{1}{16\pi} G^{\alpha\beta}\sqrt{-g} + \left(\frac{\partial \mathcal{L}_m}{\partial g_{\alpha\beta}} - \partial_\nu \frac{\partial \mathcal{L}_m}{\partial g_{\alpha\beta,\nu}} \right) \right\} \delta g_{\alpha\beta}$$
$$+ \left\{ \frac{\partial \mathcal{L}_m}{\partial \phi} - \partial_\nu \frac{\partial \mathcal{L}_m}{\partial(\partial_\nu \phi)} \right\} \delta\phi \Bigg] d^4x. \tag{276}$$

The last term in each curly brackets was obtained by an integration by parts, as follows:

$$\frac{\partial \mathcal{L}}{\partial f_{\mu}} \partial f_{,\mu} = \partial_{\mu} \left(\frac{\partial \mathcal{L}}{\partial f_{,\mu}} \delta f \right) - \partial_{\mu} \frac{\partial \mathcal{L}}{\partial f_{,\mu}} \delta f, \tag{277}$$

where f stands for either $g_{\mu\nu}$ or ϕ. The integral over the first term on the right vanishes because the f are not varied on the boundary. Because the variations are otherwise arbitrary, the vanishing of the action implies the vanishing of the coefficients of the varied fields. The variation of the matterfields yields (252) (where we have removed $\sqrt{-g}$ because it is unaffected by the ϕ variation). The variation of the gravitational fields yields

$$G^{\alpha\beta}\sqrt{-g} = 16\pi \left(\frac{\partial \mathcal{L}}{\partial g_{\alpha\beta}} - \partial_{\nu} \frac{\partial \mathcal{L}_m}{\partial g_{\alpha\beta,\nu}} \right). \tag{278}$$

This equation shows how the gravitational and matter fields are coupled. We use it to derive (253) by showing that the right side is the energy–momentum tensor in the form (254). The familiar form (256) then follows. Evaluate the first term on the right side of (278):

$$\frac{\partial \mathcal{L}_m}{\partial g_{\alpha\beta}} = L_m \frac{\partial \sqrt{-g}}{\partial g_{\alpha\beta}}. \tag{279}$$

Because

$$\frac{\partial g}{\partial g_{\mu\nu}} = g g^{\mu\nu}, \tag{280}$$

which follows by differentiating the identity $g = g_{\mu\nu} g^{\mu\nu}$, we obtain

$$\frac{\partial \mathcal{L}_m}{\partial g_{\alpha\beta}} = \left(-\frac{1}{2} g^{\alpha\beta} L_m + \frac{\partial L_m}{\partial g_{\alpha\beta}} \right) \sqrt{-g}\,. \tag{281}$$

The matter Lagrangian will not usually depend on derivatives of the metric, but only on the metric itself. Thus, with the above equation we have derived (253) with the energy momentum tensor given by (254).

Acknowledgments

This work is supported by the Director, Office of Energy Research, Office of High Energy and Nuclear Physics, Nuclear Physics Division of the U.S. Department of Energy under Contract No. DE-AC02-05-CH 1231.

References

Abramowicz, M. A. and Prasanna, A. R., *Mon. Not. R. Astron. Soc.* **245**, 720 (1990).

Baade, W. and Zwicky, F., *Phys. Rev.* **45**, p. 138 (1934).

Berry, M., *Principles of Cosmology and Gravitation* (Adam Hilgar Publ., Bristol, England, 1989, 1st ed. 1976).

Bjorken, J. D. and Drell, S. D., *Relativistic Quantum Fields* (McGraw Hill, New York, USA, 1965).

Einstein, A., *Annalen der Physik* **49**, 769 (1916).

Einstein, A., unpublished manuscript in the Pierpont Morgan Library, New York, USA, 1920.

Einstein, A., *The Meaning of Relativity* (Methuen and Co., London, England, 1951, 5th ed., A Lecture Series at Princeton University, 1921, with several revisions in later editions).

Einstein, A. and de Sitter, W., *Proc. Nat. Acad. Sci.* **18**, 213 (1932).

Eötvös, R. V., *Math. Nat. Ber. Ungarn* **8**, 65 (1890).

Glendenning, N. K., *Phys. Rep.* **342**, 393 (2001).

Glendenning, N. K., Kodama, T. and Klinkhamer, F. R., *Phys. Rev.* **D3**, 3226 (1988).

Harrison, B. K., Thorne, K. S., Wakano, M. and Wheeler, J. A., *Gravitation Theory and Gravitational Collapse* (University of Chicago Press, Chicago, USA, 1965).

Hartle, J. B., in *Relativity, Astrophysics and Cosmology*, W. Israel (ed.) (D. Riedel Publishing Co., Dordrecht, Holland, 1973).

Ishiwara, J., *Albert Einstein in J. Ishiwara, Einstein Koen-Roku.* (Tokyo-Tosho, Tokyo, Japan, 1977).

Kruskal, M. D., *Phys. Rev.* **119**, 1743 (1960).

Lee, T. D., *Phys. Rev.* **D35**, 3637 (1987); Friedberg, R., Lee, T. D. and Pang, Y., *Phys. Rev.* **D35**, 3640 (1987); *ibid.*, p. 3658 (1987); Lee, T. D. and Pang, Y., *Phys. Rev.* **D35**, 3678 (1987).

Michelson, A. A. and Morley, E. W., *Am. J. Sci.* **34**, 333 (1887).

Minkowski, H., *Annalen der Physik* (Leipzig) **47**, 927 (1915).

Nordvedt, K., *Phys. Rev.* **169**, 1017 (1968); *op. cit.* **180**, 1293 (1969); *Phys. Rev.* **D3**, 1633 (1971).

Oppenheimer, J. R. and Snyder, H., *Phys. Rev.* **56**, 455 (1939).

Pound, R. V. and Rebka, G. A., *Phys. Rev. Lett.* **4**, 337 (1960).

Schild, A., *Texas Quarterly* **3**, 42 (1960).

Schild, A., in *Evidence for Gravitational Theories*, C. Møller (ed.) (Academic Press, New York, USA, 1962).

Schild, A., in *Relativity Theory and Astrophysics*, J. Ehlers (ed.) (American Mathematical Society, Providence, Rhode Island, USA, 1967).

Swiatecki, W. J., *Phys. Scripta* **28**, 349 (1983).

Szerkeres, G., *Publ. Mat. Debrecen.* **7**, 285 (1960).

Taylor, J. H., Wolszczan, A., Damour, T. and Weisberg, J. M., *Nature* **355**, 132 (1992).
Voigt, W., *Goett. Nachr.*, p. 41 (1887).
Weinberg, S., *Gravitation and Cosmology* (John Wiley & Sons, New York, USA, 1972).
Will, C. M., *Was Einstein Right?* (Oxford University Press, England, 1995).

Chapter 2

Non-Spherical Compact Stellar Objects in Einstein's Theory of General Relativity

Omair Zubairi* and Fridolin Weber†

*Department of Sciences, Wentworth Institute of Technology
550 Huntington Avenue, Boston, MA 02115, USA
zubairio@wit.edu

†Department of Physics, San Diego State University
5500 Campanile Drive, San Diego, California 92182, USA
Center for Astrophysics and Space Sciences, University of California
San Diego, La Jolla, California 92093, USA
fweber@ucsd.edu

In this work, we derive a Tolman–Oppenheimer–Volkoff (TOV)-like stellar structure equations for deformed compact stellar objects, whose mathematical form is similar to the traditional TOV equation for spherical stars. We then solve these equations numerically for a given equation of state (EoS) and produce stellar properties such as masses and radii along with pressure and density profiles and investigate any changes from spherical models of compact stars. If rotating, deformed compact objects are among the possible astrophysical sources emitting gravitational waves which could be detected by gravitational-wave detectors.

Contents

1. Introduction

Neutron stars are generally assumed to be perfect spheres, whose properties are described in the framework of general relativity theory by the Tolman–Oppenheimer–Volkoff (TOV) equation [Oppenheimer and Volkoff (1939); Tolman (1939)]. The TOV equation is a first-order differential equation which can be solved with a given model for the equation of state (EoS) of neutron star matter. However, assuming perfect spherical symmetry may not always be correct. It is known that magnetic fields are present inside of neutron stars. In particular, if the magnetic field is strong (up to around 10^{18} Gauss in the core), such as for magnetars [Thompson and Duncan (1995, 1996); Mereghetti (2008)], and/or the pressure of the matter in the cores of neutron stars is non-isotropic, as predicted by some models of color superconducting quark matter [Ferrer *et al.* (2010)], then deformation of neutron stars can occur [Chandrasekhar and Fermi (1953); Ferraro (1954); Goossens (1972); Katz (1989); Payne and Melatos (2004); Haskell *et al.* (2008)].

In this work, we derive a TOV-like stellar structure equations for deformed neutron stars whose mathematical form is similar to the traditional TOV equation for spherical stars. We then solve these equations numerically for a given EoS and produce stellar properties such as masses and radii along with pressure and density profiles and investigate any changes from spherical models of neutron stars.

2. Non-Spherical Symmetry

In order to study deformed neutron stars, as shown in Fig. 1, we first need to investigate the symmetry of these objects. Our first assumption of deformity will lie within the geometry of these objects. As shown in Fig. 1, the axial symmetry lies in the polar axis being orthogonal with the equatorial axis. Mathematically, this is can be described by the Weyl metric [Weyl (1918); Herrera *et al.* (1999)],

$$ds^2 = e^{2\lambda} dt^2 - e^{-2\lambda}\left[e^{2\nu}\left(dr^2 + dz^2\right) + r^2 d\phi^2\right] = g_{\mu\nu} dx^\mu dx^\nu, \quad (1)$$

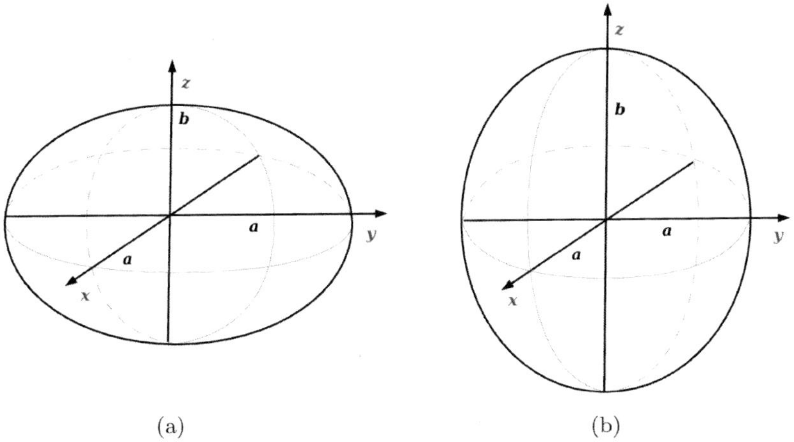

Fig. 1. Schematic illustration of the geometry of an oblate (a) and prolate (b) spheroid.

where t is time and r, z, and ϕ are the spatial components in cylindrical coordinates. The terms λ and ν are the metric functions that depend on both the radial r and polar z coordinates such that $\lambda = \lambda(r, z)$ and $\nu = \nu(r, z)$.

Since we have distinct polar and radial directions both coupled together, as described in Eq. (1), we can now start to investigate the hydrostatic equilibrium of deformed stars in two dimensions and utilize a model for the equation of state of neutron star matter that takes the pressure in both the radial and polar directions into considerations. However, we must first examine each component of Eq. (1) and find out which one makes any contributions to the stellar structure. For this purpose, we will need to make a few assumptions about the stellar configuration. Firstly, we assume that the configuration is non-rotating and not pulsating. Secondly, we assume that the star is axially symmetric, as shown in Fig. 1. Hence, the only terms that contribute to the stellar configuration will be the radial and polar components.

3. Weyl Metric Solutions

As stated just above, the only components that will make any contributions to the stellar structure will be the radial and polar components of Einstein's field equations,

$$G^{\mu}{}_{\nu} \equiv R^{\mu}{}_{\nu} - \frac{1}{2}\delta^{\mu}{}_{\nu} R = -8\pi T^{\mu}{}_{\nu}, \tag{2}$$

where $G^{\mu}{}_{\nu}$ denotes the Einstein tensor, $R^{\mu}{}_{\nu}$ is the Ricci tensor, R the curvature scalar, and $T^{\mu}{}_{\nu}$ the energy–momentum tensor. We begin with the Weyl metric of Eq. (1), and calculate the Christoffel symbols to be

$$\Gamma^{r}{}_{tt} = \left[\mathrm{e}^{-2\nu(r,z)+4\lambda(r,z)}\right] \partial_r \lambda(r,z), \tag{3}$$

$$\Gamma^{z}{}_{tt} = \left[\mathrm{e}^{-2\nu(r,z)+4\lambda(r,z)}\right] \partial_z \lambda(r,z), \tag{4}$$

$$\Gamma^{t}{}_{tr} = \partial_r \lambda(r,z),\ \Gamma^{t}{}_{tz} = \partial_z \lambda(r,z), \tag{5}$$

$$\Gamma^{r}{}_{rr} = \partial_r \nu(r,z) - \partial_r \lambda(r,z),\ \Gamma^{z}{}_{rr} = \partial_z \lambda(r,z) - \partial_z \nu(r,z), \tag{6}$$

$$\Gamma^{r}{}_{rz} = \partial_z \nu(r,z) - \partial_z \lambda(r,z),\ \Gamma^{z}{}_{rz} = \partial_r \nu(r,z) - \partial_r \lambda(r,z), \tag{7}$$

$$\Gamma^{z}{}_{rz} = \partial_r \nu(r,z) - \partial_r \lambda(r,z),\ \Gamma^{\phi}{}_{r\phi} = \left(1 - r\partial_r \lambda(r,z)\right)/r, \tag{8}$$

$$\Gamma^{r}{}_{zz} = -\Gamma^{r}{}_{rr},\ \ \Gamma^{z}{}_{zz} = -\Gamma^{z}{}_{rr},\ \ \Gamma^{\phi}{}_{z\phi} = -\Gamma^{t}{}_{tz}, \tag{9}$$

$$\Gamma^{r}{}_{\phi\phi} = \left[\mathrm{e}^{-2\nu(r,z)}\right] r\left(-1 + r\partial_r \lambda(r,z)\right), \tag{10}$$

$$\Gamma^{z}{}_{\phi\phi} = \left[\mathrm{e}^{-2\nu(r,z)}\right] r^2 \partial_z \lambda(r,z), \tag{11}$$

where ∂_r and ∂_z denote first partial derivatives with respect to r and z. Using these Christoffel symbols and from Eq. (2) we calculate the components of the Ricci tensor to be

$$R^{t}{}_{t} = \frac{1}{r}\mathrm{e}^{-2\nu+2\lambda}\left(r\partial_r^2\lambda + r\partial_z^2\lambda + \partial_r\lambda\right), \tag{12}$$

$$R^{r}{}_{r} = -\frac{1}{r}\mathrm{e}^{-2\nu+2\lambda}\left(r\partial_r^2\lambda - 2\left(\partial_r\lambda\right)^2 - r\partial_z^2\nu + r\partial_z^2\lambda - r\partial_r^2\nu + \partial_r\lambda + \partial_r\nu\right), \tag{13}$$

$$R^{z}{}_{r} = R^{r}{}_{z} = \frac{1}{r}\mathrm{e}^{-2\nu+2\lambda}\left(2r\left(\partial_r\lambda\right)^2 - \partial_z\nu\right), \tag{14}$$

$$R^{z}{}_{z} = -\frac{1}{r}\mathrm{e}^{-2\nu+2\lambda}\left(r\partial_r^2\nu - 2\left(\partial_z\lambda\right)^2 + r\partial_r^2\lambda - r\partial_z^2\nu + r\partial_z^2\lambda - \partial_r\nu + \partial_r\lambda\right), \tag{15}$$

$$R^{\phi}{}_{\phi} = \frac{1}{r}\mathrm{e}^{2\nu+2\lambda}\left(r\partial_r^2\lambda + r\partial_z^2\lambda + \partial_r\lambda\right), \tag{16}$$

where ∂_r^2 and ∂_z^2 denote second partial derivatives with respect to r and z. The Ricci scalar R is then given by

$$R = -\frac{1}{r}\Big[2\mathrm{e}^{-2\nu+2\lambda}\Big(r\partial_r^2\lambda - r\left(\partial_r\lambda\right)^2 - r\partial_z^2\nu + r\partial_z^2\lambda - r\partial_r^2\nu + \partial_r\lambda - r\left(\partial_z\lambda\right)^2\Big)\Big], \tag{17}$$

Using Eqs. (12) through (17), one obtains for the components of the Einstein tensor the following expressions,

$$G^t{}_t = \frac{1}{r}\Big[\mathrm{e}^{-2\nu+2\lambda}\Big(2r\partial_r^2\lambda + 2r\partial_z^2\lambda + 2\partial_r\lambda - r\left(\partial_r\lambda\right)^2 - r\partial_z^2\nu - r\partial_r^2\nu - r\left(\partial_z\lambda\right)^2\Big)\Big], \tag{18}$$

$$G^r{}_r = \frac{1}{r}\mathrm{e}^{-2\nu+2\lambda}\left(r\left(\partial_r\lambda\right)^2 - \partial_r\nu - r\left(\partial_z\lambda\right)^2\right), \tag{19}$$

$$G^z{}_r = G^r{}_z = \frac{1}{r}\mathrm{e}^{-2\nu+2\lambda}\left(\left(2r\partial_r\lambda\right)\left(\partial_z\lambda\right) - \partial_z\nu\right), \tag{20}$$

$$G^z{}_z = -\frac{1}{r}\mathrm{e}^{-2\nu+2\lambda}\left(r\left(\partial_r\lambda\right)^2 - \partial_r\nu - r\left(\partial_z\lambda\right)^2\right), \tag{21}$$

$$G^\phi{}_\phi = -\mathrm{e}^{-2\nu+2\lambda}\left(\left(\partial_r\lambda\right)^2 + \partial_z^2\nu + \partial_r^2\nu + \left(\partial_z\lambda\right)^2\right). \tag{22}$$

Due to the mathematical form of Eq. (1), there will be off-diagonal terms $T^r{}_z = T^z{}_r \neq 0$ in the energy–momentum tensor,

$$T^\mu{}_\lambda = \begin{pmatrix} \epsilon & 0 & 0 & 0 \\ 0 & P_\parallel & \tilde{P} & 0 \\ 0 & \tilde{P} & P_\perp & 0 \\ 0 & 0 & 0 & \tilde{P} \end{pmatrix}, \tag{23}$$

where ϵ is the energy density, $P_\parallel$ is the pressure in the radial direction, and $P_\perp$ is the pressure in the polar direction. The $\tilde{P}$ terms are the off-diagonal pressure contributions associated with the r–z direction of Eq. (1).

Before evaluating Einstein's field equation (2), we define the expressions for the total gravitational mass next. In cylindrical coordinates, the mass in the equatorial and polar directions will depend on both r and z coordinates such that the gravitational mass has the form of $m = m(r, z)$.

3.1. *Gravitational mass formalism*

Due to the axial symmetry assumptions, we define two differential masses, one which will correspond to an infinitesimal cylinder of radius r, height

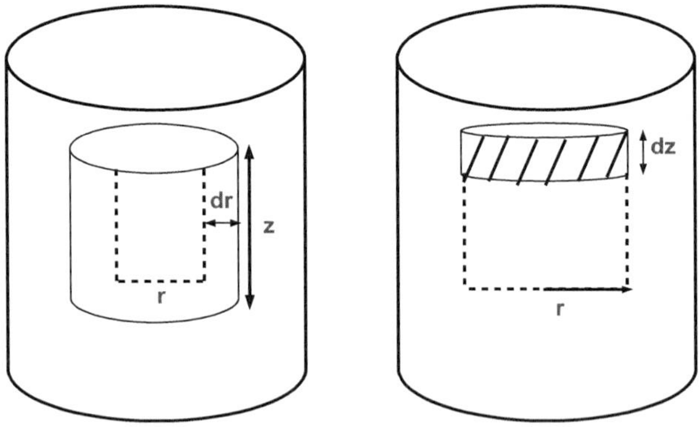

(a) (b)

Fig. 2. Schematic illustration of differential masses in cylindrical coordinates for cross-sectional masses in the equatorial direction (a) and cross-sectional masses in the polar direction (b).

z, and thickness dr, and a second differential mass for an infinitesimal slab of radius r and thickness dz. The situation is graphically illustrated in Fig. 2. Mathematically, we obtain for the cross-sectional masses the following expressions,

$$\partial_r m(r,z) = 2\pi r z \epsilon(r,z), \tag{24}$$

$$\partial_z m(r,z) = \pi r^2 \epsilon(r,z). \tag{25}$$

By making use of Einstein's field equations $G^t{}_t = -8\pi T^t{}_t$, Eq. (24) can be written as

$$\partial_r m(r,z) = -\frac{rz}{4} G^t{}_t. \tag{26}$$

Using the expressions for the Einstein tensor of Eqs. (18) through (22) along with the expression for the energy–momentum tensor given in Eq. (23) leads for Eq. (26) to

$$\partial_r m(r,z) = \partial_r \left(-\mathrm{e}^{-2\nu+2\lambda}\, z\, (\partial_r \lambda)\,(\partial_r \nu - \partial_r \lambda)\right). \tag{27}$$

Applying an integral on both sides of Eq. (27), we obtain an expression that takes the form

$$\int_0^r (\partial_r m(r,z))\, dr = -\int_0^r \partial_r \left(\mathrm{e}^{-2\nu+2\lambda}\, z\, (\partial_r \lambda)\,(\partial_r \nu - \partial_r \lambda)\right) dr. \tag{28}$$

Looking at Eq. (28), it becomes obvious that there is a serious inconsistency with the r component, as the r term has vanished from Eq. (28).

In addition to the r term vanishing, the unknown metric functions λ'_r and ν'_r are still present in Eq. (28). Therefore we can not solve for the r component $\left(e^{-2\nu+2\lambda}\right)$ directly. Having no r term implies that there can be no interior solution of Einstein's field equations that will match the exterior solution for the metric described by Eq. (1) and thus we have lost all the information about the radial coordinate. We therefore need to resort to another way to derive the hydrostatic equilibrium equations of deformed stars in cylindrical coordinates.

If the metric functions were known in Eq. (1), we could approximate a solution and derive the hydrostatic equilibrium equations. As stated in [Hernandez (1967)], one can make some assumptions of the geometry of the interior of a non-spherical mass distribution. In addition to this assumption about the geometry, we will also have to make an assumption on the off-diagonal terms in the energy–momentum tensor given in Eq. (23).

3.2. *Flat-space assumption*

The main problem with trying to directly solve for the r and z components of the Weyl metric is that both the metric functions ($\lambda = \lambda(r,z)$ and $\nu = \nu(r,z)$) are coupled together and depend on both the radial and polar directions. If we assume locally a small radial and polar step size as we construct a deformed compact star then, in the local frame of reference, the geometry may be assumed to be flat. If the space is flat, then the metric functions $\nu(r,z)$ and $\lambda(r,z)$ can be approximated by

$$\nu(r,z) = m^2 r^2 \beta^{-4}, \tag{29}$$

$$\lambda(r,z) = -m\beta^{-2}, \tag{30}$$

where $\beta^2 = \sqrt{r^2+z^2}$, as stated in [Hernandez (1967); Herrera *et al.* (2013)]. Using the expressions given in Eqs. (29) and (30), the metric functions can be written as

$$\nu(r,z) = -\frac{1}{2}\ln\left(m^2 r^2 \beta^{-4}\right), \tag{31}$$

$$\lambda(r,z) = \frac{1}{2}\ln\left(-m\beta^{-2}\right). \tag{32}$$

Now if we want to obtain expressions for $e^{2\lambda(r,z)}$ and $e^{-2\nu(r,z)}$, one can easily see that by taking exponentials on both sides, we obtain

$$e^{2\lambda} = -m^2\,\beta^{-2}, \tag{33}$$

$$e^{-2\nu} = m^2\,r^2\,\beta^{-4}. \tag{34}$$

With the expressions in Eqs. (33) and (34), one can now calculate the r and z components of the Weyl metric. Using these expressions and after some algebra, we calculate the r and z components of the Weyl metric to be

$$g_{rr} = g_{zz} = \mathrm{e}^{-2\nu+2\lambda} = \left(m^3\, r^2\right)\left(r^2+z^2\right)^{-6}. \tag{35}$$

Using Eqs. (33), (34), and (35), one can now take derivatives with respect to r and z and then utilize the Einstein equations, which are also given in terms of derivatives of ν and λ, and attempt to obtain solutions for hydrostatic equilibrium configurations.

3.3. *Toward a self-consistent description of non-spherical compact objects*

We first begin with the Einstein equations described by Eqs. (18), (19), and (21). Assuming that $\tilde{P}$ from Eq. (23) is equal to zero along with some algebra, the Einstein equations then simplify to

$$8\pi\epsilon = \mathrm{e}^{-2\nu+2\lambda}\left(-2\partial_r^2\lambda - 2\partial_z^2\lambda - \frac{2}{r}\partial_r\lambda\right), \tag{36}$$

$$8\pi P_{\parallel} = -\frac{1}{r}\mathrm{e}^{-2\nu+2\lambda}\left(r\left(\partial_r\lambda\right)^2 - \partial_r\nu - r\left(\partial_z\lambda\right)^2\right), \tag{37}$$

$$8\pi P_{\perp} = \frac{1}{r}\mathrm{e}^{-2\nu+2\lambda}\left(r\left(\partial_r\lambda\right)^2 + \partial_r\nu + r\left(\partial_z\lambda\right)^2\right). \tag{38}$$

Next, we utilize the Euler's equation of relativistic hydrodynamics [Weber (1999)] which describes the relationship of pressure and energy density,

$$0 = (\epsilon + P)u_{r\ ;\mu}u^{\mu} - P_{,r}. \tag{39}$$

Applying Eq. (39) for parallel and perpendicular pressure gradients, one obtains

$$0 = \left(\epsilon + P_{\parallel}\right) u_{r\ ;\mu}u^{\mu} - P_{\parallel\ ,r}\ , \tag{40}$$

$$0 = \left(\epsilon + P_{\perp}\right) u_{z\ ;\mu}u^{\mu} - P_{\perp\ ,z}\ , \tag{41}$$

where the semicolons represent covariant derivatives, which are defined as

$$u_{\lambda\ ;\mu} \equiv u_{\lambda\ ,\mu} - \Gamma^{\sigma}_{\lambda\mu}u_{\sigma}. \tag{42}$$

The commas in Eqs. (40), (41) and (42) are standard partial derivatives with respect to r, z and μ such that

$$P_{\parallel,r} \equiv \frac{\partial P_{\parallel}}{\partial r}, \quad P_{\perp,z} \equiv \frac{\partial P_{\perp}}{\partial z}. \tag{43}$$

For a star that is static, there is no time dependence on the pressure and energy density such that $\partial\epsilon/\partial t = \partial P/\partial t = 0$. The four velocity is given by $u^\mu = (u^t, 0, 0, 0)$. Using the normalization condition of the four velocity $u^\nu g_{\nu\mu} u^\mu = 1$ along with the term $g_{tt} = \mathrm{e}^{2\lambda(r,z)}$, we find the time-like component of the four vector u^t to be $u^t = \mathrm{e}^{-\lambda(r,z)}$.

A closer examination of the expressions of Eqs. (40) and (41), one sees that $P_{\parallel\ ,r} \neq P_{\perp\ ,z} \neq 0$. The next step is to incorporate the expressions in Eqs. (36) to (38) with the expressions described by

$$\frac{\partial P_\parallel}{\partial r} = -\left(\epsilon + P_\parallel\right)\partial_r\lambda, \tag{44}$$

$$\frac{\partial P_\perp}{\partial z} = -\left(\epsilon + P_\perp\right)\partial_z\lambda, \tag{45}$$

where the expressions of Eqs. (44) and (45) were obtained from [Zubairi (2015)]. Even though we have calculated all the expressions for $\nu(r,z)$ and $\lambda(r,z)$, we cannot just apply a simple substitution. We must first take the gravity of the deformed star into account as described in general relativity. For this, we must use the expressions given in Eqs. (36) to (38). By adding Eqs. (36) and (37) along with the use of Eq. (44), one finds the equation of hydrostatic equilibrium in the parallel direction to be

$$\frac{\partial P_\parallel}{\partial r} = -\left(\epsilon + P_\parallel\right)\frac{\left(r^2+z^2\right)^6\left(4\pi\left(\epsilon + P_\parallel\right) - 4\pi P_\parallel r\left(1 - \frac{m}{r}\right)\right)}{m^3 r^2}. \tag{46}$$

Applying the same mathematical methodology and by adding Eqs. (36) and (38) along with the utilization of the expression given in Eq. (45), one finds the equation of hydrostatic equilibrium in the perpendicular direction to be

$$\frac{\partial P_\perp}{\partial z} = -\left(\epsilon + P_\perp\right)\frac{2z}{\left(r^2+z^2\right)}\left(\frac{m^3r^2 + m^3 r}{4\pi r\left(\epsilon + P_\perp\right)\left(r^2+z^2\right)^6} - \frac{m^3 r}{4\pi P_\perp}\right). \tag{47}$$

In Eqs. (46) and (47), ϵ is the energy density, and $P_\parallel$ and $P_\perp$ are the parallel and perpendicular pressures, respectively. The terms r and z are the radial and polar radii. The gravitational mass is given by $m = m(r,z)$.

In order to see if Eqs. (46) and (47) are correct solutions to Einstein's field equations, one must examine their mathematical form and units associated with them. As mentioned earlier in this chapter, it is impossible to obtain an interior solution for the axial symmetric Weyl metric. However, if we know what those metric functions are, then we can approximate a solution that would correctly describe the space inside a non-spherical mass distribution. In Eqs. (46) and (47), the pressure gradients are negative, implying that both the parallel and perpendicular pressures decrease monotonically from the center towards the surface of the stellar configuration. However, the units that describe Eqs. (46) and (47) do not yield the expected units of $\mathrm{MeV/fm^4}$. This inconsistency in units stems directly from the assumptions we have made for the flat-space regime.

First, we have assumed a flat local geometry inside a non-spherical mass distribution. Even though the units of the flat-space assumptions of the metric functions $\nu(r,z)$ and $\lambda(r,z)$ cancel out, the internal geometry is not really flat even locally at small radial and polar steps. Secondly, we have made a mathematical assumption of the energy–momentum tensor. We have assumed that the $G^{\phi}{}_{\phi}$ and the off-diagonal terms $G^{r}{}_{z}$ and $G^{z}{}_{r}$, which are the pressure terms described by $\tilde{P}$, are zero. Due to this simplification, we were able to substitute the derivatives of Eqs. (29) and (30) for the metric functions. But at the same time, the Einstein field equations were also simplified where the r–z terms of the Einstein tensor were reduced to equaling zero. The term $\tilde{P}$ is not actually zero but needs to be taken into account when calculating the components of the Einstein tensor.

The reason we assumed that $\tilde{P} = 0$ is that there is no equation of state model to date that takes the off-diagonal pressure into consideration when calculating the pressure gradients as a function of density. By applying these simplifications, it is shown that a direct derivation of the hydrostatic equilibrium equations for parallel and perpendicular directions is nearly impossible to do.

However, this assumption of a local flat space is not a complete loss. By deriving the equations for the flat-space assumption, we are able to gain some valuable insight into the problem. By examining Eqs. (46) and (47), one sees that each equation does not depend on a coupled change of parallel and perpendicular pressures, but the total gravitational mass is incorporated into both equations. The idea that the total gravitational mass must be taken into account for both pressure gradients is feasible considering that the star as a whole must depend on the radial and polar directions. Having this insight into the gravitational mass is helpful in that we can use a spherical solution for hydrostatic equilibrium but also take the total gravitational mass of a deformed object into account. In the next section we present a solution for hydrostatic equilibrium configurations in two dimensions that will take distinct parallel and perpendicular pressure gradients into consideration by using a parameterized spherical solution which will be modified to incorporate the total gravitational mass of a deformed object.

4. An Investigative Parametric Solution

Another way of investigating deformity of neutron stars is by applying a parametrization on the polar radius in terms of the equatorial radius along with a deformation constant γ on the metric described by Eq. (1). Such a metric reads as

$$ds^2 = -e^{2\Phi(r)}dt^2 + \left(1 - \frac{2m(r)}{r}\right)^{-\gamma} dr^2 + r^2 d\theta^2 + r^2 \sin^2(\theta) d\phi^2, \quad (48)$$

as described in [Zubairi *et al.* (2015a)]. In Eq. (48), γ determines the degree of deformation either in the oblate ($\gamma < 1$) or prolate case ($\gamma > 1$) (see Fig. 1), where the parametrization is described by $z = \gamma r$. The Christoffel symbols are then calculated to be

$$\Gamma^r{}_{tt} = \beta\, e^{2\Phi(r)}\Phi'(r),\ \Gamma^t{}_{tr} = \Phi'(r), \quad (49)$$

$$\Gamma^r{}_{rr} = \frac{\gamma\left[-m'(r)r + m(r)\right]}{r[-r + 2m(r)]},\ \Gamma^\theta{}_{r\theta} = \Gamma^\Phi{}_{r\Phi} = \frac{1}{r}, \quad (50)$$

$$\Gamma^r{}_{\theta\theta} = -\beta\, r,\ \Gamma^\Phi{}_{\Phi\Phi} = \cot(\theta), \quad (51)$$

$$\Gamma^r{}_{\Phi\Phi} = -\beta\, r \sin^2(\theta),\ \Gamma^\theta{}_{\theta\theta} = -\sin(\theta)\cos(\theta), \quad (52)$$

where primes denote derivatives with respect to the radial coordinate, r, and

$$\beta \equiv \left(\frac{r - 2m(r)}{r}\right)^\gamma. \quad (53)$$

The components of the Ricci tensor $R^\mu{}_\nu$ for the metric of Eq. (48) are calculated to be

$$R^t{}_t = \frac{1}{r(r - 2m(r))}\Big[\beta\Phi'(r)m'(r)\gamma r - \Phi'(r)m(r)\gamma - (\Phi'(r))^2\, rm(r)$$
$$- \Phi''(r)r^2 + 2\Phi''(r)rm(r) - 2\Phi'(r)r + 4\Phi'(r)m(r)\Big], \quad (54)$$

$$R^r{}_r = \frac{1}{r^2(r - 2m(r))}\Big[- \beta\Phi''(r)r^3 - 2\Phi''(r)r^2m(r) + (\Phi'(r))^2\, r^3$$
$$- 2\,(\Phi'(r))^2\, r^2m(r) - \gamma\Phi'(r)m'(r)r^2 + \gamma\Phi'(r)rm(r) - 2\gamma m'(r)r$$
$$+ 2\gamma m(r)\Big], \quad (55)$$

$$R^{\theta}{}_{\theta} = \frac{1}{r^2(r-2m(r))}\Big[-\beta r^2\Phi'(r) + 2\beta\gamma r\Phi'(r)m(r) + \beta\gamma m'(r)r - \beta\gamma m(r)$$
$$+\ r - 2m(r) - \beta\gamma + 2\beta m(r)\Big], \tag{56}$$

and

$$R^{\phi}_{\phi} = R^{\theta}_{\theta}. \tag{57}$$

The Ricci scalar R is given by

$$R = \frac{2}{r^2(r-2m(r))}\Big[\beta\gamma\Phi'(r)m'(r)r^2 - \beta\gamma\Phi'(r)m(r)r - \beta\left(\Phi'(r)\right)^2 r^3$$
$$+ 2\beta\left(\Phi'(r)\right)^2 r^2 m(r) - \beta\Phi''(r)r^3 + 2\beta\Phi''(r)r^2 m(r) - 2m(r)$$
$$- 2\beta\gamma^2\Phi'(r) + 4\beta\Phi'(r)m(r) + 2\beta\gamma m'(r)r - 2\beta\gamma m(r)$$
$$+ r - 2m(r) - \beta\gamma + 2\beta m(r)\Big]. \tag{58}$$

Using Eqs. (54) through (58) and methods similar to that as described in Sec. 3.2 it was shown in [Zubairi *et al.* (2015a)] that the hydrostatic equilibrium equation is then given by

$$\frac{dP}{dr} = -\frac{(\epsilon + P)\left(\frac{1}{2}r + 4\pi r^3 P - \frac{1}{2}r\left(1 - \frac{2m}{r}\right)^{\gamma}\right)}{r^2\left(1 - \frac{2m}{r}\right)^{\gamma}}. \tag{59}$$

In the case when $\gamma = 1$, Eq. (59) simplifies to the well-known Tolman–Oppenheimer–Volkoff equation [Oppenheimer and Volkoff (1939); Tolman (1939)] which describes the stellar structure of perfectly spherically symmetric object. The gravitational mass of a deformed neutron star parameterized with the deformation constant γ is then described by

$$\frac{dm}{dr} = 4\pi r^2 \epsilon\gamma, \tag{60}$$

so that the total gravitational mass, M, of a deformed neutron star with an equatorial radius R follows as [Herrera *et al.* (1999)] $M = \gamma\, m(R)$.

For a given deformation γ, this is a very useful and easy-to-use equation for the description of deformed stars. However, if the deformation is unknown, it is difficult to derive distinct hydrostatic equilibrium equations that will describe the deformity separately. However, we can use Eq. (59) and make some assumptions about the solution and attempt to find hydrostatic equilibrium equations that will take the parallel and perpendicular pressure gradients into account separately.

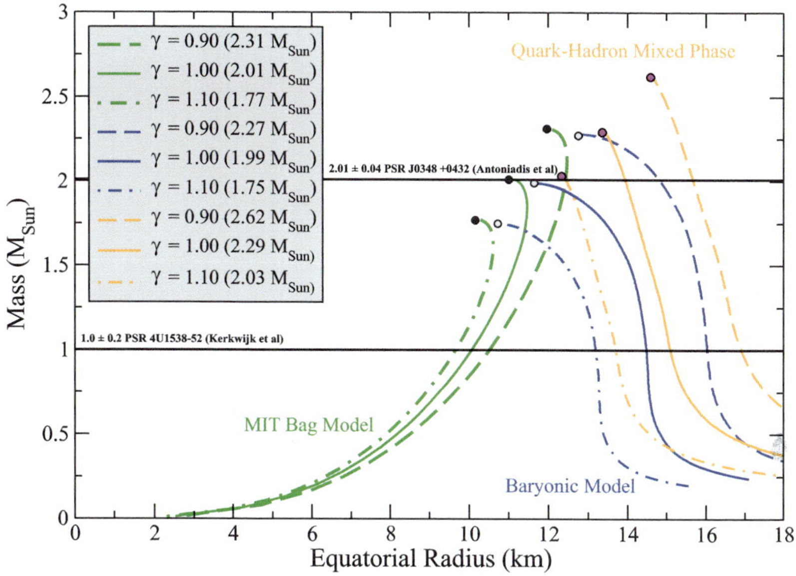

Fig. 3. Mass–radius relationships calculated from Eq. (59) for various isotropic nuclear equations of state.

What we want to accomplish is to investigate deformation due to anisotropies contained in the equation of state. It was shown in [Zubairi *et al.* (2015a)] that if the equation of state is isotropic, but we have deformity, the mass of neutron stars could either increase (oblate stars) or decrease (prolate stars) (see Fig. 1), as shown in Fig. 3. If the parallel and perpendicular pressures are distinct, then deformation can still occur, implying that the type of deformation will also depend on these anisotropies.

5. Fully Self-Consistent Models of Non-Spherical Compact Objects

Using the parametrization $z = \gamma r$ and assuming our radial step size is small (on the order of meters), we can look at the deformation constant γ as a differential ratio of the polar and radial step sizes given by

$$\gamma = \frac{dz}{dr}. \tag{61}$$

Hence, we can use Eq. (61), along with the parametrization $z = \gamma r$, and apply a transformation on Eq. (59) to obtain a relationship between pressure and the polar direction. Using Eq. (61), we find that

$$\frac{dP}{dz} = -\frac{(\epsilon + P)\left(\frac{z}{2\gamma} + 4\pi\left(\frac{z}{\gamma}\right)^3 P - \frac{z}{2\gamma}\left(1 - \frac{2m\gamma}{z}\right)^\gamma\right)}{\frac{z^2}{\gamma^3}\left(1 - \frac{2m\gamma}{z}\right)^\gamma}, \tag{62}$$

where now we can take distinct pressure gradients into account by assuming that the parallel pressure will be associated with dr and the perpendicular pressure will be associated with dz. We will also have to take a look at the total gravitational mass m in Eqs. (59) and (62). This is extremely important due to the fact that both Eqs. (59) and (62) need to be coupled together. For the total gravitational mass, we utilize our mass elements described by

$$\partial_r m(r, z) = 2\pi r z \epsilon(r, z), \tag{63}$$

$$\partial_z m(r, z) = \pi r^2 \epsilon(r, z). \tag{64}$$

If we were to integrate then add the expressions given in Eqs. (63) and (64), we would obtain some total gravitational mass described as

$$\begin{aligned} m_{\text{total}}(r, z) &= 2\pi\epsilon(r, z) r^2 z, \\ &= 2m(r, z). \end{aligned} \tag{65}$$

However this expression for $m_{\text{total}}(r, z)$ does not correctly define the gravitational mass of a oblong spheroid, which is given by

$$M(r, z) = \frac{4}{3}\pi\epsilon(r, z) r^2 z. \tag{66}$$

In order to obtain Eq. (66), we must subtract off another cross-sectional term from Eq. (65) that will ensure the gravitational confinement of the star. This leads to

$$\begin{aligned} M(r, z) &= 2\pi\epsilon(r, z) r^2 z - \frac{2}{3}\pi\epsilon(r, z) r^2 z, \\ &= 2m(r, z) - \frac{2}{3} m(r, z), \end{aligned} \tag{67}$$

however, we must break up the top expression in Eq. (67) equally to take the mass in the equatorial and polar directions separately but that are confined and result in a mass described by Eq. (66). Using the expressions from Eqs. (63) and (64) we find the gravitational mass to be

$$\mathcal{M}(r,z) = 2\pi\epsilon(r,z)rz + \pi\epsilon(r,z)r^2 - \frac{1}{3}\pi\epsilon(r,z)r^2 z,$$
$$= \partial_r m(r,z) + \partial_z m(r,z) - \frac{1}{3}\pi\epsilon(r,z)r^2 z, \tag{68}$$

where we can define

$$\partial_r m(r,z) = m_{||}(r,z),$$
$$\partial_z m(r,z) = m_{\perp}(r,z). \tag{69}$$

The total gravitational mass is then given by

$$\mathcal{M}_{\text{total}}(r,z) = 2\pi\epsilon(r,z)r^2 z - \frac{2}{3}\pi\epsilon(r,z)r^2 z. \tag{70}$$

Using the expressions given in Eq. (68), we make substitutions into Eqs. (59) and Eq. (62) along with the substitutions for the parallel and perpendicular pressures and find that the hydrostatic equilibrium equations modify to

$$\frac{\partial P_{||}}{\partial r} = -\frac{(\epsilon + P_{||})\left(\frac{1}{2}r + 4\pi r^3 P_{||} - \frac{1}{2}r\left(1 - \frac{2\mathcal{M}(r,z)}{r}\right)^{\gamma}\right)}{r^2\left(1 - \frac{2\mathcal{M}(r,z)}{r}\right)^{\gamma}}, \tag{71}$$

$$\frac{\partial P_{\perp}}{\partial z} = -\frac{(\epsilon + P_{\perp})\left(\frac{z}{2\gamma} + 4\pi\left(\frac{z}{\gamma}\right)^3 P_{\perp} - \frac{z}{2\gamma}\left(1 - \frac{2\mathcal{M}(r,z)\gamma}{z}\right)^{\gamma}\right)}{\frac{z^2}{\gamma^3}\left(1 - \frac{2\mathcal{M}(r,z)\gamma}{z}\right)^{\gamma}}, \tag{72}$$

where now the expressions given in Eqs. (71) and (72) are coupled together with the gravitational mass $\mathcal{M}(r,z)$ all governed by distinct parallel $(P_{||})$ and perpendicular $(P_{\perp})$ pressure gradients. It is also important to note that now we do not need to deform the star by γ. For all of our calculations, we fixed γ, so that $\gamma = 1$. Having this fixed constant will allow the anisotropies in the equation of state to dictate the deformation.

Hence, our two dimensional parameterized stellar model will consist of solving the expressions in Eqs. (71) and (72) in conjunction with the expressions given in Eqs. (63) and (64). The surface of the star will be defined when the pressure gradients both in the parallel and perpendicular directions vanish. That is, we will define the surface of the star by

$$P_{||}(r = R) = 0, \tag{73}$$
$$P_{\perp}(z = Z) = 0, \tag{74}$$

where the total gravitational mass of the object is given by

$$M_{\text{Total}}(R, Z) = \mathcal{M}_{\text{total}}. \tag{75}$$

In order to solve our coupled system of equations, we now need an equation of state which has distinct pressure gradients in the parallel and perpendicular directions. The next section briefly describes the equation of state used in these two dimensional calculations.

6. Non-Isotropic Equations of State

In the one dimensional parameterized deformed stellar model described in Fig. 3, three different isotropic equation of state models were used to produce stellar properties such as masses and radii. The three EoS models which produce the stellar properties in Fig. 3 are shown in Table 1 which summarizes key differences of each model. In the two dimensional model, we modify the quark–hadron mixed phase equation of state (Model III) by changing the pressure by various amounts greater than and less than the spherical case. Modifying the equation of state this way will ensure some symmetry in our calculations. We took the data presented in [Contrera *et al.* (2014)], and changed the pressure terms by increasing the pressure by 4, 8, and 12 percent. We also decreased the pressure by 4, 8, and 12 percent. Graphically, this is illustrated in Fig. 4. Using the data shown in Fig. 4, we can now solve our two dimensional system of hydrostatic equilibrium for deformed neutron stars. Hence, with distinct pressure gradients, deformity should result; however, it should not be large. We have only deviated from the isotropic case by a few percent, and therefore the deformation of spheres to oblate and prolate stars should vary slightly.

Table 1. Equation of state models studied in this work.

Model	Equation of state	Composition	Stellar object
Model I	MIT bag model	u, d, s quarks	Strange quark star
Model II	Standard baryonic	n, p, e, Σ, Λ, Ξ	Neutron star
Model III	Quark–hadron mixed phase	n, p, e, Σ, Λ, Ξ, u, d, s quarks	Neutron star

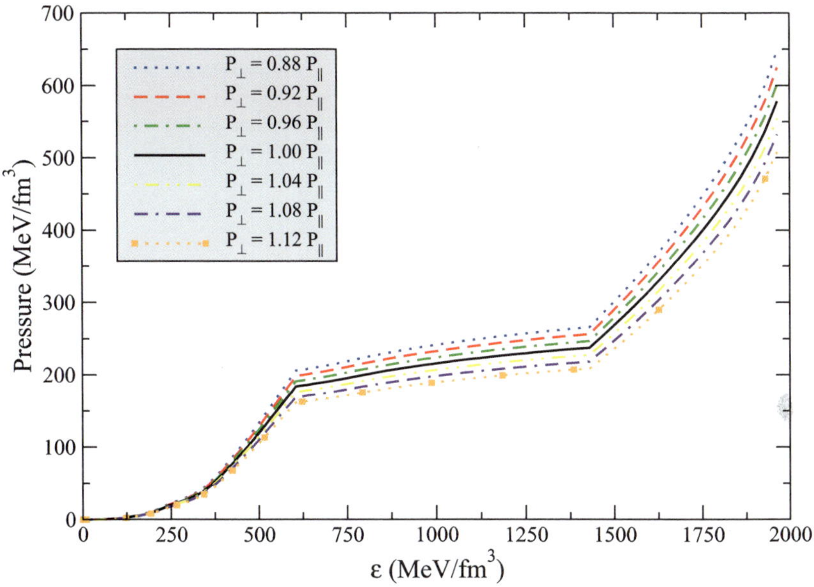

Fig. 4. (Color online) Parameterized anisotropic relationship of pressure and energy density for a quark–hadron equation of state. In this model, the core is comprised of a mixed phase of up, down and strange quarks and hadronic matter. The pressure is varied by 4, 8, and 12 percent to mimic anisotropies in the pressure gradients.

7. Results and Discussion

In contrast to standard spherical models, solving our two dimensional model requires other strategies. In our case, we simply cannot have mass–radius relationships to describe the maximum mass. We can however choose a central density and build our deformed star and calculate the pressure and energy density profiles associated with that star. From the pressure profiles, we will be able to tell the deformity. Due to the non-isotropic equation of state, we can investigate the change in mass.

We begin with choosing a central density of 933 $\mathrm{MeV/fm^3}$. This central density was chosen from the one dimensional parameterized model for a spherical object [Zubairi *et al.* (2015b)]. Using this density, we calculate the pressure and density profiles along with the total mass of the object. We first look at the pressure and density profiles associated with increased pressures

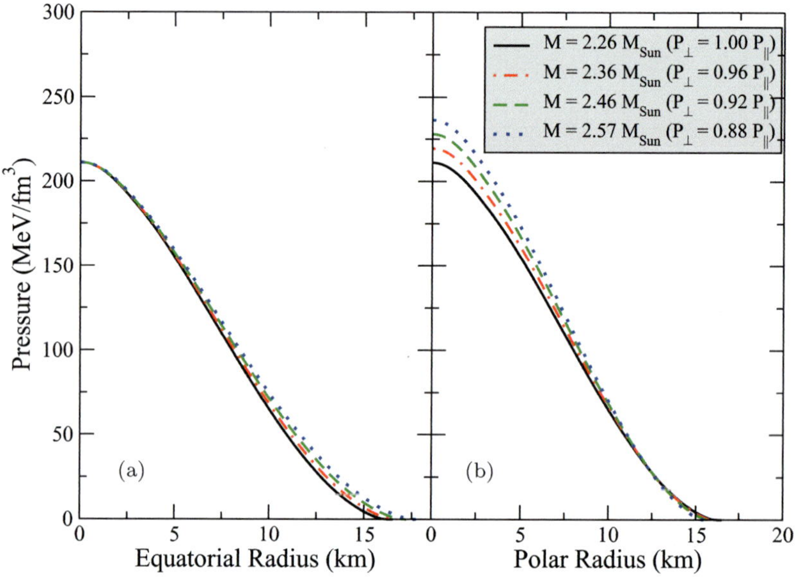

Fig. 5. (Color online) Pressure profiles for both the equatorial (a) and polar (b) directions for pressure gradients in the equatorial direction greater than 4, 8, and 12 percent. The masses of such stars increase with oblateness.

of 4, 8, and 12 percent. The results are shown in Figs. 5 and 7 respectively. The pressure profiles for each maximum mass star in Fig. 5 converge to zero at the stellar surfaces for both the equatorial and polar radii. A zoomed-in version of the pressure profiles shown in Fig. 5 is provided in Fig. 6.

As shown in Figs. 5 and 6, for no change in pressure, the mass is about 2.26 $M_{\odot}$ which is in good agreement with results from [Contrera *et al.* (2014); Zubairi *et al.* (2015b)] for the spherical case. From our calculations, the mass increases as the equatorial radius increases and the polar radius decreases. This increase in mass due to increased oblateness is consistent with our one dimensional parameterized model and validates our results presented using our parametrization. It is also important to note that from the pressure profiles shown in Figs. 5 and 6, the pressure decrease monotonically as the equatorial and polar radii increase to the surface and hence the pressure becomes zero as stated by our boundary conditions. The increase in mass is appreciable even though we have small changes in the pressure in our equation of state.

We next investigate the deformation in another way to obtain prolate spheroids and see if the masses decreases as we increase prolateness. We

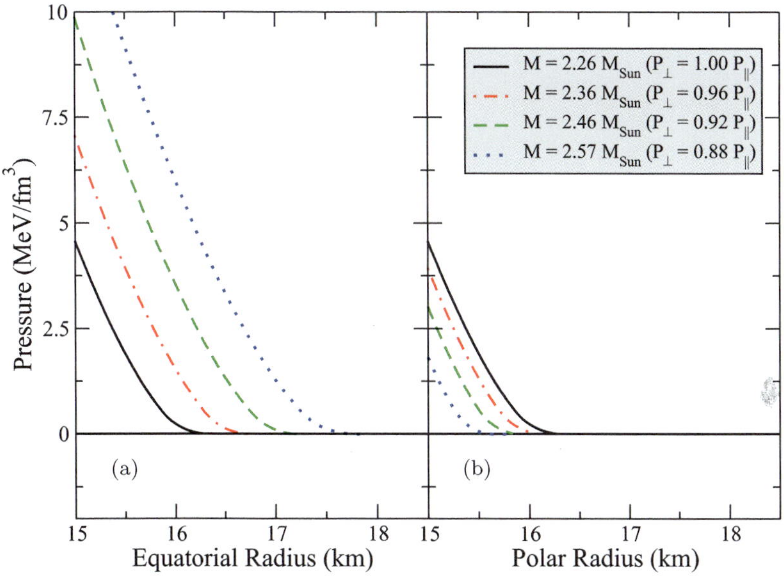

Fig. 6. (Color online) Zoomed-in version of Fig. 5. The radial locations where the pressure becomes zero define the stars' surface. The pressure profiles being larger as we increase the anisotropies in the pressure results in a decrease of pressure in the polar radius implying we get oblate spheroids.

do this by calculating the pressure profiles by decreasing the pressure by 4, 8 and 12 percent from our non-isotropic equation of state. However, we have to be mindful when applying the different pressure gradients. We do not want to rotate the star by any means. Therefore, we decrease the pressure gradients in the equatorial direction, but keep the pressure in the polar direction fixed. This will ensure we obtain prolate spheroid but not an oblate spheroid just rotated by 90 degrees. The results of the pressure and density profiles calculated this way are shown in Figs. 8 and 10 respectively. The zoomed-in version of the pressure profiles in Fig. 8 is shown in Fig. 9. Just as in the oblate case, we see clearly that the mass decreases as we increase prolateness. This is consistent with the results from [Zubairi *et al.* (2015a,b)] for the one dimensional parameterized model for prolate stars. We see significant changes in mass due to increasing prolateness. It is important to note that the mass change (either increase or decrease) is roughly the same and it deviates slightly. This implies the two dimensional calculations are more sensitive to the anisotropies in the equation of state.

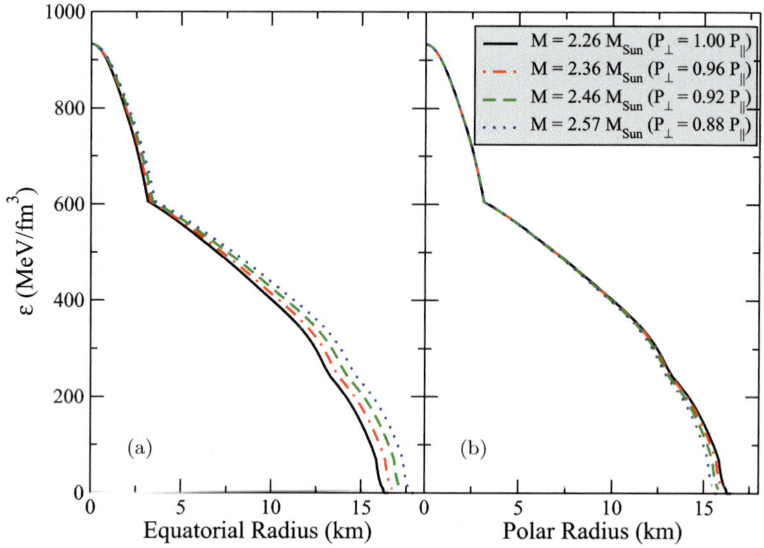

Fig. 7. (Color online) Energy density profiles for both the equatorial (a) and polar (b) directions for pressure gradients in the equatorial direction greater than the polar direction by 4, 8, and 12 percent.

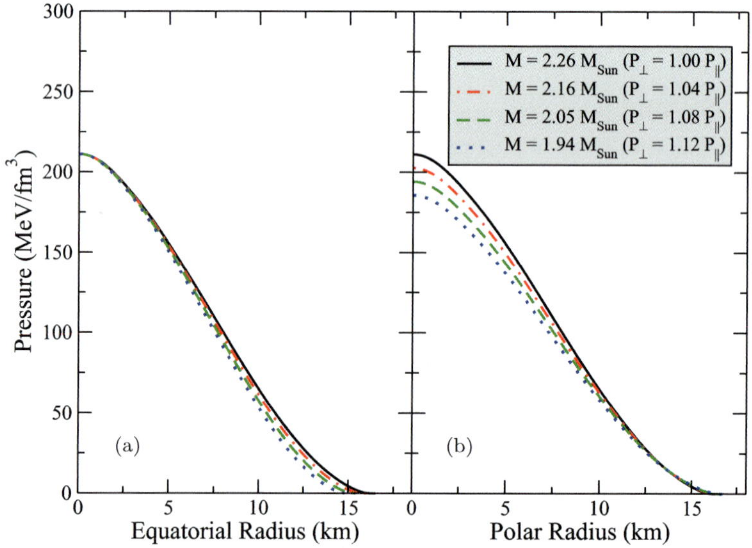

Fig. 8. (Color online) Pressure profiles for both the equatorial (a) and polar (b) directions for pressure gradients in the equatorial direction less than the polar direction by 4, 8, and 12 percent. The masses of such stars decrease for increasing prolateness.

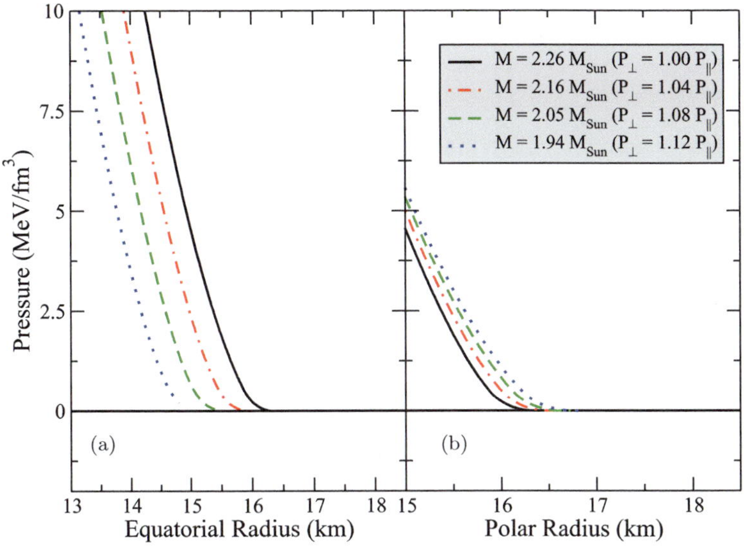

Fig. 9. (Color online) Zoomed-in version of Fig. 8. The radial locations where pressure becomes zero define the stellar surfaces.

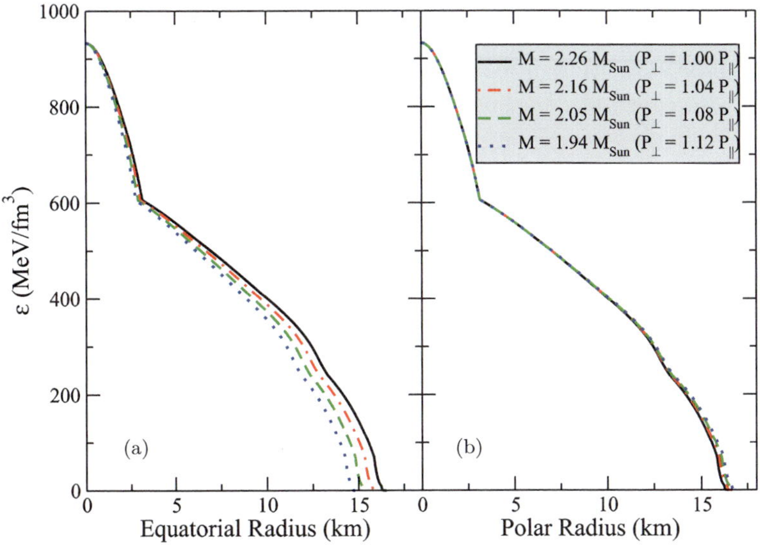

Fig. 10. (Color online) Energy density profiles for both the equatorial (a) and polar (b) directions for pressure gradients in the equatorial direction less than the polar direction by 4, 8, and 12 percent.

Table 2. Stellar properties of deformed neutron stars for the non-isotropic equations of state given in Fig. 4.

Parameterized EoS	R (km)	Z (km)	Total mass ($M_\odot$)	Shape
$P_\perp = 0.88P_\parallel$	17.92	15.77	2.57	Oblate
$P_\perp = 0.92P_\parallel$	17.21	15.83	2.46	Oblate
$P_\perp = 0.96P_\parallel$	16.73	16.06	2.36	Oblate
$P_\perp = 1.00P_\parallel$	16.51	16.51	2.26	Sphere
$P_\perp = 1.04P_\parallel$	16.00	16.67	2.16	Prolate
$P_\perp = 1.08P_\parallel$	15.39	16.73	2.05	Prolate
$P_\perp = 1.12P_\parallel$	14.80	16.82	1.94	Prolate

A summary of the stellar properties resulting from our two dimensional model is provided in Table 2.

From our results given in Figs. 5 through 10 along with the values listed in Table 2, we clearly show that deformation plays a vital role in the stellar structure of neutron stars and we cannot simply ignore this deformation. From our results, we conclude that anisotropies in the equation of state can deform the star either in the equatorial or polar direction resulting in oblate or prolate spheroids. Depending on the deformation the mass could either increase or decrease. This change in mass could explain discrepancies in various equation of state models and help us better understand the internal structure of neutron stars.

Acknowledgments

This work is supported through the National Science Foundation under grants PHY-1411708 and DUE-1259951.

References

Chandrasekhar, S. and Fermi, E. (1953), *Astrophys. J.* **118**, 116.

Contrera, G. A., Spinella, W., Orsaria, M. and Weber, F. (2014), arXiv: 1403.7415 [hep-ph].

Ferraro, V. C. A. (1954), *Astrophys. J.* **119**, 407.

Ferrer, E. J., de la Incera, V., Keith, J. P., Portillo, I. and Springsteen, P. L. (2010), *Phys. Rev. C.* **82**, 6.

Goossens, M. (1972), *Ap & SS* **16**, 286.

Haskell, B., Samuelsson, S., Glampedakis, K., and Andersson, N. (2008), *MNRAS* **385**, 531.

Hernandez, W. (1967), *Phys. Rev.* **153**, 1359.

Herrera, L., Paiva, F. M., and Santos, O. N. (1999), *J. Math Phys.* **40**, 4064.

Herrera, L., Prisco, A. D., Ibanez, J., and Ospino, J. (2013), arXiv:1301.2424v1 [gr-qc].

Katz, J. I. (1989), *MNRAS* **239**, 751.

Mereghetti, S. (2008), *Astron. Astrophys. Rev* **15**, 225.

Oppenheimer, J. R. and Volkoff, G. M. (1939), *Phys. Rev.* **55**, 374.

Payne, D. J. B. and Melatos, A. (2004), *MNRAS* **351**, 569.

Thompson, C. and Duncan, R. C. (1995), *MNRAS* **275**, 255.

Thompson, C. and Duncan, R. C. (1996), *Astrophys. J.* **473**, 322.

Tolman, R. C. (1939), *Phys. Rev.* **55**, 364.

Weber, F. (1999), *Pulsars as Astrophysical Laboratories for Nuclear and Particles Physics* (IoP Publishing, London).

Weyl, H. (1918), *Ann. Phys.* (*Leipzig*) **54**, 117.

Zubairi, O. (2015), *An Investigation of Deformation of the Stellar Structure of Neutron Stars* (Montezuma Publishing, San Diego State University).

Zubairi, O., Romero, A., and Weber, F. (2015a), *J. Phys. Conf.: Ser.* **615**, 012003.

Zubairi, O., Spinella, W., Romero, A., Mellinger, R., Weber, F., Orsaria, M., and Contrera, G. (2015b), arXiv:1504.03006 [astro-ph.SR].

Chapter 3

Pseudo-Complex General Relativity: Theory and Observational Consequences

Peter O. Hess* and Walter Greiner†

*Instituto de Ciencias Nucleares, UNAM, Circuito Exterior
C.U., A.P. 70-543, 04510, Mexico D.F., Mexico
†Frankfurt Institute for Advanced Studies, Wolfgang Goethe University
Ruth-Moufang-Strasse 1, 60438 Frankfurt am Main, Germany
hess@nucleares.unam.mx, greiner@fias.uni-frankfurt.de

General Relativity (GR) is algebraically extended to pseudo-complex (pc) coordinates. The new theory, called pc-GR, contains a minimal length and in addition requires the appearance of an energy–momentum tensor, related to the vacuum fluctuations (dark energy), provoked by the presence of the central mass. The dark energy density increases toward the central mass and avoids the appearance of an event horizon. Observational consequences are presented, related to *Quasi-Periodic Oscillations* in accretion disks around so-called *galactic black holes*, and the structure of these disks.

Contents

1. Introduction: General Relativity

The Theory of General Relativity (GR) [Misner *et al.* (1973)] had its 100th anniversary in 2015. It is not only one of the most beautiful cultural achievements of mankind but also a billion dollar enterprise, considering the *Global Positioning System* (GPS), which is unthinkable without the new concept of time, introduced by Einstein. Up until now, the theory has passed many observational tests [Will (2006)]; however, all tests were realized in weak gravitational fields. New ones have to be performed in strong gravitational fields near the Schwarzschild radius.

The standard GR predicts singularities, due to the black-hole solution, which has a singularity in the center and a coordinate singularity, called the *event horizon*, which for non-rotating black holes is at the Schwarzschild radius. No information beyond this event horizon can reach an external observer, even nearby.

In our philosophical understanding, *no theory should have a singularity*, even a coordinate singularity of the type discussed above, and the appearance of a singularity hints to the incompleteness of the theory. In addition, one has to take into account that *no direct proof of the existence of the event horizon exists* [Abramowicz *et al.* (2002)], yet.

In this contribution we present the current status of an algebraic extension of GR, involving *pseudo-complex* (pc) variables. Earlier extensions of GR to a complex metric were presented by A. Einstein [Einstein (1945, 1948)] and an algebraic extension to complex variables in [Mantz (2007); Mantz and Prokopec (2011); Moffat (1979)]. We will argue why the use of pc-variables is consistent; in fact, the extension to pc-variables is the only viable way to avoid the appearance of non-physical particles.

One consequence is the presence of an energy–momentum tensor on the right hand side of the Einstein equations, which will be associated to dark energy. The accumulation of dark energy toward the central mass affects the classical form of the metric tensor, such that the *event horizon will disappear*. A general principle emerges, namely that *mass not only curves the space* (which leads to the standard GR) *but also changes the space (vacuum) structure in its vicinity*, leading to an important deviation from the classical solution.

Because no quantized theory of gravitation exists yet, we are led to the construction of models for the distribution of the dark energy. In our theory it will be treated as an ideal anisotropic fluid in the exterior of a star. This permits to solve the Einstein equations with the modified metric and the disappearance of the event horizon. In [Hess *et al.* (2009); Schönenbach *et al.* (2012)] the details of the theory are presented and here we will assume the main ideas and the results of some calculations. This enables us to avoid to get too much into detail which could potentially obscure the line of arguments.

In [Schönenbach, Caspar *et al.* (2013)] the consequences for the case of the Kerr solution were investigated. It was shown that particles in a circular orbit exhibit a maximum in its orbital frequency, which from there on is decreasing again toward lower radial distances.

In [Schönenbach *et al.* (2014)] a model for the accretion disk around a central mass was applied and simulations for it were performed, allowing some important predictions for its observation, which can discriminate between GR and pc-GR. In this contribution we will add new results and simulate the accretion disks as seen from several angles in the line of sight to an observer.

The outline of the paper is as follows. In Section 2 the theory of pc-GR is described with a short introduction to the mathematics of pc-variables. Also arguments are presented on why only the extension to pc-coordinates is consistent and not the extensions to complex variables or quaternions and others. In addition we show that the formerly proposed modification on the variational principle is equivalent to a standard variational principle with a constraint. In Section 3 circular orbits are considered and the consequences compared to the observations of accretion disks around *galactic black holes*, where hints to deviations to standard GR are already observed. In Section 4 simulations of accretions disks are presented for various angles of view to the observer relative to the accretion disk (nearly from above, at 45° and nearly head on). In Section 5 we present further results, related to stable stars and cosmological models, including some brief thoughts on an *oscillating universe.* In Section 6 conclusions are drawn.

In this contribution we set c and κ, the gravitational constant, to 1, except in places where for illustrative reasons it is preferred to maintain c and κ. For the metric we use the signature $(- + ++)$.

2. Pseudo-Complex General Relativity (pc-GR)

An algebraic extension of GR consists of mapping the real coordinates to a different type, as for example complex or pseudo-complex (pc) variables

$$X^\mu = x^\mu + I y^\mu, \tag{1}$$

with $I^2 = \pm 1$ and where x^μ is the standard coordinate in spacetime and y^μ the complex component. When $I^2 = -1$ it denotes *complex* variables, while when $I^2 = +1$ it denotes pseudo-complex (pc) variables.

As shown in [Kelly and Mann (1986)], where algebraic extensions to more general than just complex or pseudo-complex variables were considered too, only the pseudo-complex algebraic extension is consistent (called in [Kelly and Mann (1986)] differently). The rough reasoning is as follows: when weak gravitational fields are considered

$$g_{\mu\nu} = \eta_{\mu\nu} + h_{\mu\nu}, \tag{2}$$

with $\eta_{\mu\nu}$ as the Minkowsky metric and $h_{\mu\nu}$ the complex correction, field equations are obtained for the real and imaginary part of $h_{\mu\nu}$. The propagator for the imaginary part has the factor I^2. A negative sign in the propagator implies a *ghost solution*, which is unphysical. Only if $I^2 = +1$ no ghost solutions appear.

Before we proceed further, some elementary properties of pc-variables have to be recalled:

- Instead of the division in a pseudo-real and a pseudo-complex component, there is an alternative form

$$\begin{aligned} X^\mu &= X^\mu_+ \sigma_+ + X^\mu_- \sigma_-, \\ \sigma_\pm &= \frac{1}{2}\left(1 \pm I\right). \end{aligned} \tag{3}$$

- The $\sigma_\pm$ satisfy the relations

$$\sigma_\pm^2 = \sigma_\pm, \; \sigma_+\sigma_- = 0. \tag{4}$$

- Due to the last property in (4), when multiplying one variable proportional to σ_+ by another one proportional to σ_-, the result is zero, i.e. there is a *zero divisor*. The variables, therefore, do not form a *field* but a *ring*.
- In both zero-divisor component ($\sigma_\pm$) the analysis is very similar to the standard complex analysis.

In pc-GR the metric is also pseudo-complex

$$g_{\mu\nu} = g^+_{\mu\nu}\sigma_+ + g^-_{\mu\nu}\sigma_-. \tag{5}$$

Due to $\sigma_+\sigma_- = 0$, a GR theory can be constructed in a completely independent way in each zero divisor. The problem is how to connect the two zero-divisor components in order to provide a new theory.

In [Hess *et al.* (2009); Schönenbach *et al.* (2012); Hess *et al.* (2015)] a modified variational principle was proposed and the argument is as follows: when the action $S = S_+\sigma_+ + S_-\sigma_-$ is varied, it can be done independently in each zero-divisor component (the one in σ_+ and the other one in σ_-), which leads to $\delta S_\pm = 0$. This gives two independent equations of motion with no connection at all. In order that a new theory emerges, the variation of the action is set to be proportional to a zero-divisor component, i.e., only proportional either to σ_+ or to σ_-. The pseudo-complex form of the Einstein equations has then an additional contribution on the right hand side, which can be associated to the dark energy. In order to get physical results a mapping to pseudo-real values had to be applied.

In [Hess *et al.* (2015)] an alternative way was proposed, which leads to the same result more directly and here we will explore it in more detail and complete the derivation.

The infinitesimal pc length element squared is given by

$$\begin{aligned} d\omega^2 &= g_{\mu\nu} dX^\mu dX^\nu \\ &= g^+_{\mu\nu} dX^\mu_+ dX^\nu_+ \sigma_+ + g^-_{\mu\nu} dX^\mu_- dX^\nu_- \sigma_- , \end{aligned} \tag{6}$$

as written in the zero-divisor components. In terms of the pseudo-real and pseudo-imaginary components, we have

$$\begin{aligned} d\omega^2 &= g^s_{\mu\nu}\left(dx^\mu dx^\nu + dy^\mu dy^\nu\right) + g^a_{\mu\nu}\left(dx^\mu dy^\nu + dy^\mu dx^\nu\right) \\ &\quad + I\left[g^a_{\mu\nu}\left(dx^\mu dx^\nu + dy^\mu dy^\nu\right) + g^s_{\mu\nu}\left(dx^\mu dy^\nu + dy^\mu dx^\nu\right)\right], \end{aligned} \tag{7}$$

with

$$\begin{aligned} g^s_{\mu\nu} &= \frac{1}{2}\left(g^+_{\mu\nu} + g^-_{\mu\nu}\right), \\ g^a_{\mu\nu} &= \frac{1}{2}\left(g^+_{\mu\nu} - g^-_{\mu\nu}\right). \end{aligned} \tag{8}$$

The upper indices s and a refer to a *symmetric* and *antisymmetric* combination of the metrics.

The connection between the two zero-divisor components is achieved by requiring that the infinitesimal length element squared in (7) *is real*, i.e.,

$$I\left[g^a_{\mu\nu}\left(dx^\mu dx^\nu + dy^\mu dy^\nu\right) + g^s_{\mu\nu}\left(dx^\mu dy^\nu + dy^\mu dx^\nu\right)\right] = 0, \tag{9}$$

or written in terms of the zero-divisor components

$$\left(\sigma_+ - \sigma_-\right)\left(g^+_{\mu\nu} dX^\mu_+ dX^\nu_+ - g^-_{\mu\nu} dX^\mu_- dX^\nu_-\right) = 0. \tag{10}$$

This is a constraint and has to be taken into account at the moment a *standard* variational procedure is applied, leading to an additional contribution in the Einstein equations, interpreted as an energy–momentum tensor. The $dX^\mu_\pm$ describe the status of the local motion of a fluid element, which are fixed but arbitrary. The 4-velocity components are defined by $u^\mu = \frac{dx^\mu}{ds}$.

The action which will be used is given by

$$S = \int dx^4 \sqrt{-g}\Big(\mathcal{R} + 2\alpha\Big), \tag{11}$$

where $\mathcal{R}$ is the Riemann scalar. The last term in the action integral allows to introduce the cosmological constant in cosmological models, where α has to be constant in order not to violate the Lorentz symmetry. This, however changes, when a system with a uniquely defined center is considered, which has spherical (Schwarzschild) or axial (Kerr) symmetry. In these cases, the α is allowed to be a function in r, for the Schwarzschild solution, and a function in r and ϑ, for the Kerr solution.

The variation of the action with respect to the metric is

$$\frac{\delta S}{\delta g_{\mu\nu}}\delta g_{\mu\nu} = \frac{\delta S_+}{\delta g^+_{\mu\nu}}\delta g^+_{\mu\nu}\sigma_+ + \frac{\delta S_-}{\delta g^-_{\mu\nu}}\delta g^-_{\mu\nu}\sigma_- \;=\; 0. \tag{12}$$

It is performed independently in both zero-divisor components, leading to

$$\int \left[\mathcal{R}^{\mu\nu}_{\pm} - \frac{1}{2}g^{\mu\nu}_{\pm}\mathcal{R}_{\pm} - \alpha g^{\mu\nu}_{\pm}\right]\delta g^{\pm}_{\mu\nu}\sqrt{-g}d^4x = 0, \tag{13}$$

where $\mathcal{R}^{\mu\nu}_{\pm}$ is the Ricci tensor and $\mathcal{R}_{\pm}$ the Riemann scalar in each zero-divisor component. The expression in the square bracket can not be set to zero, due to the constraint which renders the metric variables linearly dependent.

Adding the constraint leads to

$$\int \left[\mathcal{R}^{\mu\nu}_{\pm} - \frac{1}{2}g^{\mu\nu}_{\pm}\mathcal{R}_{-} - \alpha g^{\mu\nu}_{\pm} - \lambda\Big(\dot{X}^{\mu}_{\pm}\dot{X}^{\nu}_{\pm}\Big)\right]\delta g^{\pm}_{\mu\nu}\sqrt{-g}d^4x = 0. \tag{14}$$

Now the expression in the square bracket can be set to zero, leading to the modified Einstein equations

$$\begin{aligned}\mathcal{R}^{\pm}_{\mu\nu} - \frac{1}{2}g^{\pm}_{\mu\nu}\mathcal{R}_{\pm} &= \lambda\Big(\dot{X}^{\pm}_{\mu}\dot{X}^{\pm}_{\nu}\Big) + \alpha g^{\pm}_{\mu\nu}\\ &= \lambda u_{\mu}u_{\nu} + \lambda\Big(\dot{y}_{\mu}\dot{y}_{\nu} \pm u_{\mu}\dot{y}_{\nu} \pm u_{\nu}\dot{y}_{\mu}\Big) + \alpha g^{\pm}_{\mu\nu}\\ &= 8\pi T^{\Lambda}_{\pm\ \mu\nu},\end{aligned} \tag{15}$$

where the derivative of the coordinate ($X^{\mu}_{\pm}$) with respect to $s = ct = t$ was expressed in terms of the pseudo-real and pseudo-imaginary component of $X^{\mu}_{\pm}$. The right hand side of the Einstein equations is identified with an energy–momentum tensor, using the appropriate expression of λ [Hess *et al.* (2015)] (also, see further below).

Now, we will analyze in more detail the right hand side of (15), mapping it to the pseudo-real part

$$T^{\Lambda}_{\mu\nu,R} = \frac{1}{2}\left(T^{\Lambda}_{+\ \mu\nu} + T^{\Lambda}_{-\ \mu\nu}\right). \tag{16}$$

Defining

$$\begin{aligned} \lambda &= 8\pi\tilde{\lambda},\ \alpha \ = \ 8\pi\tilde{\alpha} \\ \tilde{\lambda} &= \left(p^{\Lambda}_{\vartheta} + \varrho_{\Lambda}\right),\ \tilde{\alpha} \ = \ p^{\Lambda}_{\vartheta},\ \tilde{\lambda} y_{\mu} y_{\nu} \ = \ \left(p^{\Lambda}_{r} - p^{\Lambda}_{\vartheta}\right) k_{\mu} k_{\nu}, \end{aligned} \tag{17}$$

the real part of the energy–momentum tensor acquires the structure of an ideal anisotropic fluid

$$T^{\Lambda}_{\mu\nu,R} = \left(\varrho_{\Lambda} + p^{\Lambda}_{\vartheta}\right) u_{\mu} u_{\nu} + p^{\Lambda}_{\vartheta} g_{\mu\nu} + \left(p^{\Lambda}_{r} - p^{\Lambda}_{\vartheta}\right) k_{\mu} k_{\nu}, \tag{18}$$

where p^{Λ}_{ϑ} and p^{Λ}_{r} are the tangential and radial pressures, respectively. For an isotropic fluid $p^{\Lambda}_{\vartheta} = p^{\Lambda}_{r} = p^{\Lambda}$. The u^{μ} are the components of the 4-velocity of the elements of the fluid and k^{μ} is a space-like vector ($k_{\mu}k^{\mu} = 1$) in the radial direction. It satisfies the relation $u_{\mu}k^{\mu} = 0$. The fluid is anisotropic due to the presence of y_{μ}. Finally, we apply in (15) the mapping of $T^{\Lambda}_{\pm\ \mu\nu}$, $\mathcal{R}^{\pm}_{\mu\nu}$, $\mathcal{R}^{\pm}$ and $g^{\pm}_{\mu\nu}$ to their real part, in order to obtain the final set of the Einstein equations.

Why an anisotropic fluid makes sense? This is understood by analyzing the *Tolman–Oppenheimer–Volkov* (TOV) equations for an isotropic fluid [Adler *et al.* (1975)], which relates the derivative with respect to r of the radial dark energy pressure $p^{\Lambda}_{\vartheta} = p^{\Lambda}_{r} = p^{\Lambda}$ to the same pressure and the dark energy density ϱ_{Λ}. Assuming the isotropic fluid and that the equation of state for the dark energy is $p^{\Lambda} = -\varrho_{\Lambda}$, the factor $(p^{\Lambda} + \varrho_{\Lambda})$ in the TOV equation for $\frac{dp^{\Lambda}}{dr}$ is zero, i.e., the pressure derivative is zero. As a result the pressure is constant and with the equation of state also the density is constant. This is obviously an incorrect result, because the dark energy density has to decrease with r. The consideration shows that the fluid has to be anisotropic: for an anisotropic fluid, in the equation of the radial pressure derivative, $\frac{dp^{\Lambda}_{r}}{dr}$, an additional term

$$\frac{2}{r}\Delta p^{\Lambda} = \frac{2}{r}\left(p^{\Lambda}_{\vartheta} - p^{\Lambda}_{r}\right) \tag{19}$$

is added [Hess *et al.* (2014)]. This permits a decreasing dark energy density as a function in r.

For the modeling of the density, we referred to results obtained in calculating the Casimir effect, performed in [Visser (1996)]. The semi-classical

Quantum Mechanics [Birrell and Davies (1986)] was applied, which assumes a *fixed* background metric and is thus only valid for not too strong gravitational fields. Far away from the Schwarzschild radius the vacuum energy density falls off proportional to $1/r^6$. However, near the Schwarzschild radius the field is very strong which manifested in a singularity of the vacuum fluctuations [Visser (1996)] (explosion of the dark energy density).

Because we treat the vacuum fluctuations as a classical ideal anisotropic fluid, we are free to propose a different fall-off of the negative energy density, which is finite at the Schwarzschild radius. The main result obtained in [Visser (1996)] is only used as an argument that the density has to increase toward the center.

Behind this finding is a general principle, namely that *mass not only curves the space but also changes the space (vacuum) properties, which in turn influence the metric.* As a consequence *the event horizon disappears*!

For the dark energy density we used

$$\varrho_\Lambda = \frac{1}{8\pi}\frac{B}{r^5}, \tag{20}$$

which is strong enough in order not to contribute to the known observations within the solar system and other systems with not too strong gravitational fields [Will (2006)].

With the assumed density, the metric for the Kerr solution changes to [Schönenbach *et al.* (2012)]

$$\begin{aligned}
g_{00} &= -\frac{r^2 - 2mr + a^2\cos^2\vartheta + \frac{B}{2r}}{r^2 + a^2\cos^2\vartheta}, \\
g_{11} &= \frac{r^2 + a^2\cos^2\vartheta}{r^2 - 2mr + a^2 + \frac{B}{2r}}, \\
g_{22} &= r^2 + a^2\cos^2\vartheta, \\
g_{33} &= (r^2 + a^2)\sin^2\vartheta + \frac{a^2\sin^4\vartheta\left(2mr - \frac{B}{2r}\right)}{r^2 + a^2\cos^2\vartheta}, \\
g_{03} &= \frac{-a\sin^2\vartheta\, 2mr + a\frac{B}{2r}\sin^2\vartheta}{r^2 + a^2\cos^2\vartheta},
\end{aligned} \tag{21}$$

where $0 \leq a \leq 1$ is the spin parameter of the Kerr solution, in units of m. The Schwarzschild solution is obtained, setting $a = 0$. The parameter $B = bm^3$ measures the coupling of the dark energy to the central mass.

In the Schwarzschild solution it is easier to see that b can be chosen such that *no event horizon appears*: One has

$$-g_{00} = \left(1 - \frac{2m}{r} + \frac{B}{2r^3}\right), \tag{22}$$

and a lower limit of the parameter B is determined, requiring that g_{00} is always larger than one. For the parameter b this means $b > \frac{64}{27}$. For the calculations presented in the rest of this contribution we use for simplicity $b = \frac{64}{27}$.

3. Circular Orbits and the so-called Galactic Black Holes

In the first example, circular orbits around a large mass concentration are considered. The detailed calculation is presented in [Schönenbach, Caspar *et al.* (2013)]. The derivation of the orbital frequency is trivial, starting from the Lagrangian

$$L = g_{00}c^2\dot{t}^2 + g_{11}\dot{r}^2 + g_{22}\dot{\vartheta}^2 + g_{33}\dot{\varphi}^2 + 2g_{03}c\dot{t}\dot{\varphi} = \frac{ds^2}{ds^2} = -1, \tag{23}$$

with $\dot{r} = \frac{dr}{ds}$ and using the standard Euler–Lagrange equations.

A circular orbit is defined by $\dot{r} = 0$ and $\vartheta = \frac{\pi}{2}$, i.e., the motion is in the orbital plane. The result is presented in Fig. 1.

The green (upper) curve is the result within GR while the red (lower) one is within pc-GR. The ω in pc-GR is always lower than in GR and furthermore a maximum appears, which is due to the fact that the derivatives of the metric components start to decrease from a certain point on to lower values of r.

The behavior of the curve in pc-GR can be understood easier as an effective lowering of the gravitational constant. The

$$-g_{00} = 1 - \frac{2m}{r} + \frac{bm^3}{2r^3} \tag{24}$$

can be rewritten as

$$1 - \frac{2m}{r}\left[1 - \frac{bm^2}{4r^2}\right] = 1 - \frac{2m'}{r}. \tag{25}$$

Taking into account that $m = \frac{\kappa M}{c^2} = M$ ($\kappa = c = 1$), with M as the mass of the star, the $m' = \frac{\kappa' M}{c^2} = \kappa' M$ (now, κ' is not one in general), with

$$\kappa' = \kappa\left[1 - \frac{bm^2}{4r^2}\right] = 1 - \frac{bm^2}{4r^2}, \tag{26}$$

which diminishes the effective gravitational constant for smaller radial distances and approaches the known one at infinite distance. This behavior is

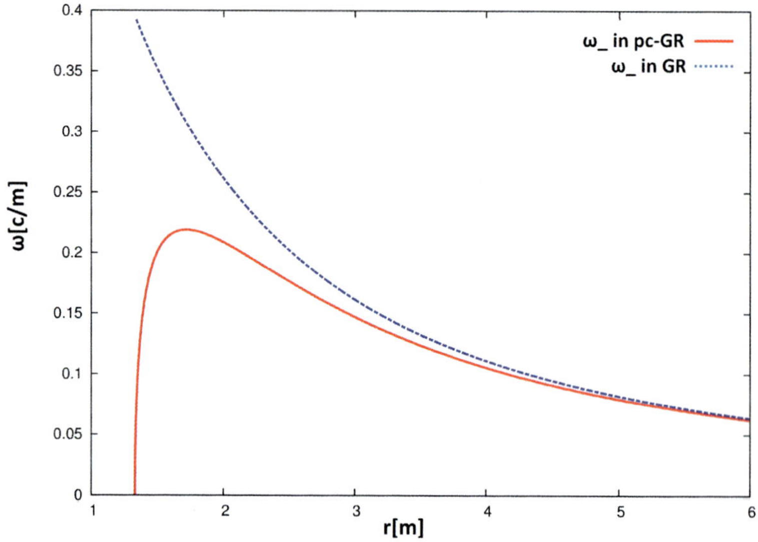

Fig. 1. Orbital frequency (in units of c/m) as a function of r, for stable geodesic prograde (rotating with the star) circular motion. The value $\omega = 0.219\frac{c}{m}$, for a mass of the star of four million suns (as in the center of our galaxy), corresponds to about 9.4 minutes for a full circle. The plot is done for parameter values of $a = 0.995m$ and $B = \frac{64}{27}m^3$.

the result of the repulsive property of the dark energy, thus diminishing the gravitational attraction. The smaller effective gravitational constant is the reason for the lower orbital frequency.

In Fig. 2 the redshift factor versus the radial distance is plotted for the case $\vartheta = \frac{\pi}{2}$, i.e. within the orbital plane. Note that in pc-GR the curve is very similar to GR but shifted to lower values of r. Also the redshift gets very large and may simulate a dark body.

We want to emphasize the main point, namely that first one can measure the orbital frequency of an object in a circular orbit and a given theory predicts a radial distance, corresponding to this orbital frequency. The same can be done for the redshift. *In a consistent theory both radial distances deduced have to be equal*, within the observational errors.

To test the theories, one possibility is observing so-called *Quasi-Periodic Objects* (QPO) [Belloni *et al.* (2012); Hynes *et al.* (2004); Reis *et al.* (2008); Steiner *et al.* (2010)], which express themselves by large brightness variations of short duration. They are observed in the accretion disks around massive mass concentrations at the center of nearly every galaxy, as in the

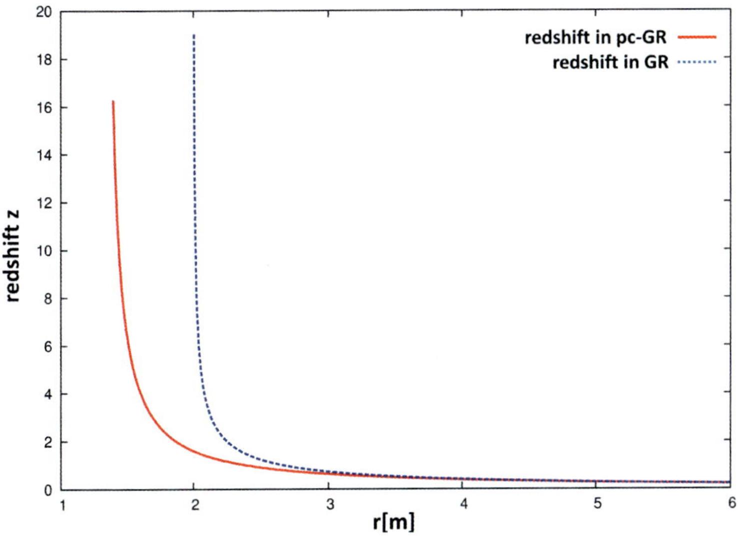

Fig. 2. Redshift for an emitter in the equatorial plane as a function of the position r, outside of a spherically symmetric, uncharged and static mass (Schwarzschild metric). For the Kerr solution this corresponds to $\vartheta = \frac{\pi}{2}$. B is set to $\frac{64}{27}m^3$.

large elliptic galaxy M87. In these cases one is sure that the QPO's are interpreted as local bright spots in the accretions disk following its motion. Unfortunately, no Fe Kα lines are observed yet, thus not permitting the measure of the redshift.

This is different in so-called *galactic black holes*, which are objects within our galaxy with a stellar partner providing mass to the accretion disk. There, the frequency of the QPO's can be measured and Fe Kα lines are observed. One case is shown in Fig. 3. The radial distance derived in both measurements *do not agree in GR* but *they do in pc-GR*! The radial distance deduced from the redshift is indicated for GR on the left side of the figure. For pc-GR it is shifted to lower values of r.

In an attempt to reconcile with the standard theory, the QPO's in galactic black holes are interpreted in a different manner, due to vibrations provoked within the accretion disk [Lai *et al.* (2012)] by the stellar partner. The definite judgment is still out, but our argument is: if QPO's in galactic centers are known to be objects which follow the accretion disk, the physics in galactic black holes should be the same, or at least very similar! Maybe,

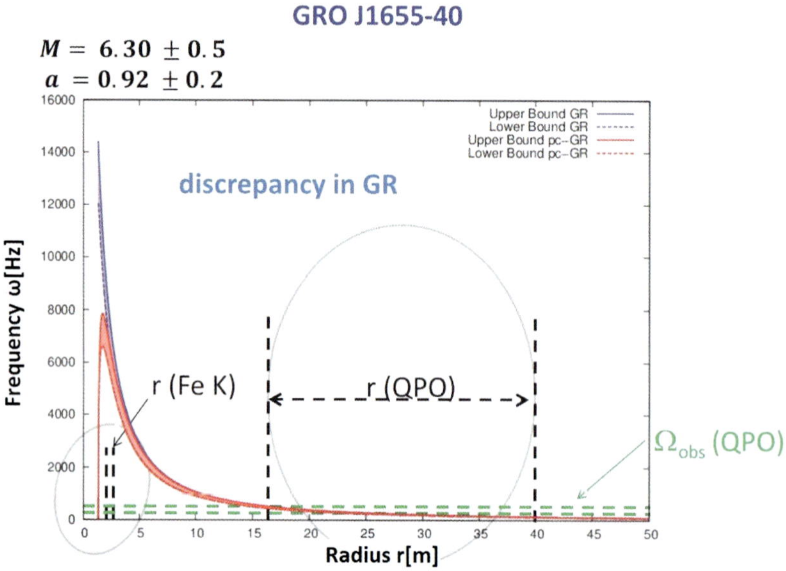

Fig. 3. Angular orbital frequency ω, in units of Hertz (Hz), versus the radial distance, in units of m. The steadyly increasing curve toward smaller r (blue curve) is the result of GR, while the other one, with the maximum, is pc-GR. The width corresponds to the errors in knowing the mass and the spin of the central object. M is in units of the mass of the Sun.

here we have the first indication that the standard Theory of General Relativity has to be modified in the regime of very strong gravitational fields!

4. Simulations of Accretion Disks

In order to connect to actual observations or those planned in near future [The Event Horizon Telescope (2015)], one possibility is to simulate accretion disks around massive objects as the one in the center of the elliptical galaxy M87. The underlying theory was published by D. N. Page and K. S. Thorne [Page and Thorne (1974)] in 1974. The basic assumptions are

- A thin, infinitely extended accretion disk.
- An energy–momentum tensor that includes all main ingredients, such as mass and electromagnetic contributions.
- Conservation laws (energy, angular momentum and mass) are imposed in order to obtain the flux function, the main result of [Page and Thorne (1974)].

- The internal energy of the disk is liberated via shears of neighboring orbitals and distributed from orbitals of higher frequency to those of lower frequency.

The flux F is given by (for more details, please consult [Page and Thorne (1974)])

$$F = \frac{f\dot{M}_0}{4\pi\sqrt{-g}}, \tag{27}$$

where $\dot{M}_0$ is the change of mass passing toward lower orbitals and g is the determinant of the metric. The f satisfies the equation

$$f = \frac{-\omega_{|r}}{\left(E - \omega L_z\right)^2} \int \left(E - \omega L_z\right) L_{z|r} dr, \tag{28}$$

with ω being the orbital frequency as a function in r, E is the energy of a particle in a circular orbit, L_z is the angular momentum around the z-axis and the lower index $|r$ refers to the derivative with respect to r. The upper limit of the integral is always r.

In [Page and Thorne (1974)] the lower limit is chosen as the position of the *Innermost Stable Circular Orbit* (ISCO), i.e., $r_0 = r_{\rm ISCO}$. The physical interpretation is that the energy is transported from lower radial distances to larger ones, i.e., *from orbitals with larger orbital frequency to those with lower values.* In GR the orbital frequency ω is continuously decreasing and justifies this choice of a lower limit. In [Schönenbach *et al.* (2014)] the integration limits of the flux function were modified in order to take into account the maximum in the orbital frequency, which is a dividing point: one possibility is that from $r_{\omega_{\rm max}}$ on, the energy is transported to larger distances, the other one corresponds to the transportation to lower distances. This is understood by examining (28). For $r > r_{\omega_{\rm max}}$ the derivative of the orbital frequency $\omega_{|r}$ is, as in GR, negative and with the negative sign the flux turns positive, as it should be. The flux at $r > r_{\omega_{\rm max}}$ receives energy from all lower orbitals starting at $r_{\omega_{\rm max}}$. For $r < r_{\omega_{\rm max}}$, however, $\omega_{|r}$ is positive and the factor in front of the integral in (28) is negative. The additional change of sign is obtained by using for the lower integral limit $r_{\omega_{\rm max}} > r$, where r is always the upper integral limit. Therefore, for a lower r the energy is transported from $r_{\omega_{\rm max}}$ to lower radial distances.

The above consideration is relevant for a larger than approximately 0.4, as can be seen from Fig. 4 (for explanations, see the figure caption) and [Schönenbach, Caspar *et al.* (2013)]. For lower values of a, in pc-GR

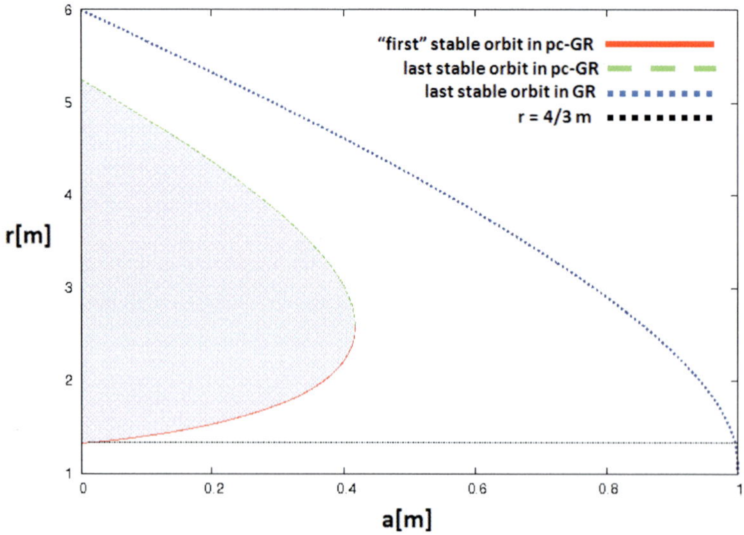

Fig. 4. The position of the *Innermost Stable Circular orbit* (ISCO) is plotted versus the rotational parameter a. The upper curve corresponds to GR and the lower curves to pc-GR. The gray shaded region corresponds to a forbidden area for circular orbits within pc-GR. For small values of a the ISCO in pc-GR follows more or less the one of GR, but at smaller values of r. For a a bit greater than 0.4, the pc-GR has no ISCO and the accretion disk reaches the surface of the star.

the last stable orbit follows the one of GR, but with lower values for the ISCO. As a consequence, the particles reach further inside and due to the decrease of the potential more energy is released, producing a brighter disk. However, the last stable orbit in pc-GR does not reach $r_{\omega_{\rm max}}$. This changes when a is a bit larger than 0.4. Now, $r_{\omega_{\rm max}}$ is crossed and the existence of the maximum of ω has to be taken into account as explained above.

Once the flux function is known, one can deduce the appearance of the accretion disk in the detector, using the *raytracing method* which is illustrated in Fig. 5, where two rays emitted by the disk are indicated, one nearer to the observer and the other one deviated significantly by the central mass. Both reach the observer screen illustrated by a black square. The raytracing method starts from the screen and follows the ray back to the accretion disk. The method used is related to the Hamilton–Jacobi theory and we refer to [Vincent *et al.* (2011)], where the program is also provided for the actual calculations. [Vincent *et al.* (2011)] permits to use different metrics, which has to be provided by the user.

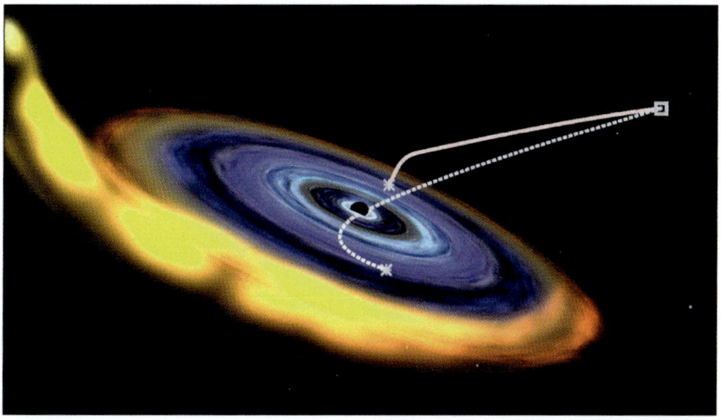

Fig. 5. Illustration of the raytracing technique: Two rays, originating from the accretion disk, are shown. The solid line represents a light path which reaches the observer at infinity on a geodesic path, having been distorted by the gravitational field. The dotted light curve represents a second order effect, where the light makes a near complete turn around the star. The raytracing method follows the ray back, starting from the observer. In each path the conservation laws and the Carter constant are verified, until one reaches a point at the accretion disk. In such a way, only light rays which reach the observer are taken into account, reducing the numerical effort enormously.

Some simulations are presented in Fig. 6. The line of sight of the observer to the accretion disk is 85° (near the edge of the accretion disk); the angle refers to the one between the axis of rotation and the line of sight. Two rotation parameters of the Kerr solution are plotted, namely $a = 0$ (no rotation of the star, corresponding to the Schwarzschild solution) and nearly the maximal rotation $a = 0.9$. The a is in units of m.

As a global feature, the accretion disk in pc-GR appears brighter, which is due to the fact that the accretion disk reaches further inside where the potential is deeper, thus releasing more gravitational energy.

Figure 7 shows the simulated accretion disk seen at an angle of 10°, i.e. nearly from the top. In the center one notes the *Einstein ring*, which is the result of light rays making a near complete turn around the central object. Barely seen is a second Einstein ring further inside, the result of nearly two turns around the central object. In Fig. 8 the same is shown for an angle of 45°. The overall feature, as at 85°, is that in pc-GR the disk is brighter and that for small a the structure of the accretion disk is similar in both theories. For a above 0.4 the difference in structure is notorious: In pc-GR a dark fringe appears, followed by a bright ring further inside.

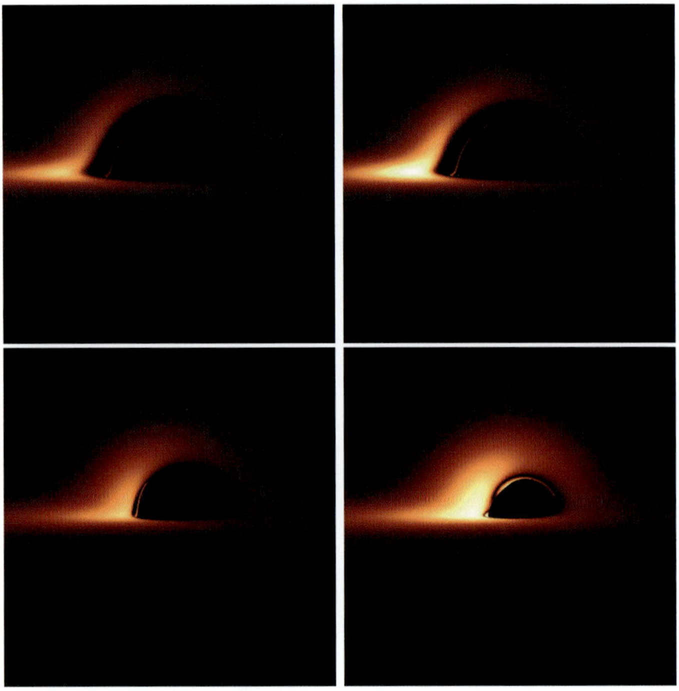

Fig. 6. Infinite, counter-clockwise rotating geometrically thin accretion disk around static and rotating compact objects viewed from an inclination of 85°. The left panels show the original disk model by [Page and Thorne (1974)]. The right panels show the modified model, including pc-GR correction terms as described in the text. Scales change between the images. The first row corresponds to the spin parameter $a = 0$, which is the Schwarzschild solution. The second row is for $a = 0.9$. a is in units of m.

A further difference compared to GR is the smaller size of the dark object in the center.

The reason for the dark fringe and bright ring is as follows: The dark ring is the position of the maximum of the orbital frequency. Neighboring orbitals have nearly the same orbital frequency, thus less friction and shear, which results locally in a lesser excitation of the disk, producing as a consequence just this dark ring. Further inside, the orbital frequency changes rapidly, producing a bright ring.

Such structures, which differ from the standard GR, represent a clear possibility to differentiate between both theories.

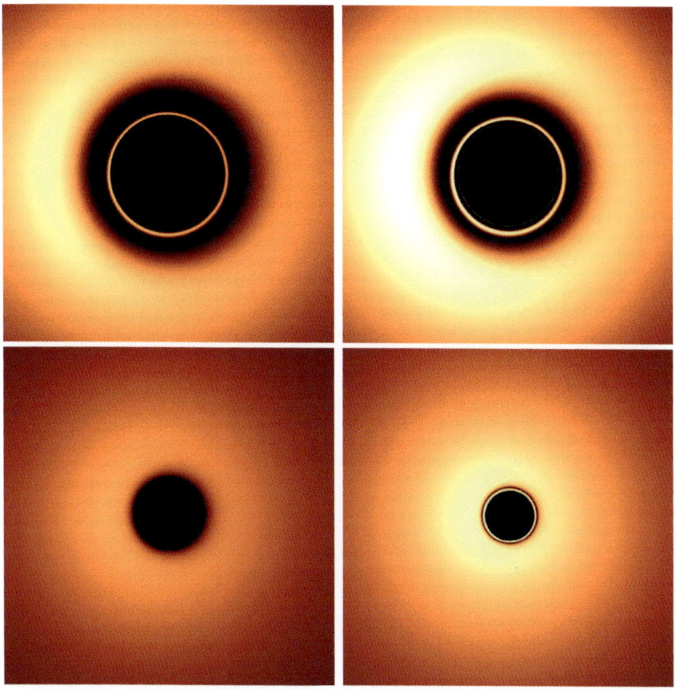

Fig. 7. Infinite, counter-clockwise rotating geometrically thin accretion disk around static and rotating compact objects viewed from an inclination of 10°. The left panels show the original disk model by [Page and Thorne (1974)]. The right panels show the modified model, including pc-GR correction terms as described in the text. Scales change between the images. The first row corresponds to the spin parameter $a = 0$, which is the Schwarzschild solution. The second row is for $a = 0.9$. a is in units of m.

5. Further Considerations and Models within pc-GR

There are more predictions of pc-GR which demonstrate the richness and the predictive power of this theory. They are related to cosmology, more precisely to versions of the *Robertson–Walker universe.* In [Hess *et al.* (2015, 2010)] it is shown that for the Robertson–Walker universe an additional contribution, interpreted as a dark energy appears, which may depend on the radius **a** of the universe. In all cases the acceleration of the universe is negative in the early epoch and is followed by a period of acceleration. Depending on this, there are universes where the acceleration increases indefinitely or reach a constant value at infinite time or even tend to zero.

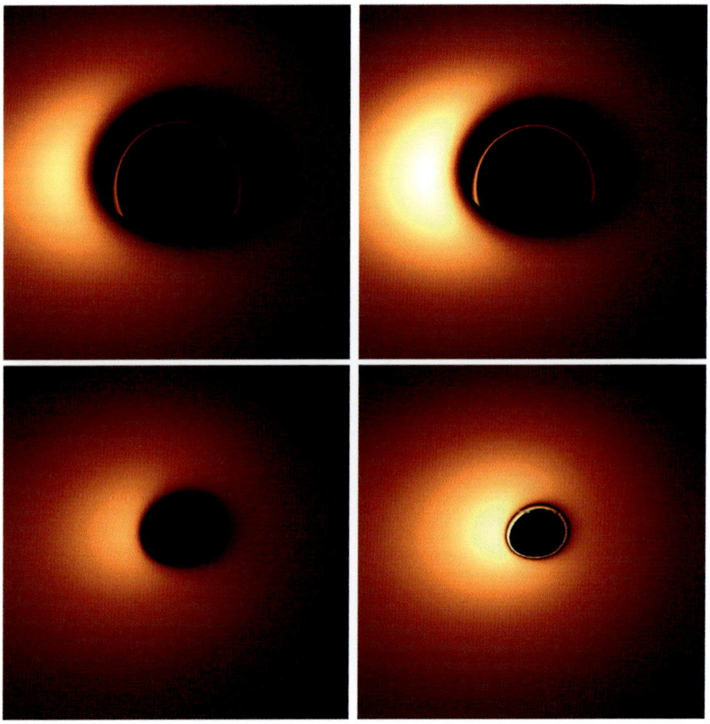

Fig. 8. Infinite, counter-clockwise rotating geometrically thin accretion disk around static and rotating compact objects viewed from an inclination of 45°. The left panels show the original disk model by [Page and Thorne (1974)]. The right panels show the modified model, including pc-GR correction terms as described in the text. Scales change between the images. The first row corresponds to the spin parameter $a = 0$, which is the Schwarzschild solution. The second row is for $a = 0.9$. a is in units of m.

In [Hess *et al.* (2015)] some thoughts are presented on *oscillating universes* and ideas similar to [Tolman (1934)] were presented. As in [Tolman (1934)] one may get solutions of an oscillating universe; however, due to an increase of the entropy in the universe; each phase of an oscillation takes a longer time. Therefore, following it back in time one reaches always a point where the universe is concentrated in a point. This seems to destroy the idea of an ever-expanding and contracting universe, a beautiful idea which answers the question why the universe is so smooth, avoiding extreme assumptions.

We are not giving up the idea of a continuously oscillating universe. The main problem is *how to deal with the entropy.* We suspect that a key role is played by the minimal length which appears in pc-GR: When all matter and light is concentrated in a phase-space volume of a cube with sides of the minimal length, then there should be no possibility for having many quantum states, in fact, there should be only one. This leads to zero entropy! Somehow the entropy is reduced in the contraction process. We have not found yet an explanation for the mechanism but we are confident that we will find it!

A further question is that an arbitrary mass of a star has to be stable in order not to contract to a singularity. In [Hess *et al.* (2014)] a simple linear coupling model was applied for the interaction of the dark energy with the mass density. Stable stars with up to six solar masses were obtained. For larger masses the coupling was too strong near the surface, such that the repulsive property of the dark energy evaporated the upper layers of the surface. This is mainly due to the ansatz and calculations are underway, with a more realistic coupling of the dark energy to the baryon density using semiclassical Quantum Mechanics [Birrell and Davies (1986)], in order to show that larger masses can be obtained, in fact arbitrarily large masses should be possible.

6. Conclusions

We presented a summary on the pseudo-complex General Relativity, showing the equivalence of a symmetric incorporation of the constraint of a real length element squared into the standard variational principle and a modified variational principle as used in [Hess *et al.* (2009); Schönenbach *et al.* (2012); Hess *et al.* (2015)].

In the discussion on circular orbits we indicated that for so-called *galactic black holes* probable deviation of GR are observed in strong gravitational fields, while pc-GR is in good agreement to observations.

Accretion disks were simulated, showing for large a the appearance of a dark ring, followed by a bright one toward smaller radial distances to the star. In pc-GR the disks are brighter because the stable orbits reach further inside releasing more gravitational energy.

Finally, some further predictions of the pc-GR were summarized, as modified cosmological models of the Robertson–Walker universes, including oscillating universes. Also the possibility on how to construct stable stars of arbitrary masses was mentioned.

Note Added in Proof

In February 2016 the first direct measurements of gravitational waves was announced [Abbott *et al.* (2016)]. It is therefore interesting to see what the predictions of pc-GR would be. In [Abbott *et al.* (2016)] a simple formula is given, relating the measured frequency to the chirping mass, which in turn gives information on the size of the black holes involved. The formula is the result of several model assumptions, i.e., both black holes are treated as point masses in a circular orbit and the gravitational field is not strong [Magiorre (2011)]. Important for our purposes is to point out that the chirping mass is inversely proportional to the gravitational constant. In spite of these assumptions, the formula serves to get a good estimate. However, numerical relativity calculations are needed [Alcubierre (2007)] to obtain a final, definitive answer.

In pc-GR the same assumptions can be applied with the difference that the gravitational constant is now dependent on the radial distance, getting smaller with lower r. This results in a greater deduced chirping mass with consequently larger objects.

Due to the complicated r-dependence of the gravitational constant toward very small distances, numerical relativity calculations have to be applied, which our group is unable to do for the moment. In the simple calculation, using (21), the gravitational constant gets even zero at the distance $r = (\frac{2}{3\sqrt{3}})r_s$, with $B = \frac{64}{27}m^3$. Thus, we cannot obtain a good estimate for the final value of the chirping mass, which may vary from an increase by a factor of the order of 2–3 to millions, in case of a very small effective gravitational constant, implying a huge build up of the vacuum energy.

In the latter case, the observed system in [Abbott *et al.* (2016)] would suggest the fusion of two black holes which belonged to two former galaxies which merged.

Acknowledgment

Peter O. Hess acknowledges the financial support from DGAPA-PAPIIT (IN100315). Peter O. Hess also acknowledges financial support from the *Frankfurt Institute for Advanced Studies* (FIAS) and the fruitful working atmosphere at this institute which led to many new ideas and products. We also thank T. Boller (MPI, Garching) for the many discussions, contributions and for providing to us Fig. 3. We thank G. Caspar, M. Schäfer and T. Schönenbach for many helpful discussions.

References

Abbott, B. P. *et al.* (2016). *Phys. Rev. Lett.* **116**, 061102.

Abramowicz, M. A., Kluzniak, W. and Lasota, J. P. (2002). *Astron. Astrophys.* **396**, L31.

Adler, R., Bazin, M. and Schiffer, M. (1975). *Introduction to General Relativity* (McGraw-Hill, N.Y.).

Alcubierre, M. (2007). *Introduction to Numerical Relativity* (Oxford University Press, Oxford).

Belloni, T. M., Sanna, A. and Mendz, M. (2012). *MNRAS* **426**, 1701.

Birrell, N. D. and Davies, P. C. W. (1986). *Quantum Field in Curved Space* (Cambridge University Press, Cambridge).

Caspar, G., Schönenbach, T., Hess, P. O., Schäfer, M. and Greiner, W. (2012). *Int. J. Mod. Phys. E* **21**, 1250015.

Einstein, A. (1945). *Ann. Math.* **46**, 578.

Einstein, A. (1948). *Rev. Mod. Phys.* **20**, 35.

Hess, P. O. and Greiner, W. (2009). *Int. J. Mod. Phys. E* **18**, 51.

Hess, P. O., Maghlaoui L. and Greiner, W. (2010). *Int. J. Mod. Phys. E* **19**, 1315.

Hess, P. O., Schäfer M. and Greiner, W. (2015), *Pseudo-Complex General Relativity* (Springer, Heidelberg), DOI 10.1007/978-3-319-25061-8.

Hynes, R. I., Steeghs, D., Casares, J., Charles, P. A. and O'Brian, K. (2004). *ApJ* **609**, 317.

Kelly, P. F. and Mann, R. B. (1986) *Class. Quant. Grav.* **3**, 705.

Lai, D., Fu, W., Tsang, D., Horak, J. and Ya, C. (2012). *Proceedings of the International Astronomical Union* **8S** (290), p. 57.

Magiorre, M. (2011). *Gravitational Waves* (Oxford University Press, Oxford).

Mantz, C. L. M. (2007). Hermitian graity and cosmology, PhD thesis, Univ. Utrecht.

Mantz, C. L. M. and Prokopec, T. (2011) *Found. Phys.* **41**, 1597.

Misner, C. W., Thorne, K. S. and Wheeler, J. A. (1973). *Gravitation*, (W. H. Freeman and Company, San Francisco).

Moffat, W. (1979). *Phys. Rev. D* **19**, 3554.

Page, D. N. and Thorne, K. S. (1974). *The Astrophys. Jour.* **191**, 499.

Reis, R. C., Fabian, A. C., Ross, R. R., Miniutti, G., Miller. J. M. and Reynolds, C. (2008). *MNRAS* **387**, 1489.

Rodríguez, I., Hess, P. O., Schramm, S. and Greiner, W. (2014). *J. Phys. G* **41**, 105201.

Schönenbach, T., Caspar, G., Hess, P. O., Boller, T., Müller, A. and Greiner, W. (2014). *MNRAS* **442**, 121.

Schönenbach, T., Caspar, G., Hess, P. O., Boller, T., Müller, A., Schäfer, M. and Greiner, W. (2013). *MNRAS* **430**, 2999.

Steiner, J., McClintock, J. Jeffrey, G. *et al.* (2010). *38th COSPAR Scientific Assembly*, 18–15 July, Bremen, Germany.

The Event Horizon Telescope Project, (2015). www.eventhorizontelescope.org.

Tolman, R. C. (1934). *Relativity, Thermodynamics and Cosmology,* (Oxford at the Clarendon Press, Oxford).

Vincent, F. H., Paumard, T., Gourgoulhon, E. and Perrin, G. (2011). *Class. Quant. Grav.* **28**, 225011.

Visser, M. (1996). *Phys. Rev. D* **54**, 5116.

Will, C. M. (2006). *Living Rev. Relativ.* **9**, 3.

Chapter 4

Strange Matter: A State before Black Hole

Renxin Xu and Yanjun Guo

School of Physics and KIAA, Peking University
Beijing 100871, P. R. China
r.x.xu@pku.edu.cn, guoyj10@pku.edu.cn

Normal baryonic matter inside an evolved massive star can be intensely compressed by gravity after a supernova. General relativity predicts formation of a black hole if the core material is compressed into a singularity, but the real state of such compressed baryonic matter (CBM) before an event horizon of black hole appears is not yet well understood because of the non-perturbative nature of the fundamental strong interaction. Certainly, the rump left behind after a supernova explosion could manifest as a pulsar if its mass is less than the unknown maximum mass, $M_{\rm max}$. In this contribution, it is conjectured that pulsar-like compact stars are made of strange matter (i.e., with 3-flavour symmetry), where quarks are still localized as in the case of nuclear matter. In principle, different manifestations of pulsar-like objects could be explained in the regime of this conjecture. Besides compact stars, strange matter could also manifest in the form of cosmic rays and even dark matter.

Contents

1. Introduction

The baryonic part of the Universe is well-understood in the Standard Model of particle physics (consolidated enormously by the discovery of Higgs boson), where quark masses are key parameters to make a judgment on the quark-flavour degrees of freedom at a certain energy scale. Unlike the leptons, quarks could be described with mass parameters to be measured indirectly through their influence on hadronic properties since they are confined inside hadrons rather than free particles. The masses of both up and down quarks are only a few MeV while the strange quark is a little bit heavier, with an averaged mass of up and down quarks, $m_{ud} = (3.40 \pm 0.25)$ MeV, as well as the strange quark mass of $m_s = (93.5 \pm 2.5)$ MeV obtained from lattice QCD (quantum chromodynamics) simulations [Olive *et al.* (2014)]. For nuclei or nuclear matter, the separation between quarks is $\Delta x \sim 0.5$ fm, and the energy scale is then in the order of $E_{\rm nucl} \sim 400$ MeV according to Heisenberg's relation $\Delta x \cdot pc \sim \hbar c \simeq 200$ MeV·fm. One may then superficially understand why nuclei are of two (i.e., u and d) flavours as these two flavours of quarks are the lightest. However, because the nuclear energy scale is much larger than the mass differences between strange and up/down quarks, $E_{\rm nucl} \gg (m_s - m_{ud})$, why is the valence strangeness degree of freedom absolutely missing in stable nuclei?

We argue and explain in this paper that 3-flavour (u, d and s) symmetry would be restored if the strong-interaction matter at low temperature is very big, with a length scale much larger than the electron Compton wavelength $\lambda_{\rm e} = h/(m_e c) \simeq 0.024$ Å. We call this kind of matter *strange matter* too, but it is worth noting that quarks are still localized with this definition (in analogy to 2-flavour symmetric nuclei) because the energy scale here (larger than but still around $E_{\rm nucl}$) is still much smaller than the perturbative scale of QCD dynamics, $\Lambda_\chi > 1$ GeV. We know that normal nuclei are relatively small, with length scale $(1 \sim 10)$ fm $\ll \lambda_{\rm e}$, and it is very difficult for us to gather up huge numbers ($> 10^9$) of nuclei together because of the Coulomb barrier between them in laboratory. Then, where could one find a large nucleus with possible 3-flavour symmetry (i.e., strange matter)?

Such kind of strange matter can only be created through extremely astrophysical events. A good candidate of strange matter could be the supernova-produced rump left behind after core-collapsing of an evolved massive star, where normal *micro*-nuclei are intensely compressed by gravity to form a single *gigantic* nucleus (also known as compressed baryonic matter, CBM), the prototype of which was speculated and discussed firstly by Lev Landau [Landau (1932)]. The strange matter object could manifest the behaviors of pulsar-like compact stars if its mass is less than $M_{\max}$, the maximum mass being dependent on the equation of state of strange matter, but it could soon collapse further into a black hole if its mass is larger than $M_{\max}$. We may then conclude that strange matter could be the state of gravity-controlled CBM before an event horizon comes out (i.e., a black hole forms).

This paper is organized as follows. In Sec. 2, the gravity-compressed dense matter (a particular form of CBM), a topic relevant to Einstein's general relativity, is introduced in order to make sense of realistic CBM/strange matter in astrophysics. We try to convince the reader that such kind of astrophysical CBM should be in a state of strange matter, which would be distinguished significantly from the previous version of strange quark matter, in Sec. 3. Cold strange matter would be in a solid state due to strong colour interaction there, but the solution of a solid star with sufficient rigidity is still a challenge in general relativity. Nevertheless, the structure of solid strange star is presented (Sec. 4) in the very simple case for static and spherically symmetric objects. Different manifestations and astrophysical implications of strange matter are broadly discussed in Sec. 5. Finally, Sec. 6 is a brief summary.

2. Dense Matter Compressed by Gravity

As the first force recognized among the four fundamental interactions, gravity is mysterious and fascinating because of its unique feature. Gravity is universal, which is well known from the epoch of Newton's theory. Nothing could escape the control of gravity, from the falling of apple towards the Earth, to the motion of the Moon in the sky. In Einstein's general relativity, gravity is related to the geometry of curved spacetime. This beautiful and elegant idea significantly influences our world view. Spacetime is curved by matter/energy, while the motion of object is along the "straight" line (geodesic) of the curved spacetime. General relativity has passed all experimental tests up to now. However, there is intrinsic conflict between quantum theory and general relativity. Lots of efforts have been made to quantize gravity, but no success has been achieved yet.

Gravity is extremely weak compared to the other fundamental forces, so it is usually ignored in micro-physics. Nonetheless, on the scale of the Universe, things are mostly controlled by gravity because it is long-range and has no screening effect. One century has passed since Einstein established general relativity, but only a few solutions to the field equation have been found, among which three solutions are most famous and useful. The simplest case is for static and spherical spacetime, and the solution was derived by Schwarzschild just one month after Einstein's field equation. The Schwarzschild solution indicates also the existence of black hole, where everything is doomed to fall towards the center after passing through the event horizon. Consider a non-vacuum case with ideal fluid as source, the field equation could be transformed to Tolman–Oppenheimer–Volkoff equation [Oppenheimer & Volkoff (1939)], which could be applied to the interior of pulsar-like compact stars. Based on the so-called cosmological Copernicus principle, Friedmann equation can be derived with the assumption of homogenous and isotropic Universe, which sets the foundation of cosmology. These three solutions of Einstein's field equation represent the most frontier topics in modern astrophysics.

At the late stage of stellar evolution, how does the core of massive star collapse to a black hole? Or equivalently, how is normal baryonic matter squeezed into the singularity? What's the state of compressed baryonic matter (CBM) before collapsing into a black hole?

We are focusing on these questions in this chapter. In the Standard Model of particle physics, there are in total six flavours of quarks. Among them, three (u, d and s) are light, with masses $< 10^2$ MeV, while the other three flavours (c, t and b), with masses $> 10^3$ MeV, are too heavy to be excited in the nuclear energy scale, $E_{\rm nucl} \simeq 400$ MeV. However, the ordinary matter in our world is built from u and d quarks only, and the numbers of these two flavours tend to be balanced in a stable nucleus. It is then interesting to think philosophically about the fact that our baryonic matter is 2-flavour symmetric.

An explanation could be: micro-nuclei are too small to have 3-flavour symmetry, but bigger is different. In fact, rational thinking about stable strangeness dates back to 1970s [Bodmer (1971)], in which Bodmer speculated that the so-called "collapsed nuclei" with strangeness could be energetically favoured if baryon number $A > A_{\rm min}$, but without quantitative estimation of the minimal number $A_{\rm min}$. Bulk matter composed of almost *free* quarks (u, d, and s) was then focused on [Itoh (1970); Witten (1984)], even for astrophysical manifestations [Alcock, Farhi & Olinto (1986); Haensel, Zdunik & Schaeffer (1986)]. CBM can manifest as a pulsar if the mass is not large enough to form a black hole, and strange quark matter could possibly exist in compact stars, either in the core of neutron star (i.e., mixed or hybrid stars [Ivanenko & Kurdgelaidze (1969)]) or as the whole star (strange quark star [Itoh (1970); Alcock, Farhi & Olinto (1986); Haensel, Zdunik & Schaeffer (1986)]). Although the asymptotic freedom is well recognized, one essential point is whether the colour coupling between quarks is still perturbative in astrophysical CBM so that quarks are itinerant there. In case of non-perturbative coupling, the strong force there might render quarks grouped in so-called *quark-clusters* (or simply as strangeon, an abbreviation of "strange nucleon"), forming a nucleus-like strange object [Xu (2003)] with 3-flavour symmetry, when CBM is big enough that relativistic electrons are inside (i.e., $A > A_{\rm min} \simeq 10^9$). Anyway, we could simply call 3-flavour baryonic matter as *strange matter*, in which the constituent quarks could be either itinerant or localized.

Why is big CBM strange? This is actually a conjecture to be extensively discussed in the next section, but it could be reasonable.

3. A Bodmer–Witten's Conjecture Generalized

Besides being meaningful for understanding the nature of sub-nucleon at a deeper level, strangeness would also have a consequence on the physics of super-dense matter. The discovery of strangeness (a general introduction to Murray Gell-Mann and his strangeness could be found in the biography by Johnson [Johnson (1999)]) is known as a milestone in particle physics since our normal baryons are non-strange. Nonetheless, condensed matter with strangeness should be worth exploring as the energy scale $E_{\rm nucl} \gg m_s$ at the nuclear and supra-nuclear densities.

Previously, bulk strange object (suggested to be 3-flavour *quark* matter) is speculated to be the absolutely stable ground state of strong-interaction matter, which is known as the Bodmer–Witten's conjecture [Bodmer (1971); Witten (1984)]. But we are discussing a general conjecture in next subsections, arguing that quarks might not be necessarily free in stable strange matter, and would still be hadron-like localized as in a nucleus if non-perturbative QCD effects are significant (i.e., $E_{\rm nucl} < \Lambda_\chi$) and the repulsive core keeps to work in both cases of 2-flavour (non-strange) nuclear matter and 3-flavour (strange) matter. In this sense, protons and neutrons are of 2-flavour quark-clusters (i.e., nucleon), while strange matter could be condensed matter of quark-clusters with strangeness (i.e., strange quark-clusters or simply strangeon).

Summarily, it is well known that micro-nuclei are non-strange, but macro-nuclei in the form of CBM could be strange. Therefore, astrophysical CBM and nucleus could be very similar, but only with a simple change from non-strange to strange: "2" → "3". We are explaining two approaches to this strange quark-cluster matter state, bottom-up and top-down, respectively as following.

3.1. *Macro-nuclei with 3-flavour symmetry: Bottom-up*

Micro-nucleus is made up of protons and neutrons, and there is an observed tendency to have equal numbers of protons (Z) and neutrons (N). In liquid-drop model, the mass formula of a nucleus with atomic number $A(= Z+N)$ consists five terms,

$$E(Z,N) = a_{\rm v}A - a_{\rm s}A^{2/3} - a_{\rm sym}\frac{(N-Z)^2}{A} - a_{\rm c}\frac{Z(Z-1)}{A^{1/3}} + a_{\rm p}\frac{\Delta(N,Z)}{A^{1/2}}, \quad (1)$$

where the third term is for the symmetry energy, which vanishes with equal number of protons and neutrons. This nuclear symmetry energy represents a symmetry between proton and neutron in the nucleon degree of freedom, and is actually that of u and d quarks in the quark degree [Li, Chen & Ko (2008)]. The underlying physics of symmetry energy is not well understood yet. If the nucleons are treated as Fermi gas, there is a term with the same form as symmetry energy in the formula of Fermi energy, known as the kinetic term of nuclear symmetry energy. But the interaction is not negligible, and the potential term of symmetry energy would be significant. Recent scattering experiments show that, because of short-range interactions, the neutron–proton pairs are nearly 20 times as prevalent as proton–proton (and neutron–neutron by inference) pairs [Subedi *et al.* (2008); Hen *et al.* (2014)], which hints that the potential term would dominate in the symmetry energy.

Since the electric charges of u and d quarks are $+2/3$ and $-1/3$ respectively, 2-flavour symmetric strong-interaction matter should be positively charged, and electrons are needed to maintain electric neutrality. The possibility of electrons inside a nucleus is negligible because the nuclear radius is much smaller than the Compton wavelength $\lambda_{\rm e} \sim 10^3$ fm, and the lepton degree of freedom would then be not significant for nucleus.

Therefore, electrons contribute negligible energy for micro-nuclei, as the coupling constant of electromagnetic interaction ($\alpha_{\rm em}$) is much less than that of strong interaction ($\alpha_{\rm s}$). The kinematic motion of electrons is bound by electromagnetic interaction, so $p^2/m_{\rm e} \sim e^2/l$. From Heisenberg's relation, $p \cdot l \sim \hbar$, combining the above two equations, we have $l \sim \frac{1}{\alpha_{\rm em}} \frac{\hbar c}{m_{\rm e}c^2}$, and the interaction energy is order of $e^2/l \sim \alpha_{\rm em}^2 m_{\rm e}c^2 \sim 10^{-5}$ MeV.

However, bigger is different, and there might be 3-flavour symmetry in gigantic/macro-nuclei, as electrons are inside a gigantic nucleus. With the number of nucleons $A > 10^9$, the scale of macro-nuclei should be larger than the Compton wavelength of electrons $\lambda_{\rm e} \sim 10^3$ fm. If the 2-flavour symmetry stays, macro-nucleus will become a huge Thomson atom with electrons evenly distributed there. Though Coulomb energy could not be significant, the Fermi energy of electrons are not negligible, being $E_{\rm F} \sim \hbar c n^{1/3} \sim 10^2$ MeV. However, the situation becomes different if strangeness is included: no electrons exist if the matter is composed of equal numbers of light quarks u, d, and s in chemical equilibrium. In this case, the 3-flavour symmetry, an analogy of the symmetry of u and d in the nucleus, may result in a ground state of matter for gigantic nuclei. Certainly the mass difference between u/d and s quarks would also break the 3-flavour symmetry, but the interaction between quarks could lower the effect of mass differences and favour the restoration of 3-flavour symmetry. If macro-nuclei are almost 3-flavour symmetric, the contribution of electrons would be negligible, with $n_{\rm e} \ll n_{\rm q}$ and $E_{\rm F} \sim 10$ MeV.

The new degree of freedom (strangeness) is also possible to be excited, according to an order-of-magnitude estimation from either Heisenberg's relation (localized quarks) or Fermi energy (free quarks). For quarks localized with length scale l, from Heisenberg's uncertainty relation, the kinetic energy would be $\sim p^2/m_{\rm q} \sim \hbar^2/(m_{\rm q}l^2)$, which has to be comparable with the colour interaction energy of $E \sim \alpha_{\rm s}\hbar c/l$ in order to have a bound state, with an assumption of Coulomb-like strong interaction. One can then have, if quarks are dressed,

$$l \sim \frac{1}{\alpha_{\rm s}} \frac{\hbar c}{m_{\rm q}c^2} \simeq \frac{1}{\alpha_{\rm s}} \,\text{fm}, \quad E \sim \alpha_{\rm s}^2 m_{\rm q}c^2 \simeq 300\alpha_{\rm s}^2 \,\text{MeV}. \tag{2}$$

As $\alpha_{\rm s}$ may well be close to or even greater than 1 at several times the nuclear density, the energy scale would be approaching or even larger than ~ 400 MeV. A further calculation of Fermi energy also gives

$$E_{\rm F}^{\rm NR} \approx \frac{\hbar^2}{2m_{\rm q}}(3\pi^2)^{2/3} \cdot n^{2/3} = 380\,\text{MeV}, \tag{3}$$

if quarks are considered to be moving non-relativistically, or

$$E_{\rm F}^{\rm ER} \approx \hbar c(3\pi^2)^{1/3} \cdot n^{1/3} = 480\,{\rm MeV}, \tag{4}$$

if quarks are considered to be moving extremely relativistically. So we have the energy scale ~400 MeV by either Heisenberg's relation or Fermi energy, which could certainly be larger than the mass difference (~100 MeV) between s and u/d quarks. However, for micro-nuclei where electron contributes negligible energy, there could be 2-flavour (rather than 3-flavour) symmetry because s quark mass is larger than u/d quark masses. We now understand that it is more economical to have 2-flavour micro-nuclei because of massive s quark and negligible electron kinematic energy, whereas macro/gigantic-nuclei might be 3-flavour symmetric.

The 2-flavour micro-nucleus consists u and d quarks grouped in nucleons, while the 3-flavour macro-nucleus is made up of u, d and s quarks grouped in so-called strange quark-clusters. Such macro-nucleus with 3-flavour symmetry can be named strange quark-cluster matter, or simply strange matter.

3.2. *Macro-nuclei with 3-flavour symmetry: Top-down*

Besides this bottom-up scenario (an approach from the hadronic state), we could also start from deconfined quark state with the inclusion of stronger and stronger interaction between quarks (a top-down scenario).

The underlying theory of the elementary strong interaction is believed to be quantum chromodynamics (QCD), a non-Abelian $SU(3)$ gauge theory. In QCD, the effective coupling between quarks decreases with energy (the *asymptotic freedom*) [Gross & Wilczek (1973); Politzer (1973)]. Quark matter (or quark–gluon plasma), the soup of deconfined quarks and gluons, is a direct consequence of asymptotic freedom when temperature or baryon density are extremely high. Hot quark matter could be reproduced in the experiments of relativistic heavy ion collisions. Ultra-high chemical potential is required to create cold quark matter, and it can only exist in rare astrophysical conditions — the compact stars.

What kind of cold matter can we expect from QCD theory, in effective models, or even based on phenomenology? This is a question too hard to answer because of (i) the non-perturbative effects of strong interaction between quarks at low energy scales and (ii) the many-body problem due to vast assemblies of interacting particles. A colour-superconductivity

(CSC) state is focused on in QCD-based models, as well as in phenomenological ones [Alford *et al.* (2008)]. The ground state of extremely dense quark matter could certainly be that of an ideal Fermi gas at an extremely high density. Nevertheless, it has been found that the highly degenerate Fermi surface could be unstable against the formation of quark Cooper pairs, which condense near the Fermi surface due to the existence of colour-attractive channels between the quarks. A BCS-like colour superconductivity, similar to electric superconductivity, has been formulated within perturbative QCD at ultra-high baryon densities. It has been argued, based on QCD-like effective models, that colour superconductivity could also occur even at the more realistic baryon densities of pulsar-like compact stars [Alford *et al.* (2008)].

Can the realistic stellar densities be high enough to justify the use of perturbative QCD? It is surely a challenge to calculate the coupling constant, α_s, from first principles. Nevertheless, there are some approaches to the non-perturbative effects of QCD, one of which uses the Dyson–Schwinger equations tried by Fischer *et al.* [Fischer and Alkofer (2002); Fischer (2006)], who formulated

$$\alpha_s(x) = \frac{\alpha_s(0)}{\ln(e + a_1 x^{a_2} + b_1 x^{b_2})}, \tag{5}$$

where $a_1 = 5.292\ \mathrm{GeV}^{-2a_2}$, $a_2 = 2.324$, $b_1 = 0.034\ \mathrm{GeV}^{-2b_2}$, $b_2 = 3.169$, $x = p^2$ with p the typical momentum in GeV, and that α_s freezes at $\alpha_s(0) = 2.972$. For our case of assumed dense quark matter at $\sim 3\rho_0$, the chemical potential is ~ 0.4 GeV, and then $p^2 \simeq 0.16\ \mathrm{GeV}^2$. Thus, it appears that the coupling in realistic dense quark matter should be greater than 2, being close to 3 in Fischer's estimate presented in Eq. (5). Therefore, this surely means that a weakly coupling treatment could be dangerous for realistic cold quark matter (the interaction energy $\sim 300\alpha_s^2$ MeV could even be much larger than the Fermi energy), i.e., the non-perturbative effect in QCD is not negligible if we try to know the real state of compact stars. It is also worth noting that the dimensionless electromagnetic coupling constant (i.e., the fine-structure constant) is $1/137 < 0.01$, which makes QED tractable. That is to say, a weakly coupling strength comparable with that of QED is possible in QCD only if the density is unbelievably and unrealistically high ($n_{\mathrm{B}} > 10^{123} n_0$! with $n_0 = 0.16\ \mathrm{fm}^{-3}$, the baryon density of nuclear matter).

Quark-clusters may form in relatively low temperature quark matter at only a few nuclear density due to the strong interaction (i.e., large α_s), and the clusters could locate in periodic lattices (normal solid) when temperature becomes sufficiently low. Although it is hitherto impossible to know if quark-clusters could form in cold quark matter via calculation from first principles, there could be a few points that favour clustering. *Experimentally*, though quark matter is argued to be weakly coupled at high energy and thus deconfined, it is worth noting that, as revealed by the recent achievements in relativistic heavy ion collision experiments, the interaction between quarks in a fireball of quarks and gluons is still very strong (i.e. the strongly coupled quark–gluon plasma, sQGP [Shuryak (2009)]). The strong coupling between quarks may naturally render quarks grouped in clusters, i.e., a condensation in position space rather than in momentum space. *Theoretically*, the hadron-like particles in quarkyonic matter [McLerran & Pisarski (2007)] might be grouped further due to residual colour interaction if the baryon density is not extremely high, and quark-clusters would form then at only a few nuclear density. Certainly, more elaborate research work is necessary.

For cold quark matter at $3n_0$ density, the distance between quarks is $\sim$ fm $\gg$ the Planck scale $\sim 10^{-20}$ fm, so quarks and electrons can well be approximated as point-like particles. If Q_α-like clusters are created in the quark matter [Xu (2003)], the distance between clusters are ~ 2 fm. The length scale l and colour interaction energy of quark-clusters have been estimated by the uncertainty relation, assuming quarks are dressed (the constituent quark mass is $m_{\rm q} \sim 300$ MeV) and move non-relativistically in a cluster. We have $l \sim \hbar c/(\alpha_{\rm s} m_{\rm q} c^2) \simeq 1$ fm if $\alpha_{\rm s} \sim 1$, and the colour interaction energy $\sim \alpha_{\rm s}^2 m_{\rm q} c^2$ could be greater than the baryon Fermi energy if $\alpha_{\rm s} \gtrsim 1$. The strong coupling could render quarks grouped in position space to form clusters, forming a nucleus-like strange object [Xu (2003)] with 3-flavour symmetry, if it is big enough that relativistic electrons are inside (i.e., $A > A_{\rm min} \simeq 10^9$). Quark-clusters could be considered as classical particles in cold quark-cluster matter and would be in lattices at a lower temperature.

In conclusion, quark-clusters could emerge in cold dense matter because of the strong coupling between quarks. The quark-clustering phase has high density and the non-perturbative interaction is still dominant, so it is different from the usual hadron phase, and on the other hand, the quark-clustering phase is also different from the conventional quark matter phase which is composed of relativistic and weakly interacting quarks. The quark-clustering phase could be considered as an intermediate state between hadron phase and free-quark phase, with deconfined quarks grouped into quark-clusters.

3.3. *Comparison of micro-nuclei and macro-nuclei*

In summary, there could be some similarities and differences between micro-nuclei and macro-nuclei, listed as follows.

Similarity 1: Both micro-nuclei and macro-nuclei are self-bound by the strong colour interaction, in which quarks are localized in groups called generally quark-clusters. We are sure that there are two kinds of quark-clusters inside micro-nuclei, the proton (with structure uud) and neutron (udd), but we don't know well the clusters in macro-nuclei due to the lack of detailed experiments on the subject.

Similarity 2: Since the strong interaction might not be very sensitive to flavour, the interaction between general quark-clusters should be similar to that of nucleon, which is found to be Lennard–Jones-like for the case of two flavours by both experiment and modeling. Especially, one could then expect a hard core [Wilczek (2007)] (or repulsive core) of the interaction potential between strange quark-clusters though no direct experiment now hints this existence.

Difference 1: The most crucial difference is the change of flavour degree of freedom, from two (u and d) in micro-nuclei to three (u, d and s) in macro/gigantic-nuclei. We could thus have following different aspects derived.

Difference 2: The number of quarks in a quark-cluster is 3 for micro-nuclei, but could be 6, 9, 12, or even 18 for macro-nuclei, since the interaction between Λ-particles could be attractive [Beane *et al.* (2011); Inoue *et al.* (2011)] so that no positive pressure can support a gravitational star of Λ-cluster matter. We therefore call proton/neutron light quark-clusters, while the strange quark-clusters heavy clusters because of (1) the massive s-quark and (2) more quarks inside.

Difference 3: A micro-nucleus could be considered as a quantum system so that one could apply quantum mean-field theory, whereas the heavy clusters in strange matter may be classical particles since the quantum wavelength of massive clusters can be even smaller than the mean distance between them.

Difference 4: The equation of state (EoS) of strange matter would be stiffer [Lai, Gao, & Xu (2013)] than that of nuclear matter because the clusters in the former should be non-relativistic but relativistic in the latter. The kinematic energy of a cluster in both micro- and macro-nuclei could be $\sim 0.5\,\text{GeV}$, which is much smaller than the rest mass (generally $\gtrsim 2\,\text{GeV}$) of a strange quark-cluster.

Difference 5: Condensed matter of strange quark-clusters could be in a solid state at low temperature much smaller than the interaction energy between clusters. We could then expect solid pulsars [Xu (2003)] in nature although an idea of solid nucleus was also addressed [Bertsch (1974)] a long time ago.

3.4. *A general conjecture of flavour symmetry*

The 3-flavour symmetry may hint at the nature of strong interaction at low energy scale. Let's tell a story of science fiction about flavour symmetry. Our protagonist is a fairy who is an expert in QCD at high energy scale (i.e., perturbative QCD) but knows little about spectacular non-perturbative effects. There is a conversation between the fairy and God about strong-interaction matter.

God: "I know six flavours of quarks, but how many flavours could there exist in stable strong-interaction matter?"

Fairy: "It depends ... how dense is the matter? (aside: the nuclear saturation density arises from the short-distance repulsive core, a consequence of non-perturbative QCD effect she may not know much.)"

God: "Hmm ... I am told that quark number density is about $0.48\ \text{fm}^{-3}$ ($3n_0$) there."

Fairy: "Ah, in this energy scale of ~ 0.5 GeV, there could only be light flavours (i.e., u, d and s) in stable matter if quarks are free."

God: "Two flavours (u and d) or three flavours?"

Fairy: "There could be two flavours of free quarks if strong-interaction matter is very small ($\ll \lambda_e$), but would be three flavours for bulk strong-interaction matter (aside: the Bodmer–Witten's conjecture)."

God: "Small 2-flavour strong-interaction matter is very useful, and I can make life and mankind with huge numbers of these pieces. We can call them atoms."

Fairy: "Thanks, God! I can also help mankind to have a better life."

God: "But ... are quarks really free there?"

Fairy: "Hmm ... there could be clustered quarks in both two and three flavour (aside: a Bodmer–Witten's conjecture generalized) cases if the interaction between quarks is so strong that quarks are grouped together. You name a piece of small two-flavour matter *atom*, what would we call a 3-flavour body in bulk?"

God: "Oh ... simply a *strange* object because of its strangeness."

4. Solid Strange Star in General Relativity

Cold strange matter with 3-flavour symmetry could be in a solid state because of (1) a relatively small quantum wave packet of quark-cluster (wavelength $\lambda_{\rm q} < a$, where a is the separation between quark-clusters) and (2) low temperature $T < (10^{-1}–10^{-2})U$ (U is the interaction energy between quark-clusters). The packet scales $\lambda_{\rm q} \sim h/(m_{\rm q}c)$ for free quark-cluster, with m_q the rest mass of a quark-cluster, but could be much smaller if it is constrained in a potential with depth of $> \hbar c/a \simeq 100$ MeV (2 fm$/a$). A star made of strange matter would then be a solid star.

It is very fundamental to study static and spherically symmetric gravitational sources in general relativity, especially for the interior solutions. The TOV solution [Oppenheimer & Volkoff (1939)] is only for perfect fluid. However, for solid strange stars, since the local pressure could be *anisotropic* in elastic matter, the radial pressure gradient could be partially balanced by the tangential shear force, although a general understanding of relativistic, elastic bodies has unfortunately not been achieved [Karlovini & Samuelsson (2004)]. The origin of this local anisotropic force in solid quark stars could be from the development of elastic energy as a star (i) spins down (its ellipticity decreases) and (ii) cools (it may shrink). Release of the elastic as well as the gravitational energies would be not negligible, and may have significant astrophysical implications.

The structure of solid quark stars can be numerically calculated as follows. For the sake of simplicity, only spherically symmetric sources are dealt with, in order to make sense of possible astrophysical consequence of solid quark stars. By introducing respectively radial and tangential pressures, P and $P_\perp$, the stellar equilibrium equation of static anisotropic matter in Newtonian gravity is [Herrera & Santobs (1997)]: $\mathrm{d}P/\mathrm{d}r = -Gm(r)\rho/r^2 + 2(P_\perp - P)/r$, where ρ and G denote mass density and the gravitational constant, respectively, and $m(r) = \int_0^r 4\pi x^2\rho(x)\mathrm{d}x$. However, in Einstein's gravity, this equilibrium equation is modified to be [Xu *et al.* (2006)]

$$\frac{\mathrm{d}P}{\mathrm{d}r} = -\frac{Gm(r)\rho}{r^2}\frac{(1+\frac{P}{\rho c^2})(1+\frac{4\pi r^3 P}{m(r)c^2})}{1-\frac{2Gm(r)}{rc^2}} + \frac{2\epsilon}{r}P, \tag{6}$$

where $P_\perp = (1+\epsilon)P$ is introduced. In case of isotropic pressure, $\epsilon = 0$, Eq. (6) turns out to be the TOV equation. It is evident from Eq. (6) that the radial pressure gradient, $|\mathrm{d}P/\mathrm{d}r|$, decreases if $P_\perp > P$, which may result in a higher maximum mass of compact stars. One can also see that a sudden decrease of $P_\perp$ (equivalently of elastic force) in a star may cause substantial energy release, since the star's radius decreases and the absolute gravitational energy increases.

Starquakes may result in a sudden change of ϵ, with release of the gravitational energy as well as the tangential strain energy. Generally, it is evident that the differences of radius, gravitational energy, and moment of inertia increase proportionally to stellar mass and the parameter ϵ. This means that an event should be more important for a bigger change of ϵ in a quark star with higher mass. Typical energy of $10^{44\sim47}$ erg is released during superflares of SGRs, and a giant starquake with $\epsilon \lesssim 10^{-4}$ could produce such a flare [Xu *et al.* (2006)]. A sudden change of ϵ can also result in a jump of spin frequency, $\Delta\Omega/\Omega = -\Delta I/I$. Glitches with $\Delta\Omega/\Omega \sim 10^{-10\sim-4}$ could occur for parameters of $M = (0.1 \sim 1.4)M_\odot$ and $\epsilon = 10^{-9\sim-4}$. It is suggestive that a giant flare may accompany a high-amplitude glitch.

5. Astrophysical Manifestations of Strange Matter

How to create macro-nuclei (even gigantic) in the Universe? Besides a collapse event where normal baryonic matter is intensely compressed by gravity, strange matter could also be produced after cosmic hadronization [Witten (1984)]. Strange matter may manifest itself as a variety of objects with a broad mass spectrum, including compact objects, cosmic rays and even dark matter.

5.1. *Pulsar-like compact star: Compressed baryonic matter after supernova*

In 1932, soon after Chandrasekhar found a unique mass (the mass limit of white dwarfs), Landau speculated a state of matter, the density of which "becomes so great that atomic nuclei come in close contact, forming one *gigantic nucleus*" [Landau (1932)]. A star composed mostly of such matter is called a "neutron" star, and Baade and Zwicky even suggested in 1934 that neutron stars (NSs) could be born after supernovae. NSs theoretically predicted were finally *discovered* when Hewish and his collaborators detected radio pulsars in 1967 [Hewish and Bell *et al.* (1968)]. More kinds of pulsar-like stars, such as X-ray pulsars and X-ray bursts in binary systems, were also discovered later, and all of them are suggested to be NSs.

In a gigantic nucleus, protons and electrons combined to form the neutronic state, which involves weak equilibrium between protons and neutrons. However, the simple and beautiful idea proposed by Landau and others had one flaw at least: nucleons (neutrons and protons) are in fact *not* structureless point-like particles although they were thought to be elementary particles in 1930s. A success in the classification of hadrons discovered in cosmic rays and in accelerators leaded Gell-Mann to coin "*quark*" with fraction charges $(\pm 1/3, \mp 2/3)$ in mathematical description, rather than in reality [Gell-Mann (1964)]. All the six flavors of quarks (u, d, c, s, t, b) have experimental evidence (the evidence for the last one, top quark, was reported in 1995). Is weak equilibrium among u, d and s quarks possible, instead of simply that between u and d quarks?

At the late stage of stellar evolution, normal baryonic matter is intensely compressed by gravity in the core of massive star during supernova. The Fermi energy of electrons are significant in CBM, and it is very essential to cancel the electrons by weak interaction in order to make lower energy state. There are two ways to kill electrons as shown in Fig. 1: one is via *neutronization*, $e^- + p \to n + \nu_e$, where the fundamental degrees of freedom could be nucleons; the other is through *strangenization*, where the degrees of freedom are quarks. While neutronization works for removing electrons, strangenization has both the advantages of minimizing the electron's contribution of kinetic energy and maximizing the flavour number, the latter could be related to the flavour symmetry of strong-interaction matter. These two ways to kill electrons are relevant to the nature of pulsar, to be neutron star or strange star, as summarized in Fig. 1.

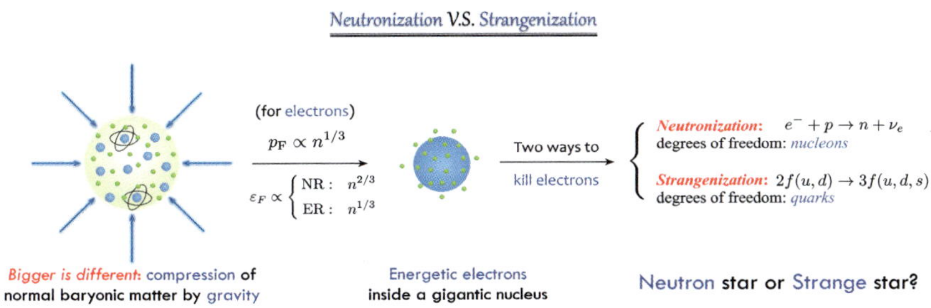

Fig. 1. Neutronization and strangenization are two competing ways to cancel energetic electrons.

There are many speculations about the nature of pulsar due to unknown non-perturbative QCD at low energy. Among different pulsar models, hadron star and hybrid/mixed star are conventional neutron stars, while quark star and quark-cluster star are strange stars with light flavour symmetry. In hadron star model, quarks are confined in hadrons such as neutron/proton and hyperon, while a quark star is dominated by de-confined free quarks. A hybrid/mixed star, with quark matter in its cores, is a mixture of hadronic and quark states. However, a quark-cluster star, in which strong coupling causes individual quarks to group in clusters, is neither a hadron star nor a quark star. As an analog of neutrons, quark-clusters are bound states of several quarks, so in this point of view a quark-cluster star is more similar to a *real* giant nucleus of self-bound (not that of Landau), rather than a "giant hadron" which describes traditional quark stars. Different models of pulsar's inner structure are illustrated in Fig. 2.

It is shown in Fig. 2 that conventional neutron stars (hadron star and hybrid/mixed star) are gravity-bound, while strange stars (strange quark star and strange quark-cluster star) are self-bound on surface by strong force. This feature difference is very useful to identify observationally. In the neutron star picture, the inner and outer cores and the crust keep chemical equilibrium at each boundary, so neutron star is bound by gravity. The core should have a boundary and is in equilibrium with the ordinary matter because the star has a surface composed of ordinary matter. There is, however, no clear observational evidence for a neutron star's surface, although most authors still take it for granted that there should be ordinary matter on the surface, and consequently a neutron star has different components from inner to outer parts. Being similar to traditional quark

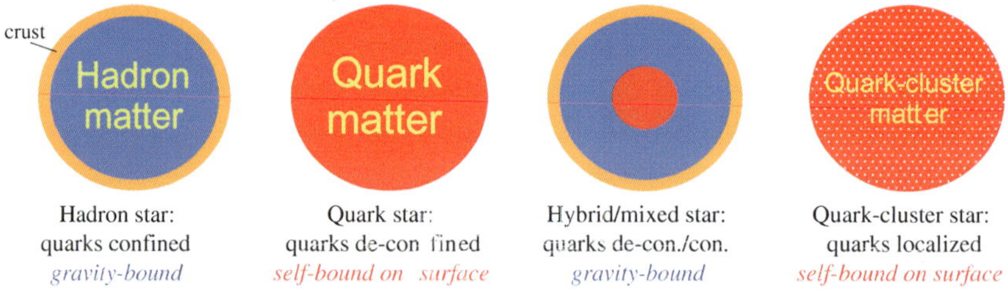

Fig. 2. Different models of pulsar's nature. Hadron star and hybrid/mixed star are of conventional neutron stars, while strangeness plays an important role for quark star and quark-cluster star (or simply *strange star*) as a result of three-light-flavour (u, d and s) symmetry.

stars, quark-cluster stars have almost the same composition from the center to the surface, and the quark matter surface could be natural for understanding some different observations. It is also worth noting that, although composed of quark-clusters, quark-cluster stars are self-bound by the residual interaction between quark-clusters. This is different from but similar to the traditional MIT bag scenario. The interaction between quark-clusters could be strong enough to make condensed matter, and on the surface, the quark-clusters are just in the potential well of the interaction, leading to non-vanishing density but vanishing pressure.

Observations of pulsar-like compact stars, including surface and global properties, could provide hints for the state of CBM, as discussed in the following.

5.1.1. *Surface properties*

Drifting subpulses. Although pulsar-like stars have many different manifestations, they are populated by radio pulsars. Among the magnetospheric emission models for pulsar radio radiative process, the user-friendly nature of Ruderman–Sutherland [Ruderman & Sutherland (1975)] model is a virtue not shared by others, and clear drifting subpulses were first explained. In the seminal paper, a vacuum gap was suggested above the polar cap of a pulsar. The sparks produced by the inner-gap breakdown result in the subpulses, and the observed drifting feature is caused by $\mathbf{E} \times \mathbf{B}$. However, that model can only work in strict conditions for conventional neutron stars — strong magnetic field and low temperature on surfaces of pulsars with $\mathbf{\Omega} \cdot \mathbf{B} < 0$ — while calculations showed, unfortunately, that these conditions usually cannot be satisfied there. The above model encounters the so-called "binding energy problem". Calculations have shown that the binding energy of Fe at the neutron star surface is < 1 keV [Flowers *et al.* (1977); Lai (2001)], which is not sufficient to reproduce the vacuum gap. These problems might be alleviated within a partially screened inner-gap model [Gil, Melikidze & Zhang (2006)] for NSs with $\mathbf{\Omega} \cdot \mathbf{B} < 0$, but could be naturally solved for any $\mathbf{\Omega} \cdot \mathbf{B}$ in the bare strange (quark-cluster) star scenario.

The magnetospheric activity of bare quark-cluster star was investigated in quantitative details [Yu & Xu (2011)]. Since quarks on the surface are confined by strong colour interaction, the binding energy of quarks can be even considered as infinity compared to electromagnetic interaction. As for electrons on the surface, on one hand the potential barrier of the vacuum gap prevents electrons from streaming into the magnetosphere, on the other hand the total energy of electrons on the Fermi surface is non-zero. Therefore, the binding energy of electrons is determined by the difference between the height of the potential barrier in the vacuum gap and the total energy of electrons. Calculations have shown that the huge potential barrier built by the electric field in the vacuum gap above the polar cap can usually prevent electrons from streaming into the magnetosphere, unless the electric potential of a pulsar is sufficiently lower than that at the infinite interstellar medium. In the bare quark-cluster star model, both positively and negatively charged particles on the surface are usually bound strongly enough to form a vacuum gap above its polar cap, and the drifting (even bi-drifting) subpulses can be understood naturally [Xu *et al.* (1999); Qiao *et al.* (2004)].

X-ray spectral lines. In conventional neutron star (NS)/crusted strange star models, an atmosphere exists above the surface of a central star. Many theoretical calculations, first developed by Romani [Romani (1987)], predicted the existence of atomic features in the thermal X-ray emission of NS (also for crusted strange star) atmospheres, and advanced facilities of *Chandra* and *XMM-Newton* were then proposed to be constructed for detecting those lines. One expects to know the chemical composition and magnetic field of the atmosphere through such observations, and eventually to constrain stellar mass and radius according to the redshift and pressure broadening of spectral lines.

However, unfortunately, none of the expected spectral features has been detected with certainty up to now, and this negative test may hint at a fundamental weakness of the NS models. Although conventional NS models cannot be completely ruled out by only non-atomic thermal spectra since modified NS atmospheric models with very strong surface magnetic fields [Ho & Lai (2003); Turolla *et al.* (2004)] might reproduce a featureless spectrum too, a natural suggestion to understand the general observation could be that pulsars are actually bare strange (quark or quark-cluster) star [Xu (2002)], almost without atoms there on the surfaces.

More observations, however, did show absorption lines of PSR-like stars, and the best absorption features were detected for the central compact object (CCO) 1E 1207.4-5209 in the center of supernova remnant PKS 1209-51/52, at ~ 0.7 keV and ~ 1.4 keV [Sanwal *et al.* (2002); Mereghetti *et al.* (2002); Bignami *et al.* (2003)]. Although initially these features were thought to be due to atomic transitions of ionized helium in an atmosphere with a strong magnetic field, soon thereafter it was noted that these lines might be of electron-cyclotron origin, and 1E 1207 could be a bare strange star with surface field of $\sim 10^{11}$ G [Xu, Wang & Qiao (2003)]. Further observations of both spectra feature [Bignami *et al.* (2003)] and precise timing [Gotthelf & Halpern (2007)] favour the electron-cyclotron model of 1E 1207.

But this simple single-particle approximation might not be reliable due to high electron density in strange stars, and Xu *et al.* investigated the global motion of the electron seas on the magnetized surfaces [Xu *et al.* (2012)]. It is found that hydrodynamic surface fluctuations of the electron sea would be greatly affected by the magnetic field, and an analysis shows that the seas may undergo hydrocyclotron oscillations whose eigen frequencies are given by $\omega(l) = \omega_c/[l(l+1)]$, where $l = 1, 2, 3, \ldots$ and $\omega_\mathrm{c} = eB/mc$ is the cyclotron frequency. The fact that the absorption feature of 1E 1207.4-5209 at 0.7 keV is not much stronger than that at 1.4 keV could be understood in this hydrocyclotron oscillations model, because these two lines with l and $l+1$ could have nearly equal intensity, while the strength of the first harmonic is much smaller than that of the fundamental in the electron-cyclotron model. Besides the absorption in 1E 1207.4-5209, the detected lines around $(17.5, 11.2, 7.5, 5.0)$ keV in the burst spectrum of SGR 1806-20 and those in other dead pulsars (e.g., radio quiet compact objects) would also be of hydrocyclotron origin [Xu *et al.* (2012)].

Planck-like continue spectra. The X-ray spectra from some sources (e.g., RX J1856) are well fitted by blackbody, especially with high-energy tails surprisingly close to Wien's formula: decreasing exponentially ($\propto e^{-\nu}$). Because there is an atmosphere above the surface of neutron star/crusted strange stars, the spectrum determined by the radiative transfer in atmosphere should differ substantially from Planck-like one, depending on the chemical composition, magnetic field, etc. [Zavlin *et al.* (1996)]. Can the thermal spectrum of quark-cluster star be well described by Planck's radiation law? In bag models where quarks are nonlocal, one limitation is that bare strange stars are generally supposed to be poor radiators in thermal X-ray as a result of their high plasma frequency, ~ 10 MeV. Nonetheless, if quarks are localized to form quark-clusters in cold quark matter due to very strong interactions, a regular lattice of the clusters (i.e., similar to a classical *solid* state) emerges as a consequence of the residual interaction between clusters [Xu (2003)]. In this latter case, the metal-like solid quark matter would induce a metal-like radiative spectrum, with which the observed thermal X-ray data of RX J1856 can be fitted [Zhang, Xu & Zhang (2004)]. Alternatively, other radiative mechanism in the electrosphere (e.g., electron bremsstrahlung in the strong electric field [Zakharo (2010)] and even of negligible ions above the sharp surface) may also reproduce a Planck-like continue spectrum.

Supernova and gamma-ray bursts. It is well known that the radiation fireballs of gamma-ray bursts (GRBs) and supernovae as a whole move towards the observer with a high Lorentz factor [Paczyǹski (1986)]. The bulk Lorentz factor of the ultra-relativistic fireball of GRBs is estimated to be in the order [Mészáros, Rees & Wijers (1988)] of $\Gamma \sim 10^2 - 10^3$. For such an ultra-relativistic fireball, the total mass of baryons cannot be too high, otherwise baryons would carry too much energy out of the central engine, the so-called "baryon contamination". For conventional neutron stars as the central engine, the number of baryons loaded with the fireball is unlikely to be small, since neutron stars are gravity-confined and the luminosity of fireball is extremely high. However, the baryon contamination problem can be solved naturally if the central compact objects are strange quark-cluster stars. The bare and chromatically confined surface of quark-cluster stars separates baryonic matter from the photon and lepton dominated fireball. Inside the star, baryons are in quark-cluster phase and cannot escape due to strong colour interaction, but $e^{\pm}$-pairs, photons and neutrino pairs can escape from the surface. Thus, the surface of quark-cluster stars automatically generates a low baryon condition for GRBs as well as supernovae [Ouyed, Rapp & Vogt (2005); Paczyǹski & Haensel (2005); Cheng & Dai (1996)].

It is still an unsolved problem to simulate supernovae successfully in the neutrino-driven explosion models of neutron stars. Nevertheless, in the quark-cluster star scenario, the *bare* quark surfaces could be essential for successful explosions of both core and accretion-induced collapses [Xu (2005)]. A nascent quark-cluster star born in the center of GRB or supernova would radiate thermal emission due to its ultrahigh surface temperature [Haensel, Paczyǹski & Amsterdamski (1991)], and the photon luminosity is not constrained by the Eddington limit since the surface of quark-cluster stars could be bare and chromatically confined. Therefore, in this photon-driven scenario [Chen, Yu & Xu (2007)] the strong radiation pressure caused by enormous thermal emissions from quark-cluster stars might play an important role in promoting core-collapse supernovae. Calculations have shown that the radiation pressure due to such strong thermal emission can push the overlying mantle away through photon–electron scattering with energy as much as $\sim 10^{51}$ ergs. Such photon-driven mechanism in core-collapse supernovae by forming a quark-cluster star inside the collapsing core is promising to alleviate the current difficulty in core-collapse supernovae. The recent discovery of highly super-luminous supernova ASASSN-15lh, with a total observed energy $(1.1 \pm 0.2) \times 10^{52}$ ergs, [Dong *et al.* (2016)] might also be understood in this regime if a very massive strange quark-cluster star, with mass smaller than but approaching $M_{\rm max}$, forms.

5.1.2. *Global properties*

Free or torque-induced precession. Rigid body precesses naturally when spinning, either freely or by torque, but fluid one can hardly. The observation of possible precession or even free precession of B1821-11 [Stairs, Lyne & Shemar (2000)] and others could suggest a global solid structure for pulsar-like stars. Low-mass quark stars with masses of $\lesssim 10^{-2} M_{\odot}$ and radii of a few kilometers are gravitationally force-free, and their surfaces could then be irregular (i.e., asteroid-like). Therefore, free or torque-induced precession may easily be excited and expected with larger amplitude in low-mass quark stars. The masses of AXPs/SGRs (anomalous X-ray pulsars/soft gamma-ray repeaters) could be approaching the mass-limit ($> 1.5 M_{\odot}$) in the AIQ (accretion-induced quake) model [Xu (2007)]; these objects could then manifest no or weak precession as observed, though they are more likely than CCOs/DTNs (eg., RX J1856) to be surrounded by dust disks because of their higher masses (thus stronger gravity).

Normal and slow glitches. A big disadvantage in pulsars being strange quark stars lies in the fact that the observation of pulsar glitches conflicts with the hypothesis of conventional quark stars in fluid states [Alpar (1987); Benvenuto, Horvath & Vucetich (1990)] (e.g., in MIT bag models). That problem could be solved in a solid quark-cluster star model since a solid stellar object would inevitably result in starquakes when strain energy develops to a critical value. Huge energy should be released, and thus large spin-change occurs, after a quake of a solid quark star. Starquakes could then be a simple and intuitional mechanism for pulsars to have glitches frequently with large amplitudes. In the regime of solid quark star, by extending the model for normal glitches [Zhou *et al.* (2004)], one can also model pulsar's slow glitches [Peng & Xu (2008)] not well understood in NS models. In addition, both types of glitches without (Vela-like, Type I) and with (AXP/SGR-like, Type II) X-ray enhancement could be naturally understood in the starquake model of solid strange star [Zhou *et al.* (2014)], since the energy release during a type I (for fast rotators) and a type II (for slow rotators) starquake are very different.

Energy budget. The substantial free energy released after starquakes, both elastic and gravitational, would power some extreme events detected in AXPs/SGRs and during GRBs. Besides persistent pulsed X-ray emission with luminosity well in excess of the spin-down power, AXPs/SGRs show occasional bursts (associated possibly with glitches), even superflares with isotropic energy $\sim 10^{44-46}$ erg and initial peak luminosity $\sim 10^{6-9}$ times of the Eddington one. They are speculated to be magnetars, with the energy reservoir of magnetic fields $\gtrsim 10^{14}$ G (the origin still debatable since the dynamo action might not be so effective and the strong magnetic field could decay effectively), but failed predictions are challenges to the model [Tong & Xu (2011)]. However, AXPs/SGRs could also be solid quark stars with surface magnetic fields similar to that of radio pulsars. Starquakes are responsible for both bursts/flares and glitches in the latter scenario [Xu (2007)], and kinematic oscillation energy could effectively power the magnetospheric activity [Lin, Xu & Zhang (2015)].

The most conspicuous asteroseismic manifestation of solid phase of quark stars is their capability of sustaining torsional shear oscillations induced by SGR's starquake [Bastrukov, Chen & Chang (2009)]. In addition, there are more and more authors who are trying to connect the GRB central engines to SGRs' flares, in order to understand different GRB light-curves observed, especially the internal-plateau X-ray emission [Xu & Liang (2009); Dai, Li & Xu (2011)].

Mass and radius of compact star. The EoS of quark-cluster matter would be stiffer than that of nuclear matter, because (1) quark-cluster should be non-relativistic particle for its large mass, and (2) there could be strong short-distance repulsion between quark-clusters. Besides, both the problems of hyperon puzzle and quark-confinement do not exist in quark-cluster star. Stiff EoS implies high maximum mass, while low mass is a direct consequence of self-bound surface.

It has been addressed that quark-cluster stars could have high maximum masses ($> 2M_\odot$) as well as very low masses ($< 10^{-2}M_\odot$) [Lai & Xu (2009)]. Later radio observations of PSR J1614-2230, a binary millisecond pulsar with a strong Shapiro delay signature, imply that the pulsar mass is 1.97±0.04 $M_\odot$ [Demorest *et al.* (2010)], which indicates a stiff EoS for CBM. Another $2M_\odot$ pulsar is also discovered afterwards [Antoniadis *et al.* (2013)]. It is conventionally thought that the state of dense matter softens and thus cannot result in high maximum mass if pulsars are quark stars, and that the discovery of massive $2M_\odot$ pulsar may make pulsars unlikely to be quark stars. However, quark-cluster star could not be ruled out by massive pulsars, and the observations of pulsars with higher mass (e.g. $> 2.5M_\odot$), would even be a strong support to quark-cluster star model, and give further constraints to the parameters. The mass and radius of 4U 1746-37 could be constrained by PRE (photospheric radius expansion) bursts, on the assumption that the touchdown flux corresponds to Eddington luminosity and the obscure effect is included [Li *et al.* (2015)]. It turns out that 4U 1746-37 could be a strange star with small radius.

There could be other observational hints of low-mass strange stars. Thermal radiation components from some PSR-like stars are detected, the radii of which are usually much smaller than 10 km in blackbody models where one fits spectral data by Planck spectrum [Pavlov, Sanwal & Teter (2004)] and Pavlov and Luna [Pavlov & Luna (2009)] find no pulsations with periods longer than ~ 0.68 s in the CCO of Cas A, and constrain stellar radius and mass to be $R = (4 \sim 5.5)\,\text{km}$ and $M \lesssim 0.8M_\odot$ in hydrogen NS atmosphere models. Two kinds of efforts are made toward an understanding of these facts in conventional NS models. (1) The emissivity of NS's surface isn't simply of blackbody or of hydrogen-like atmospheres. The CCO in Cas A is suggested to be covered by a carbon atmosphere [Ho & Heinke (2009)]. However, the spectra from some sources (e.g., RX J1856) are still puzzling, being well fitted by blackbody. (2) The small emission areas would represent hot spots on NS's surfaces, i.e., to fit the X-ray spectra with at least two blackbodies, but this has three points of weakness in NS models. *a*, about P and $\dot{P}$. No or very weak pulsation has been detected in some of thermal component-dominated sources (e.g., the Cas A CCO [Pavlov & Luna (2009)]), and the inferred magnetic field from $\dot{P}$ seems not to be consistent with the atmosphere models at least for RX J1856 [Kerkwijk & Kaplan (2008)]. *b*, fitting of thermal X-ray spectra (e.g., PSR J1852+0040) with two blackbodies finds two small emitting radii (significantly smaller than 10 km), which are not yet understood [Halpern & Gotthelf (2010)]. *c*, the blackbody temperature of the entire surface of some PSR-like stars are much lower than those predicted by the standard NS cooling models [Li,

Lu & Li (2005)], even provided that hot spots exist. Nevertheless, besides those two above, a *natural* idea could be that the detected small thermal regions (*if* being global) of CCOs and others may reflect their small radii (and thus low masses in quark-cluster star scenario) [Xu (2005)].

Another low-mass strange (quark-cluster) star could be 4U 1700+24. Because of strangeness barrier existing above a quark-cluster surface, a strange star may be surrounded by a hot corona or an atmosphere, or even a crust for different accretion rates. Both the redshifted O VIII Ly-α emission line (only $z = 0.009$) and the change in the blackbody radiation area (with an inferred scale of $\sim$ (10–10^2) m) could naturally be understood if 4U 1700+24 is a low-mass quark-cluster star which exhibits weak wind accretion [Xu (2014)]. Additionally, the mass function via observing the G-type red giant company is only $f_o = (1.8 \pm 0.9) \times 10^{-5} M_\odot$ [Galloway, Sokoloski & Kenyon (2002)], from which the derived mass of compact star should be much lower than $1 M_\odot$ unless there is geometrical fine-tuning (inclination angle $i < 2°$, see Fig. 3). All these three independent observations (redshift, hot spot and mass function) suggest that 4U 1700+24 could be a low-mass strange quark-cluster star.

Future observations with more advanced facilities, such as FAST and SKA, could provide more observational hints for the nature of CBM. Pulsar mass measurement could help us find more massive pulsar, while measurement of the momentum of inertia may give information on the radius. Searching sub-millisecond pulsars could be an expected way to provide clear

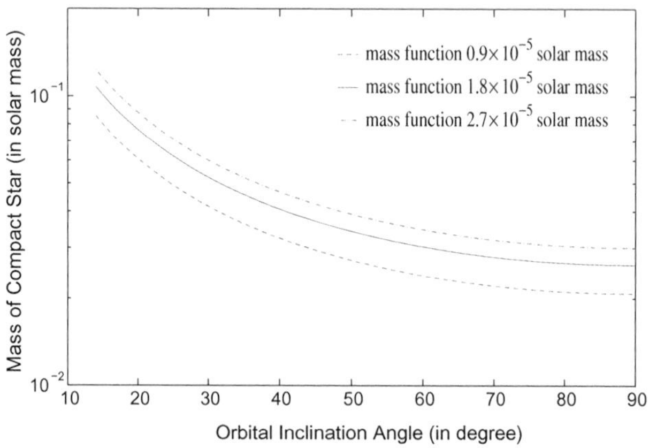

Fig. 3. The compact star mass as a function of orbital inclination for different values of mass function (4U 1700+24).

evidence for (low-mass) quark stars. Normal neutron stars cannot spin with periods less than $\sim 0.5 M_1^{1/2} R_6^{-3/2}$ ms ($R_6 = R/10^6$ cm), as the rotation is limited by Kepler frequency. But low-mass bare strange stars has no such limitation on the spin period, which could be even less than 1 ms. We need thus a much shorter sampling time, and would deal with then a huge amount of data in order to find sub-millisecond pulsars. Besides, the pulse profile of pulsar is helpful for the understanding of its magnetospheric activity.

5.2. *Strange matter in cosmic rays and as dark matter candidate*

Strange quark-nuggets, in the form of cosmic rays, could be ejected during the birth of central compact star [Xu & Wu (2003)], or during collision of strange stars in a binary system spiraling towards each other due to loss of orbital energy via gravitational waves [Madsen (2005)].

A strangelet with mass per baryon < 940 MeV (i.e., binding energy per baryon $\gtrsim 100$ MeV) could be stable in cosmic rays, and would decay finally into nucleons when collision-induced decrease of baryon number makes it unstable due to the increase of surface energy. When a stable strangelet bombards the atmosphere of the Earth, its fragmented nuggets may decay quickly into Λ particles by strong interaction and further into nucleons by weak interaction. What if a strange nugget made of quark clusters bombards the Earth? It is interesting and necessary to investigate.

In the early Universe ($\sim 10\,\mu$s), quark–gluon plasma condenses to form hadron gas during the QCD phase transition. If the cosmological QCD transition is first-order, bubbles of hadron gas are nucleated and grow until they merge and fill up the whole Universe. A separation of phases during the coexistence of the hadronic and the quark phase could gather a large number of baryons in strange nuggets [Witten (1984)]. If quark clustering occurs, evaporation and boiling may be suppressed, and strange nuggets may survive and contribute to the dark matter today. Strange nuggets as cold quark matter may favour the formation of seed black holes in primordial halos, alleviating the current difficulty of quasars at redshift as high as $z \sim 6$ [Lai & Xu (2010)], and the small pulsar glitches detected may hint the role of strange nuggets [Lai & Xu (2016)].

6. Conclusions

Although normal micro-nuclei are 2-flavour symmetric, we argue that 3-flavour symmetry would be restored in macro/gigantic-nuclei compressed by gravity during a supernova. Strange matter is conjectured to be condensed matter of 3-flavour quark-clusters, and future advanced facilities (e.g., FAST, SKA) would provide clear evidence for strange stars. Strange nuggets manifested in the form of cosmic rays and even dark matter has significant astrophysical consequences, to be tested observationally.

Note added in proof

After submission of this chapter, the discovery of the gravitational waves is announced (Abbott *et al.* *PRL* **116**, 061102 (2016)). The proposed model of strange star with rigidity (i.e., solid strange quark-cluster star) is quite likely to be tested further by kilo-Hz gravitational wave observations of two kinds of events as follow, at least. (1) Merger of pulsar–pulsar/pulsar–black hole binary. The predicted waveform depends on the state equation of supra-nuclear matter, and the tidal effects during inspiral should be much weaker for solid strange star than for normal neutron star. (2) Starquake of pulsar-like compact star. Sensitive detectors may discover starquake-induced gravitational waves of compact stars, and then show very different signatures of neutron and strange stars.

Acknowledgments

This work is supported by the National Basic Research Program of China (No. 2012CB821801) and NNSFC (No. 11225314). The FAST FELLOWSHIP is supported by the Special Funding for Advanced Users, budgeted and administrated by Center for Astronomical Mega-Science, Chinese Academy of Sciences (CAS). We would like to thank Ms. Yong Su for reading and checking the story of science fiction about flavour symmetry.

References

Alcock, C., Farhi R. and Olinto, A., *ApJ* **310**, 261 (1986).
Alford, M. J. *et al.*, *Rev. Mod. Phys.* **80**, 1455 (2008).
Alpar, M. A., *Phys. Rev. Lett.* **58**, 2152 (1987).
Antoniadis, J., Freire, P. C. C., Wex, N. *et al.*, *Science* **340**, 448 (2013).
Bastrukov, S. I., Chen, G. T., Chang, H. K. *et al.*, *ApJ* **690**, 998 (2009).
Beane, S. R., Chang, E., Detmold, W. *et al.*, *Phys. Rev. Lett.* **106**, 162001 (2011).
Benvenuto, O. G., Horvath, J. E., and Vucetich, H., *Phys. Rev. Lett.* **64**, 713 (1990).
Bertsch, G. F., *Ann. Phys.* **86**, 138 (1974).
Bignami, G. F., Caraveo, P. A., Luca, A. De, and Mereghetti, S., *Nature* **423**, 725 (2003).

Bodmer, A. R., *Phys. Rev.* **D4**, 16 (1971).
Chen, A. B., Yu, T. H., and Xu, R. X., *ApJ* **668**, L55 (2007).
Cheng, K. S. and Dai, Z. G., *Phys. Rev. Lett.* **77**, 1210 (1996).
Dai, S., Li, L. X. and Xu, R. X., *Sci. Chin. Ser. G: Phys., Mech. Astron.* **54**, 1514 (2011).
Demorest, P., Pennucci, T., Ransom, S., Roberts, M., and Hessels, J., *Nature* **467**, 1081 (2010).
Dong, S., Shappee, B. J., Prieto, J. L. *et al.*, *Science* **351**, 257 (2016).
Fischer, C. S., *J. Phys. G: Part. Nucl. Phys* **32**, R253 (2006).
Fischer, C. S. and Alkofer, R., *Phys. Lett.* **B536**, 177 (2002).
Flowers, E. G., Ruderman, M. A., Lee, J. F., Sutherland, P. G., Hillebrandt, W. and Mueller, E., *ApJ* **215**, 291 (1977).
Galloway, D. K., Sokoloski, J. L., and Kenyon, S. J., *ApJ* **580**, 1065 (2002).
Gell-Mann, M., *Phys. Lett.* **8**, 214 (1964).
Gil, J., Melikidze, G. and Zhang, G., *ApJ* **650**, 1048 (2006).
Gotthelf, E. V., and Halpern, J. P., *ApJ* **664**, 35 (2007).
Gross, D. J. and Wilczek, F., *Phys. Rev. Lett.* **30**, 1343 (1973).
Haensel, P., Paczyǹski, B., and Amsterdamski, P., *ApJ* **375**, 209 (1991).
Haensel, P., Zdunik, J. L. and Schaeffer, R., *A&A* **160**, 121 (1986).
Halpern, J. P. and Gotthelf, E. V., *ApJ* **709**, 436 (2010).
Hen, O. *et al.*, *Science* **346**, 614 (2014).
Herrera, L. and Santobs, N. O., *Phys. Rep.* **286**, 53 (1997).
Hewish, A. and Bell, J. *et al.*, *Nature* **217**, 709 (1968).
Ho, W. C. G. and Heinke, C. O, *Nature* **462**, 71 (2009).
Ho, W. C. G. and Lai, D., *MNRAS* **338**, 233 (2003).
Inoue, T. *et al.*, *Phys. Rev. Lett.* **106**, 162002 (2011).
Itoh, N., *Prog. Theor. Phys.* **44**, 291 (1970).
Ivanenko, D. and Kurdgelaidze, D. F., *Lett. Nuovo Cimento* **2**, 13 (1969).
Johnson, G., *Strange Beauty* (A Division of Random House, Inc., New York, 1999).
Karlovini, M. and Samuelsson, L., *Class. Quant. Grav.* **21**, 4531 (2004).
Kerkwijk, M. H. van and Kaplan, D. L., *ApJ* **673**, L163 (2008).
Lai, D., *Rev. Mod. Phys.* **73**, 629 (2001).
Lai, X. Y., Gao, C. Y., and Xu, R. X., *MNRAS* **431**, 3282 (2013).
Lai, X. Y. and Xu, R. X., *MNRAS* **398**, L31 (2009).
Lai, X. Y. and Xu, R. X., *J. Cos. & Astropart. Phys.* **5**, 28 (2010).
Lai, X. Y. and Xu, R. X., *RAA* **16**, (2016) in press (arXiv:1506.04172).
Landau, L. Phys. *Z. Sowjetunion* **1**, 285 (1932).
Li, B. A., Chen, L. W. and Ko, C. M., *Phys. Rep.* **464**, 113 (2008).
Li, X. H., Lu, F. J. and Li, T. P., *ApJ* **628**, 931 (2005).
Li, Z. S., Qu, Z. J. and Chen, L. *et al.*, *ApJ* **798**, 56 (2015).
Lin, M. X., Xu, R. X. and Zhang, B., *ApJ* **799**, 152 (2015).
Madsen, J., *Phys. Rev.* **D71**, 014026 (2005).
McLerran, L. and Pisarski, R. D., *Nucl. Phys.* **A796**, 83 (2007).
Mereghetti, S., Luca, A. De, Caraveo, P., Becker, W., Mignani, R., and Bignami, G. F., *ApJ* **581**, 1280 (2002).

Mészáros, P., Rees, M. J., and Wijers, R. A. M. J., *ApJ* **499**, 301 (1998).
Olive, K. A. *et al.*, (Particle Data Group), *Chin. Phys.* **C38**, 090001, (2014) .
Oppenheimer J. R. and Volkoff, G.B., *Phys. Rev.* **55**, 374 (1939).
Ouyed, R., Rapp, R., and Vogt, C., *ApJ* **632**, 1001 (2005).
Paczyǹski, B., *ApJ* **308**, 43 (1986).
Paczyǹski, B., and Haensel, P., *MNRAS* **362**, 4 (2005).
Pavlov, G. G. and Luna, G. J. M., *ApJ* **703**, 910 (2009).
Pavlov, G. G., Sanwal, D., and Teter, M. A., in *Young Neutron Stars and Their Environments, IAU Symposium* **218**, 239 (2004).
Peng, C. and Xu, R. X., *MNRAS* **384**, 1034 (2008).
Politzer, H. D., *Phys. Rev. Lett.* **30**, 1346 (1973).
Qiao, G. J., Lee, K. J., Zhang, B., Xu, R. X. and Wang, H. G., *ApJ* **616**, L127 (2004).
Romani, R. W., *ApJ* **313**, 718 (1987).
Ruderman, M. A. and Sutherland, P. G., *ApJ* **196**, 51 (1975).
Sanwal, D., Pavlov, G. G, Zavlin, V. E., and Teter, M., *ApJ* **574**, 61 (2002).
Shuryak, E. V., *Prog. Part. & Nucl. Phys.* **62**, 48 (2009).
Stairs, I. H., Lyne, A. G., and Shemar, S. L., *Nature* **406**, 484 (2000).
Subedi, R. *et al.*, *Science* **320**, 1476 (2008).
Tong, H. and Xu, R. X., *IJMP* **E20** (S2), 15 (2011).
Turolla, R., Zane, S. and Drake, J. J., *ApJ* **603**, 265 (2004).
Wilczek, F., *Nature* **445**, 156 (2007).
Witten, E., *Phys. Rev.* **D30**, 272 (1984).
Xu, R. X., *ApJ* **570**, L65 (2002).
Xu, R. X., *ApJ* **596**, L59 (2003).
Xu, R. X., *MNRAS* **356**, 359 (2005).
Xu, R. X., *Adv. Space Res.* **40**, 1453 (2007).
Xu, R. X., *RAA* **14**, 617 (2014).
Xu, R. X., Bastrukov, S. I., Weber, F., Yu, J. W., and Molodtsova, I. V., *Phys. Rev.* **D85**, 023008 (2012).
Xu, R. X. and Liang, E. W., *Sci. Chin. Ser. G: Phys., Mech. Astron.* **52**, 315 (2009).
Xu, R. X., Qiao, G. J., and Zhang, B., *ApJ* **522**, L109 (1999).
Xu, R, X., Tao, D. J. and Yang, Y., *MNRAS* **373**, L85 (2006).
Xu, R. X., Wang, H. G., and Qiao, G. J., *Chin. Phys. Lett.* **20**, 314 (2003).
Xu, R. X. and Wu, F., *Chin. Phys. Lett.* **20**, 80 (2003).
Yu, J. W. and Xu, R. X., *MNRAS* **414**, 489 (2011).
Zakharov, B. G., *Phys. Lett.* **B690**, 250 (2010).
Zavlin, V. E., Pavlov, G. G., and Shibanov, Y. A., *A&A* **315**, 141 (1996).
Zhang, X. L., Xu, R. X., and Zhang, S. N., A strange star with solid quark surface? in *Young Neutron Stars and their Environments*, eds. F. Camilo and B. M. Gaensler (San Francisco, 2004) p. 303.
Zhou, A. Z., Xu, R. X., Wu, X. J. and Wang, N., *Astropart. Phys.* **22**, 73 (2004).
Zhou, E. P., Lu, J. G., Tong, H., and Xu, R. X., *MNRAS* **443**, 2705 (2014).

Chapter 5

Building Non-Spherical Cosmic Structures

Roberto A. Sussman

Instituto de Ciencias Nucleares
Universidad Nacional Autónoma de México (ICN-UNAM)
A. P. 70–543, 04510 México D. F., México
sussman@nucleares.unam.mx

We describe how an exact solution of Einstein's equations (the Szekeres models) can be used to construct assorted configurations of multiple non-spherical self-gravitating cold dark matter structures (over-densities and voids) that evolve from realistic initial data. Since the dynamical freedom of the Szekeres models allow (under certain restrictions) for placing these structures in previously specified locations, we are able to provide a fully relativistic non-perturbative coarse-grained description of actually existing cosmic structure at various scales. We believe that there is an enormous range of potential applications of this work to current astrophysical and cosmological problems.

Contents

1. Introduction

The theoretical interpretation of a large amount of high quality and precise cosmological observations at all astrophysical and cosmic scales requires a robust modelling of self-gravitating systems. Conventionally, the large cosmic scale dynamics of these sources is examined through linear perturbations on a ΛCDM background, while Newtonian gravity (perturbative, non-perturbative and numerical simulations) is often employed for self-gravitational systems at smaller galactic and galactic cluster scales [Ellis, Maartens & MacCallum (2012)].

A non-perturbative approach by means of analytic or numeric solutions of Einstein's equations [Ellis, Maartens & MacCallum (2012); Krasiński (1997); Plebański & Krasiński (2006); Bolejko, Krasiński, Hellaby & Célérier (2009)] is less favoured because (unless powerful numerical methods are employed) the high non-linear complexity of Einstein's equations renders realistic models mathematically untractable. As a consequence, extremely idealised toy models are used in most cosmological applications following a fully relativistic non-perturbative approach. The most prominent examples are the spherically symmetric Lemaître–Tolman (LT) models [Lemaître (1933); Tolman (1934); Bondi (1947)] (see extensive reviews in [Krasiński (1997); Plebański & Krasiński (2006); Bolejko, Krasiński, Hellaby & Célérier (2009)]), which were used to construct large scale CDM void models in the recent effort to explore the possibility of fitting cosmological observations (supernovas, CMB, etc.) without the assumption of a dark energy source or a cosmological constant [Bolejko, Célérier & Krasiński (2011); Marra & Notari (2011); Biswas, Notari & Valkenburg (2010)].

Even if the usage of non-linear inhomogeneity based on LT void models to fit large scale observations has apparently failed [Redlich *et al.*

(2014); Zibin & Moss (2011a,b); Bull, Clifton & Ferreira (2012)]), which seems to reafirm the current ΛCDM paradigm, non-perturbative general relativistic models are still needed and can be useful to probe structure formation scenarios and to provide theoretical support to cosmological observations. However, less idealised models that are not restricted by spherical symmetry are needed for this purpose, as the CDM structures we observe in all scales (from galactic surveys) are far from spherically symmetric. In this context, the well known Szekeres solutions [Szekeres (1975a,b); Goode & Wainwright (1982)] (see details and classification of them in [Krasiński (1997); Plebański & Krasiński (2006); Bolejko, Krasiński, Hellaby & Célérier (2009); Bolejko, Célérier & Krasiński (2011)]), which provide a straightforward non-spherical generalisation of LT models, are often considered as convenient tool for an improved study of cosmic structure. In fact, there is an extensive literature on the application of Szekeres models to describe cosmological structures and to fit observations [Nwankwo, Ishak & Thompson (2011); Bolejko (2007); Peel, Ishak & Troxel (2012); Krasiński & Bolejko (2010); Bolejko (2009); Ishak *et al.* (2008); Bolejko & Célérier (2010); Ishak, Peel & Troxel (2013)], with recent attempts to explore the relevance of their theoretical properties for these applications [Bolejko & Sussman (2011); Sussman & Bolejko (2012); Walters & Hellaby (2012); Vrba & Svitek (2014); Bolejko, Ahsan Nazer & Wiltshire (2016)].

The CDM structures currently observed typically consist of spatial distributions of elongated filamentary over-dense regions surrounding spheroidal under-dense regions (voids) of typical 30–50 Mpc size [Raccanelli *et al.* (2008)]. Therefore, we aim in this work at extending and improving previous efforts to describe the dynamics of this type of sources with Szekeres models that have been undertaken, specially in [Bolejko & Sussman (2011); Sussman & Bolejko (2012); Walters & Hellaby (2012); Vrba & Svitek (2014)] (see review in [Bolejko, Krasiński, Hellaby & Célérier (2009)]). The result is a theoretically robust procedure to select the free parameters of the models in order to describe the full time evolution of multiple CDM structures: over-densities and density voids, whose spatial location can be roughly prescribed as desired (an extended version of this work is found in [Sussman, Delgado Gaspar & Hidalgo (2016); Sussman & Delgado Gaspar (2015)]). We believe that Szekeres models constructed along these lines provide an effective (even if still coarse-grained) description of the evolution of observed structures in a wide range of astrophysical and cosmological scales.

2. Szekeres Models in Spherical Coordinates

The metric of Szekeres models in spherical coordinates is given by[1]:

$$\begin{aligned} g_{tt} &= -1, \\ g_{rr} &= a^2 \left\{ \frac{(\Gamma - \mathbf{W})^2}{1 - \mathcal{K}_{q0} r^2} + \frac{\sin^4\theta}{(1+\cos\theta)^2} \left[\mathcal{W}^2 - 2\frac{1+\cos\theta}{\sin^2\theta}\, Z\,\mathbf{W} \right] \right\}, \\ g_{r\theta} &= \frac{a^2\, r\, \sin\theta}{1+\cos\theta} \left(\mathbf{W} - Z\right), \\ g_{r\phi} &= -\frac{a^2\, r\, \sin^2\theta}{1+\cos\theta} \mathbf{W}_{,\phi}, \\ g_{\theta\theta} &= a^2 r^2, \qquad g_{\phi\phi} = a^2 r^2 \sin^2\theta, \end{aligned} \tag{1}$$

where $\mathcal{K}_{q0} = \mathcal{K}_{q0}(r)$. The time dependence is contained in the scale factors:

$$a = a(t,r), \qquad \Gamma = \Gamma(t,r) = 1 + \frac{ra'}{a}, \qquad a' = \frac{\partial a}{\partial r}, \tag{2}$$

while $\mathbf{W}$ and its magnitude $\mathcal{W}$ are given by [Sussman & Delgado Gaspar (2015)]

$$\mathbf{W} = -X\,\sin\theta\,\cos\phi - Y\,\sin\theta\,\sin\phi - Z\,\cos\theta, \tag{3}$$

$$\mathcal{W}^2 \equiv X^2 + Y^2 + Z^2 = \frac{3}{4\pi} \int_0^{2\pi}\!\!\int_0^{\pi} \mathbf{W}^2\,\sin\theta d\theta\, d\phi, \tag{4}$$

where $X = X(r)$, $Y = Y(r)$, $Z = Z(r)$ are arbitrary functions satisfying the regularity conditions $X(0) = Y(0) = Z(0) = 0$ and $X'(0) = Y'(0) = Z'(0) = 0$.

The main covariant scalars associated with Szekeres models are: the density ρ; the Hubble scalar $\mathcal{H} \equiv \Theta/3$ with $\Theta = \nabla_a u^a$; the spatial curvature scalar $\mathcal{K} = (1/6)^{(3)}\mathcal{R}$ with ${}^{(3)}\mathcal{R}$ the Ricci scalar of the hypersurfaces t constant; the eigenvalues Σ, Ψ_2 of the shear and electric Weyl tensors (Ψ_2 is the nonzero conformal invariant of Petrov type D spacetimes). These

[1]By "Szekeres models" we refer henceforth to quasi-spherical models of class I with a dust source, see details of the obtention and classification of these models in [Krasiński (1997); Plebański & Krasiński (2006)]. These spherical coordinates are defined as a "stereographic" projection in [Plebański & Krasiński (2006)]. The standard diagonal representation of the Szekeres metric and the transformations relating it to Eq. (1) are given in [Sussman & Delgado Gaspar (2015)].

scalars can be given by the following concise and elegant form as

$$A = A_q + \mathbf{D}^{(A)}, \qquad A = \rho,\, \mathcal{H},\, \mathcal{K}, \tag{5}$$

$$\Sigma = -\mathbf{D}^{(\mathcal{H})} = -(\mathcal{H} - \mathcal{H}_q), \qquad \Psi_2 = \frac{4\pi}{3}\mathbf{D}^{(\rho)} = \frac{4\pi}{3}(\rho - \rho_q), \tag{6}$$

where we have introduced the "q-scalars" A_q and their exact fluctuations $\mathbf{D}^{(A)}$ (for $A = \rho,\, \mathcal{H},\, \mathcal{K}$) defined as [Sussman & Bolejko (2012)]:

$$A_q = \frac{\int\limits_r\int\limits_\theta\int\limits_\phi A\,\mathcal{F}\sqrt{\mathcal{J}}\,\mathrm{d}r\,\mathrm{d}\theta\,\mathrm{d}\phi}{\int\limits_r\int\limits_\theta\int\limits_\phi \mathcal{F}\sqrt{\mathcal{J}}\,\mathrm{d}r\,\mathrm{d}\theta\,\mathrm{d}\phi}, \qquad \mathbf{D}^{(A)} = A - A_q = \frac{r\,A_q'}{3(\Gamma - \mathbf{W})}, \tag{7}$$

where $\mathcal{F} \equiv \sqrt{1 - \mathcal{K}_{q0}\, r^2}$ and $\mathcal{J}$ is the determinant of the spatial part of the metric (1). Evaluating this integral for each scalar A yields the following scaling laws and constraints

$$\rho_q = \frac{\rho_{q0}}{a^3}, \qquad \mathcal{K}_q = \frac{\mathcal{K}_{q0}}{a^2}, \qquad \mathcal{H}_q = \frac{\dot{a}}{a}, \tag{8}$$

$$\mathcal{H}_q^2 = \frac{\dot{a}^2}{a^2} = \frac{8\pi}{3}\rho_q - \mathcal{K}_q + \frac{8\pi}{3}\Lambda, \tag{9}$$

$$2\mathcal{H}_q\,\mathbf{D}^{(\mathcal{H})} = \frac{8\pi}{3}\mathbf{D}^{(\rho)} - \mathbf{D}^{(\mathcal{K})}, \tag{10}$$

$$\frac{\mathbf{D}^{(\rho)}}{\rho_q} - \frac{3}{2}\frac{\mathbf{D}^{(\mathcal{K})}}{\mathcal{K}_q} = \frac{1 - \mathbf{W}}{\Gamma - \mathbf{W}}\left[\frac{\mathbf{D}_0^{(\rho)}}{\rho_{q0}} - \frac{3}{2}\frac{\mathbf{D}_0^{(\mathcal{K})}}{\mathcal{K}_{q0}}\right], \tag{11}$$

where the subindex $_0$ denotes evaluation at an arbitrary fixed $t = t_0$. Every Szekeres model becomes fully specified by five free parameters: two of ρ_{q0}, $\mathcal{H}_{q0}$, $\mathcal{K}_{q0}$ plus the dipole parameters X, Y, Z. The models become determined either by solving the Friedmann equation (9) or by integrating the evolution equations for the A_q, $\mathbf{D}^{(A)}$ (see Eqs. (39)–(44) in Sec. 10, see also [Sussman & Bolejko (2012)]).

It is evident that surfaces of constant t and constant r are 2-spheres, since setting $\mathrm{d}t = \mathrm{d}r = 0$ in Eq. (1) yields the metric of a 2-sphere with surface areas $4\pi a^2 r^2$. These surfaces constitute a smooth foliation of any time slice by non-concentric 2-spheres. This follows from the fact that g_{rr} depends on the angular coordinates, hence the proper radial lengths along radial rays from the origin worldline[2] to points in any 2-sphere $\ell = \int_0^r \sqrt{g_{rr}}\mathrm{d}r$ smoothly depend on the angular coordinates of the points.

[2]While the origin worldline is not a symmetry centre (fixed point of SO(3)), it is still the locus of a special observer [Bolejko & Sussman (2011)], see Sec. 5.

3. Angular Orientation: The Szekeres Dipole

It is evident from Eqs. (1), (5), (6) and (7) that the angular dependence of the metric and of all covariant scalars is entirely contained in the Szekeres dipole $\mathbf{W}$. This angular dependence can be characterised by a unique angular orientation at each 2-sphere of constant r determined by the "**angular extrema**" defined by the conditions:

$$\mathbf{W}_{,\theta} = \mathbf{W}_{,\phi} = 0, \tag{12}$$

which yield two antipodal positions in the (r, θ, ϕ) coordinates:

$$\phi_- = \arctan\left(\frac{Y}{X}\right), \qquad \theta_- = \arctan\left(\frac{\sqrt{X^2+Y^2}}{Z}\right), \tag{13}$$

$$\phi_+ = \pi + \phi_-, \qquad \theta_+ = \pi - \theta_-. \tag{14}$$

Since r varies smoothly along the 2-spheres, the solutions (13)–(14) define in the spherical coordinate system the following two curves parameterised by r:

$$\mathcal{B}_\pm(r) = [r,\, \theta_\pm(r),\, \phi_\pm(r)], \tag{15}$$

which cross the origin. The magnitude of $\mathbf{W}$ at curves $\mathcal{B}_\pm(r)$ is

$$[\mathbf{W}]_{\pm(r)} \equiv \mathbf{W}(r, \theta_\pm(r), \phi_\pm(r)) = \pm\mathcal{W}(r), \tag{16}$$

where $\mathcal{W}$ is the dipole magnitude defined in Eq. (4). Therefore, angular extrema necessarily lie on the curves $\mathcal{B}_\pm(r)$. Since $\mathbf{W} > 0$ along $\mathcal{B}_+(r)$ and $\mathbf{W} < 0$ along $\mathcal{B}_-(r)$ we have at each 2-sphere of constant r:

$$\text{Angular maxima of } \mathbf{W} \text{ lie along } \mathcal{B}_+(r), \tag{17}$$

$$\text{Angular minima of } \mathbf{W} \text{ lie along } \mathcal{B}_-(r), \tag{18}$$

while allowing for the radial dependence of $\mathbf{W}$ the full extrema of $\mathbf{W}$ located on the curves $\mathcal{B}_\pm(r)$ at some $r - r_e$ such that $\mathcal{W}' = 0$. Hence maxima and minima are located at $[r_e, \theta_\pm(r_e), \phi_\pm(r_e)]$ and $r = 0$ is a saddle point of $\mathbf{W}$.

Restrictions on the orientation of the dipole in the coordinates (r, θ, ϕ) correspond to particular cases of Szekeres models that follow from restrictions on the parameters X, Y, Z.

4. The "LT Seed Model"

We shall denote by "**LT seed model**" the unique spherically symmetric LT model that follows by setting $\mathbf{W} = 0$ in the metric (1) and in the covariant scalars (5)–(6). Evidently, every Szekeres model can be constructed from its LT seed model by defining the dipole parameters X, Y, Z that characterise the Szekeres dipole. Therefore, Szekeres models inherit some of the properties of their LT seed model: the scale factors a, Γ and the q-scalars in Eq. (8) are common to both. In fact, the volume integral (7) evaluated for LT scalars yields the same q-scalars and Friedmann equations (8) and (9) [Sussman (2013a,b); Sussman *et al.* (2015)]. Hence, the LT scalars $A_{(\mathrm{lt})} = \rho_{(\mathrm{lt})}, \mathcal{H}_{(\mathrm{lt})}, \mathcal{K}_{(\mathrm{lt})}, \Sigma_{(\mathrm{lt})}, \Psi_2^{(\mathrm{lt})}$ satisfy the same relations (5)–(6), with the exact fluctuations replaced by analogous LT fluctuations whose relation with $\mathbf{D}^{(A)}$ is given by:

$$\mathbf{D}^{(A)}_{(\mathrm{lt})} = A_{(\mathrm{lt})} - A_q = \frac{r\,A'_q}{3\Gamma} \quad \Rightarrow \quad \mathbf{D}^{(A)} = \frac{\Gamma}{\Gamma - \mathbf{W}}\,\mathbf{D}^{(A)}_{(\mathrm{lt})}, \tag{19}$$

where we assume henceforth that

$$\Gamma > 0 \quad \text{and} \quad \Gamma - \mathbf{W} > 0, \qquad \mathbf{D}^{(A)} = 0 \quad \Leftrightarrow \quad \mathbf{D}^{(A)}_{(\mathrm{lt})} = 0, \tag{20}$$

hold everywhere in order to prevent shell-crossing singularities [Sussman & Bolejko (2012)]. Since the LT fluctuations $\mathbf{D}^{(A)}_{(\mathrm{lt})}$ only depend on (t, r), it is useful to express the Szekeres scalars $A = \rho$, $\mathcal{H}$, $\mathcal{K}$ and Σ, Ψ_2 exclusively in terms of LT objects and $\mathbf{W}$:

$$A = A_q + \frac{\Gamma\,\mathbf{D}^{(A)}_{(\mathrm{lt})}}{\Gamma - \mathbf{W}}, \qquad \Sigma = -\frac{\Gamma\,\mathbf{D}^{(\mathcal{H})}_{(\mathrm{lt})}}{\Gamma - \mathbf{W}}, \qquad \Psi_2 = \frac{4\pi}{3}\,\frac{\Gamma\,\mathbf{D}^{(\rho)}_{(\mathrm{lt})}}{\Gamma - \mathbf{W}}, \tag{21}$$

so that all angular dependence (i.e. non-sphericity) is contained in the dipole function $\mathbf{W}$.

5. Local Homogeneity Conditions

The condition $\Sigma = \Psi_2 = 0$ holding everywhere is the covariant coordinate independent characterisation of the FLRW limit of Szekeres models [Krasiński (1997); Plebański & Krasiński (2006); Sussman & Bolejko (2012)]. From Eqs. (5)–(6) and (10)–(11), this condition yields $\mathbf{D}^{(A)} = 0$ for $A = \rho, \mathcal{H}, \mathcal{K}$, and from Eqs. (7) and (8) we also have $A'_q = 0$ and thus $A'_{q0} = 0$ hold for all r (the FLRW limit is also obtained with these conditions if the dipole parameters X, Y, Z are not zero). It is important to remark that the origin worldline of a Szekeres model[3] can be identified with an observer complying with "local homogeneity" for all t (the same

[3]We assume henceforth that the LT seed model admits a symmetry centre at $r = 0$, and thus their associated Szekeres models admit an origin worldline whose

remark applies to the symmetry centre worldline $r = 0$ of an LT model), as standard regularity conditions at $r = 0$ are the same as the homogeneity conditions $\Sigma = \Psi_2 = \mathbf{D}^{(A)} = 0$ [Plebański & Krasiński (2006)]. However, the same type of local homogeneity at $r = 0$ can be defined for a discrete set of fixed radial comoving coordinate values $r > 0$.

Consider any finite sequence of $n+1$ arbitrary increasing radial comoving coordinate values and n open intervals between them

$$r_*^i = 0,\, r_*^1,\, ..,\, r_*^n, \qquad \Delta_*^i = 0 < r < r_*^1,\, ..,\, r_*^{n-1} < r < r_*^n. \tag{22}$$

Initial conditions that define local homogeneity at each r_*^i are given by

$$A'_{q0} = 0 \quad \Rightarrow \quad [\mathbf{D}^{(A)}_{(\mathrm{lt})}]_0 = 0 \quad \Rightarrow \quad \mathbf{D}^{(A)}_0 = 0 \quad \text{hold at all } r_*^i, \tag{23}$$

$$A'_{q0},\ [\mathbf{D}^{(A)}_{(\mathrm{lt})}]_0,\ \mathbf{D}^{(A)}_0 \quad \text{alternate signs at each } \Delta_*^i. \tag{24}$$

Since Eqs. (10)–(11) and (19) are preserved throughout the full evolution, then initial conditions (23)–(24) imply that local homogeneity is preserved for all t

$$[\mathbf{D}^{(A)}]_* = [\mathbf{D}^{(A)}_{(\mathrm{lt})}]_* = 0,\ A_* = A_{q*} = [A_{(\mathrm{lt})}]_* = \bar{A}_*(t),\ \Sigma_* = [\Psi_2]_* = 0, \tag{25}$$

$$A'_q, [\mathbf{D}^{(A)}_{(\mathrm{lt})}], \mathbf{D}^{(A)}, \Sigma, \Psi_2 \text{ alternate signs at each } \Delta_*^i, \tag{26}$$

where the subindex $_*$ denotes evaluation at the $n+1$ fixed comoving radii (22). The following points are worth commenting: (i) As a consequence of Eqs. (23)–(26) the radial coordinates r_*^i that define local homogeneity also define the time evolution of 2-spheres with surface area is $4\pi a(t, r_*)\, r_*^2$ (i.e. **local homogeneity spheres**) that are common to the Szekeres model and its LT seed model; (ii) the origin $r = 0$ is effectively the local homogeneity sphere of zero area; (iii) the dipole parameters X, Y, Z are (in general) nonzero at $r = r_*^1, \ldots, r_*^n$, hence the local homogeneity spheres are (in general) not concentric with respect to the origin; (iv) condition $\mathbf{D}^{(A)} = \mathbf{D}^{(A)}_{(\mathrm{lt})} = 0$ may hold only for one of the scalars $A = \rho, \mathcal{H}, \mathcal{K}$ (see examples for generic LT models in [Sussman (2010)]).

regularity conditions are given in [Plebański & Krasiński (2006)]. This is a physically motivated assumption, but it is not absolutely necessary, as regular LT models exist that either admit two symmetry centres or none (for time slices with the topology of a 3-sphere or of a "wormhole" [Plebański & Krasiński (2006)]).

6. Extrema of the Radial Profiles of Szekeres Scalars

Scalars like $A_q(t,r)$ and $A_{(\text{lt})}(t,r)$ evaluated at any arbitrary fixed t define "radial profiles" that can be treated as real-valued one-variable functions of r. We can also define radial profiles of Szekeres scalars at a fixed t, either by evaluating them for fixed (θ,ϕ) or by evaluating them along the curves of angular extremal $\mathcal{B}_\pm(r)$:

$$A_\pm = [A]_{\pm(r)} = A_q + \frac{\Gamma\,\mathbf{D}^{(A)}_{(\text{lt})}}{\Gamma \mp \mathcal{W}} = A_{(\text{lt})} \pm \frac{\mathcal{W}\,\mathbf{D}^{(A)}_{(\text{lt})}}{\Gamma \mp \mathcal{W}}, \tag{27}$$

$$\Sigma_\pm = -\frac{\Gamma\,\mathbf{D}^{(\mathcal{H})}_{(\text{lt})}}{\Gamma \mp \mathcal{W}}, \qquad [\Psi_2]_\pm = \frac{4\pi}{3}\,\frac{\Gamma\,\mathbf{D}^{(\rho)}_{(\text{lt})}}{\Gamma \mp \mathcal{W}}, \tag{28}$$

where we used Eqs. (16) and (21).

The choice of initial conditions (23)–(24) fully determines the concavity pattern (i.e. the extrema) of the radial profile of q-scalars A_q and of the scalars Σ and Ψ_2 in all hypersurfaces of fixed t. As a direct consequence of Eqs. (25)–(26), we have the following patterns in the radial intervals Δ^i_* defined in Eq. (22) between the local homogeneity spheres:

- The sign of A'_q alternates from positive to negative or from negative to positive.
- The signs of $\Sigma_\pm$ and $[\Psi_2]_\pm$ (and also of the radial profiles of $\Sigma_{(\text{lt})}$ and $\Psi_2^{(\text{lt})}$) alternate from positive to negative or from negative to positive.

Therefore, the radial profiles of A_q, Σ, Ψ_2 (and of $\Sigma_{(\text{lt})}$, $\Psi_2^{(\text{lt})}$) in every hypersurface of fixed t oscillate between a sequence of $n+1$ maxima and minima (depending on the alternating pattern) at the radial coordinates (22) of the local homogeneity spheres (notice that Eq. (24) prevents the possibility that the r^i_* are saddle points of these radial profiles).

The above mentioned concavity pattern of A_q, Σ, Ψ_2 leads to a similar pattern for the radial profiles of the remaining scalars $A_{(\text{lt})}$, $A_\pm$ in the radial intervals Δ^i_* between the local homogeneity spheres. We have the following result:

Proposition 1 Initial conditions (23)–(24) associated with n local homogeneity spheres marked by Eq. (22) constitute a sufficient condition for the existence, at all t, of n extrema of the radial profiles $A_\pm$, $\Sigma_\pm$, $[\Psi_2]_\pm$ of each of the Szekeres scalars A, Σ, Ψ_2. Each extremum is located in radial coordinates $r^i_{\text{tv}} > 0$ in one of the n radial intervals Δ^i_*, i.e. $0 < r^1_{\text{tv}} < r^1_*, \ldots, r^{n-1}_* < r^n_{\text{tv}} < r^n_*$.

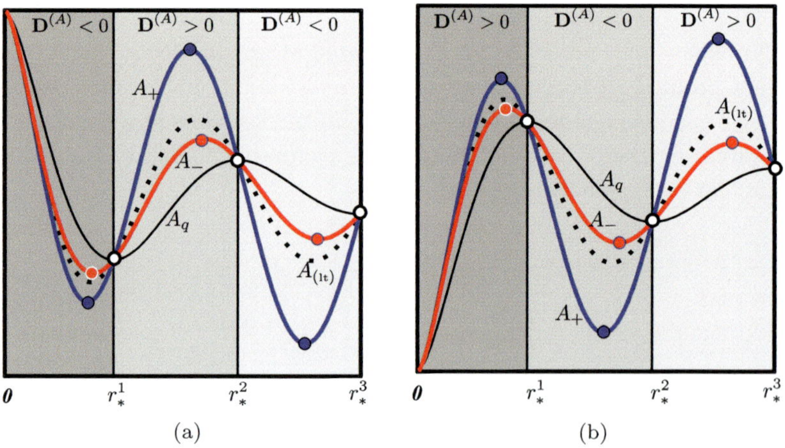

Fig. 1. **Concavity patterns of radial profiles.** The shapes of the curves necessarily follow from the initial conditions (23)–(24), their implications (25)–(26) and the inequalities (*). The profiles at arbitrary fixed t of A_q, $A_{(\mathrm{lt})}$, A_+, A_- (see Eq. (27)) are depicted by the solid black, dashed, blue and red curves, with the three radial intervals Δ_*^1, Δ_*^2, Δ_*^3 between four local homogeneity spheres marked by $r_*^i = 0, r_*^1, r_*^2, r_*^3$ displayed by shaded rectangles. The extrema of A_q (white dots) coincide with the r_*^i, while the extrema of $A_{(\mathrm{lt})}$ and $A_\pm$ occur in the three intervals Δ_*^i between them (Proposition 1) with an alternating pattern for the sign of $\mathbf{D}^{(A)}$. Similar patterns occur for the radial profiles of Σ and Ψ_2.

Proposition 1 can be proven by qualitative arguments assisted by Fig. 1 (a rigorous proof is given in [Sussman & Delgado Gaspar (2015)]). Considering initial conditions (23)–(24) and their implications for all t in Eqs. (25)–(26), as well as Eqs. (27)–(28), we have at all t an alternating pattern of the following inequalities in the radial intervals $r_*^i < r < r_*^{i+1}$ between them:

$$\begin{aligned}
&A_+ < A_{(\mathrm{lt})} < A_q, \quad \text{and} \quad A_{(\mathrm{lt})} < A_- < A_q \quad \text{if}\, \mathbf{D}^{(A)} < 0,\\
&A_+ > A_{(\mathrm{lt})} > A_q, \quad \text{and} \quad A_{(\mathrm{lt})} > A_- > A_q \quad \text{if}\, \mathbf{D}^{(A)} > 0, \qquad (*)\\
&\Sigma > 0,\ \Sigma_{(\mathrm{lt})} > 0 \quad \text{if } \mathbf{D}^{(\mathcal{H})} < 0 \quad \text{and} \quad \Sigma < 0,\ \Sigma_{(\mathrm{lt})} < 0 \quad \text{if } \mathbf{D}^{(\mathcal{H})} > 0,\\
&\Psi_2 < 0,\ \Psi_2^{(\mathrm{lt})} < 0 \quad \text{if } \mathbf{D}^{(\rho)} < 0 \quad \text{and} \quad \Psi_2 > 0,\ \Psi_2^{(\mathrm{lt})} > 0 \quad \text{if } \mathbf{D}^{(\rho)} > 0,
\end{aligned}$$

which lead to a qualitative but robust inference of the concavity patterns given by a sequence of maxima and minima of the radial profiles of all scalars at all t. This is illustrated in Fig. 1, which depicts the radial profiles of A_q, $A_{(\mathrm{lt})}$ and $A_\pm$ for a radial range containing the comoving radii of four local homogeneity spheres marked by $r = 0, r_*^1, r_*^2, r_*^3$. It is evident that the radial profiles of the scalars $A_{(\rm lt)}$ and A (i.e. $A_\pm$) displayed in the figure closely follow the concavity pattern of A_q: (i) the origin $r = 0$ is a common extremum for all the profiles whose type (maximum or minimum) follows from the sign of A'_q (or $\mathbf{D}^{(A)}$) in the first interval $\Delta^1_* = 0 < r < r^1_*$; (ii) since all the scalars coincide at $r = 0$ and $r = r^1_*$ when $A'_q = 0$, then there is necessarily an extremum of opposite type in $A_\pm$ and in the radial profile of $A_{(\rm lt)}$ at some $r = r_{\rm tv}$ such that $0 < r_{\rm tv} < r^1_*$; (iii) the same pattern goes on for the next interval and the following ones, as all scalars coincide at every extremum of A_q at the r^i_*, the radial profiles $A_\pm$ and $A_{(\rm lt)}$ must have extremum of alternating type in values $r^i_{\rm tv}$ in the intervals $\Delta^i_* = r^i_* < r < r^{i+1}_*$. The same arguments apply to Σ and Ψ_2, which alternate from positive to negative signs in the intervals Δ^i_* and vanish at the r^i_*.

7. Location of the Spatial Extrema of Szekeres Scalars

The conditions for an extremum of $A = \rho, \mathcal{H}, \mathcal{K}$ at any time slice (i.e. spatial extremum) are given by

$$A' = 0, \qquad A_{,\theta} = 0, \qquad A_{,\phi} = 0, \tag{29}$$

with similar conditions for Σ and Ψ_2. To examine the fulfilment of these conditions we will use the forms (21), (27)–(28) and will assume, conditions (20) to avoid shell crossings hold everywhere.

7.1. *Angular extrema*

We obtain from Eq. (21) the conditions for the angular extrema of $A = \rho, \mathcal{H}, \mathcal{K}$ at any fixed t:

$$\begin{aligned} A_{,\theta} &= \frac{\mathbf{D}^{(A)}_{(\rm lt)}\,\Gamma\,\mathbf{W}_{,\theta}}{(\Gamma - \mathbf{W})^2} = \frac{r\,A'_q\,\mathbf{W}_{,\theta}}{3(\Gamma - \mathbf{W})^2} = 0,\\ A_{,\phi} &= \frac{\mathbf{D}^{(A)}_{(\rm lt)}\,\Gamma\,\mathbf{W}_{,\phi}}{(\Gamma - \mathbf{W})^2} = \frac{r\,A'_q\,\mathbf{W}_{,\phi}}{3(\Gamma - \mathbf{W})^2} = 0, \end{aligned} \tag{30}$$

with analogous expressions for Σ and Ψ_2. Since the vanishing of $\mathbf{D}^{(A)}_{(\rm lt)}$ does not define an angular extremum of A, but a local homogeneity sphere, we have the following important result:

> The angular extrema of all Szekeres scalars $A = \rho, \mathcal{H}, \mathcal{K}$, as well as Σ and Ψ_2, coincide at all t with the angular extrema of the Szekeres

dipole $\mathbf{W}$, i.e. $\mathbf{W}_{,\theta} = \mathbf{W}_{,\phi} = 0$ implies $A_{,\theta} = A_{,\phi} = 0$, which is a necessary (not sufficient) condition for Eq. (29). Therefore, the extrema of all Szekeres scalars are necessarily located along the curves $\mathcal{B}_{\pm}(r)$.

7.2. *Location of the spatial extrema*

The missing condition in Eq. (29) for a spatial extremum of $A = \rho, \mathcal{H}, \mathcal{K}$ is the radial equation $A' = 0$ obtained from Eq. (21). Since the angular conditions $A_{,\theta} = A_{,\phi} = 0$ must hold, the location of all spatial extrema of Szekeres scalars in the spherical coordinates is given by

$$\left(t,\, r_{e\pm},\, \theta_{\pm}(r_{e\pm}),\, \phi_{\pm}(r_{e\pm}) \right), \tag{31}$$

where $r = r_{e\pm}$ are solutions of the radial condition $A' = 0$ evaluated along the curves $\mathcal{B}_{\pm}(r)$:

$$[A']_{+(r)} = 0 \quad \text{and} \quad [A']_{-(r)} = 0, \tag{32}$$

with similar pairs of conditions $[\Sigma']_{\pm(r)} = 0$ and $[\Psi_2']_{\pm(r)} = 0$. We can ascertain the following points:

- **The regular origin is a spatial extremum.** If standard regularity conditions hold, Eqs. (30) and (32) (and thus Eq. (29)) hold at $r = 0$ for all t.
- **The spatial extrema are not comoving.** The constraints (32) depend on time, hence their solutions are different for different values of t and define constraints of the form $r_{e\pm} = r_{e\pm}(t)$. However, to simplify the notation we will denote these radial coordinates simply by $r_{e\pm}$. This is the same situation as the extrema $r^i_{\rm tv}$ of the radial profiles.
- **Spatial extrema of different scalars.** Evidently, the radial conditions (32) are different constraints for different scalars, hence they must yield different solutions for different scalars.

8. Sufficient Conditions for the Existence of Spatial Extrema of Szekeres Scalars

Evidently, finding solutions $r_{e\pm}$ of conditions (32) is practically impossible without numerical methods. However, these are basically conditions on the radial profiles of the scalars at arbitrary fixed t. Hence, Proposition 1 discussed in Sec. 5 provides sufficient conditions for the existence of such solutions. We have then the following result:

Proposition 2. Initial conditions (23)–(24) associated with $n+1$ local homogeneity spheres marked by coordinate values (22) constitute a sufficient condition for the existence of $2n+1$ spatial extrema of the Szekeres scalars $A = \rho$, $\mathcal{H}$, $\mathcal{K}$ and of Σ, Ψ_2 for all t. One extremum is located at the origin $r = 0$. The remaining $2n$ extrema are marked by coordinate values (31) with radial coordinates $r^1_{e\pm}, \ldots, r^n_{e\pm}$ and distributed in n pairs as follows:

- the radial coordinates of each pair $r^i_{e\pm}$ are in the interval Δ^i_*, hence they satisfy $0 < r^1_{e\pm} < r^1_*, \ldots, r^{n-1}_* < r^n_{e\pm} < r^n_*$ for all t,
- every spatial extremum at r^i_{e+} is located on the curve $\mathcal{B}_+(r)$ and the spatial extremum at r^i_{e-} on the curve $\mathcal{B}_-(r)$.

Proof. It is evident that the radial coordinates $r^i_{e\pm}$ of the $2n$ pairs of spatial extrema with $r > 0$ correspond to the radial coordinates r^i_{tv} of the extrema with $r > 0$ of $A_\pm$ and $\Sigma_\pm$, $[\Psi_2]_\pm$ given in Eqs. (27)–(28), as these are the radial profiles of the scalars $A = \rho$, $\mathcal{H}$, $\mathcal{K}$ and Σ, Ψ_2 along the curves $\mathcal{B}_\pm(r)$ (hence they satisfy Eq. (30)). From Proposition 1, initial conditions (23)–(24) are sufficient for the extrema of these radial profiles to exist in the intervals Δ^i_* for all t, which means that (depending on the involved scalar) they are solutions of the radial constraints Eq. (32) for all t. Hence, (29) holds.

The following points readily emerge and are worth commenting:

(1) Proposition 2 provides sufficient conditions, not necessary ones, hence spatial extrema of A may exist even if initial conditions (23)–(24) are not assumed, and thus A_q has a monotonic radial profile without extrema in the full radial range (see qualitative arguments on their existence in [Sussman & Delgado Gaspar (2015)]).

(2) Proposition 2 does not provide the precise location of the spatial extrema in a given radial range: this must be obtained from numerical solutions of Eq. (32). The coordinate location of spatial extrema of different scalars will be (in general) different.

9. Classification of the Spatial Extrema of Szekeres Scalars

The type of extremum of the radial profile of a scalar (i.e. the maxima and minima in Fig. 1) does not (necessarily) determine the type of the spatial extremum (maximum/minimum or saddle point) of the same scalar as a full 3-dimensional object. To properly classify the extrema of A and Σ, Ψ_2 we need to compute their second derivatives evaluated at the location of generic extrema, which necessarily lie at the origin or at $\mathcal{B}_\pm(r_{e\pm})$, with $r_{e\pm}$ solutions of Eq. (32). We assume that such solutions exist and, in particular, we consider models defined by initial conditions (23)–(24), so that Proposition 2 guarantees the existence of spatial extrema of all Szekeres scalars in the n intervals Δ_*^i between radial coordinates (22) of $n+1$ local homogeneity spheres.

9.1. *Extrema at the origin*

It is straightforward to show (see [Sussman & Delgado Gaspar (2015)]) that all second derivatives of A vanish as $r \to 0$, save for the second radial derivative in this limit:

$$[A'']_{r=0} = \left[A_q(\delta^{(A)}_{(\mathrm{lt})})''\right]_{r=0} \approx \frac{2}{3}\left[A_q''\right]_{r=0} = \frac{2}{3}\left[A''_{(\mathrm{lt})}\right]_{r=0}, \tag{33}$$

where $\delta^{(A)}_{(\mathrm{lt})} = \mathbf{D}^{(A)}_{(\mathrm{lt})}/A_q$ and we used Eq. (19) to expand $A_q(\delta^{(A)}_{(\mathrm{lt})})''$, bearing in mind that the regularity conditions at the origin imply that $A_q'(t,0) = 0$ and $\Gamma = 1$ hold for all t at $r = 0$ (see [Sussman (2010)]).

As a consequence of Eq. (33), the extrema of a Szekeres scalar A at the origin $r = 0$ are of the same type as the extrema of scalar $A_{(\mathrm{lt})}$ of the seed LT model at the symmetry centre $r = 0$:

$$\left[A''_{(\mathrm{lt})}\right]_{r=0} < 0 \quad \text{local maximum (or central clump) of } A, \tag{34}$$

$$\left[A''_{(\mathrm{lt})}\right]_{r=0} > 0 \quad \text{local minimum (or central void) of } A, \tag{35}$$

which implies that either $[A]_{r=0} > A$ (central clump) or $[A]_{r=0} < A$ (central void) hold for all points in any small neighbourhood around $r = 0$. If $A'' = 0$ at $r = 0$, it may be necessary to expand A at $r \approx 0$ to higher order.

9.2. *Spatial extrema at* $r > 0$

Since we have assumed the existence of spatial extrema by virtue of choosing local homogeneity initial conditions (23)–(24), a classification of these extrema can be obtained by means of qualitative arguments that consider the angular extrema of A and the extrema of the radial profiles $A_\pm$ plotted in Fig. 1 (see [Sussman & Delgado Gaspar (2015)] for a more rigorous approach). For this purpose, we remark that the type (maxima/minima) of the angular extrema follow from the signs of the angular derivatives

$$[A_{,\theta\theta}]_{\pm(e)} = \mp\frac{\Gamma_e \mathcal{W}_e [\mathbf{D}^{(A)}_{(\mathrm{lt})}]_e}{(\Gamma_e \mp \mathcal{W}_e)^2},$$
$$[A_{,\phi\phi}]_{\pm(e)} = \mp\frac{\Gamma_e\,(X_e^2 + Y_e^2)[\mathbf{D}^{(A)}_{(\mathrm{lt})}]_e}{(\Gamma_e \mp \mathcal{W}_e)^2\,\mathcal{W}_e},$$
$$[A_{,\theta\phi}]_{\pm(e)} = 0, \tag{36}$$

while the type of radial extrema follows from the signs of the profile derivatives $A''_{\pm(e)}$. Fortunately, we can obtain the signs of $A''_{\pm(e)}$ by qualitative arguments from the curves displayed in Fig. 1 without having to evaluate this derivative.

Consider the concavity pattern illustrated in the left panel of Fig. 1 (we obtain analogous results for the right panel). By looking at the signs of $[A_{,\theta\theta}]_{\pm(e)}$, $[A_{,\phi\phi}]_{\pm(e)}$ and $A''_{\pm(e)}$ at the extrema in each interval Δ^1_*, Δ^2_*, Δ^3_* of this panel (excluding the extrema at $r = 0$) we obtain along each curve $\mathcal{B}_\pm(r)$ the type of extrema of $A_\pm(r)$:

- Extrema along the curve $B_+(r)$ are located in each interval at $r^i_{+(e)} = r^i_{\mathrm{tv}}$, the extrema of A_+ are given by

$$\begin{aligned}
&\Delta^1_*: \quad A''_{+(e)} > 0, (\mathrm{rmin})\\
&\left[\mathbf{D}^{(A)}_{(\mathrm{lt})}\right]_{+(e)} < 0 \Rightarrow [A_{,\theta\theta}]_{+(e)}, [A_{,\phi\phi}]_{+(e)} > 0, (\mathrm{amin})\\
&\Delta^2_*: \quad A''_{+(e)} < 0, (\mathrm{rmax})\\
&\left[\mathbf{D}^{(A)}_{(\mathrm{lt})}\right]_{+(e)} > 0 \Rightarrow [A_{,\theta\theta}]_{+(e)}, [A_{,\phi\phi}]_{+(e)} < 0, (\mathrm{amax})\\
&\Delta^3_*: \quad A''_{+(e)} > 0, (\mathrm{rmin})\\
&\left[\mathbf{D}^{(A)}_{(\mathrm{lt})}\right]_{+(e)} < 0 \Rightarrow [A_{,\theta\theta}]_{+(e)}, [A_{,\phi\phi}]_{+(e)} > 0, (\mathrm{amin})
\end{aligned} \tag{37}$$

  where (rmin), (rmax), (amin) and (amax) stand for radial/angular maxima and minima.
- Extrema along the curve $B_-(r)$ are located in each interval at $r^i_{-(e)} = r^i_{\mathrm{tv}}$, the extrema of A_- are given by

$$\begin{aligned}
&\Delta^1_*: \quad A''_{+(e)} > 0, (\mathrm{rmin})\\
&\left[\mathbf{D}^{(A)}_{(\mathrm{lt})}\right]_{+(e)} < 0 \Rightarrow [A_{,\theta\theta}]_{+(e)}, [A_{,\phi\phi}]_{+(e)} < 0, (\mathrm{amax})
\end{aligned}$$

$$\begin{aligned}
\Delta_*^2 : \quad & A''_{+(e)} < 0, \text{ (rmax)} \\
& \left[\mathbf{D}^{(A)}_{(\mathrm{lt})}\right]_{+(e)} > 0 \Rightarrow [A_{,\theta\theta}]_{+(e)}\,, [A_{,\phi\phi}]_{+(e)} > 0, \text{ (amin)} \\
\Delta_*^3 : \quad & A''_{+(e)} > 0, \text{ (rmin)} \\
& \left[\mathbf{D}^{(A)}_{(\mathrm{lt})}\right]_{+(e)} < 0 \Rightarrow [A_{,\theta\theta}]_{+(e)}\,, [A_{,\phi\phi}]_{+(e)} < 0. \text{ (amax)}
\end{aligned} \tag{38}$$

It is evident from the signs of the second derivatives in Eq. (37) that along $\mathcal{B}_+(r)$ the angular maxima and minima coincide with radial maxima and minima, whereas along $\mathcal{B}_-(r)$ the signs of Eq. (38) reveal the opposite pattern: radial maxima coincide with angular minima and vice versa. The sign pattern shown in Eqs. (37) and (38) goes on for maxima and minima in n intervals Δ_*^i following the concavity pattern of the left panel of Fig. 1 (the right panel simply switches the sequence of maxima and minima). As a consequence, we can now state that:

> The $2n$ spatial extrema of Szekeres scalars located at $r > 0$ that follow from the choice of initial conditions (23)–(24) are of the following type:
>
> - an alternating pattern of n local spatial maxima and minima in the radial intervals Δ_*^i along the curve $\mathcal{B}_+(r)$. If the extremum at $r = 0$ is a local maximum (central clump) then the pattern is maximum, minimum, maximum and so on. If there is a local minimum at $r = 0$ (central void) the pattern is the opposite.
> - n spatial saddle points in the radial intervals Δ_*^i along the curve $\mathcal{B}_-(r)$.

A rigorous classification of the spatial extrema of A is undertaken in [Sussman & Delgado Gaspar (2015)] in terms of the determinants of the Hessian matrix and its minors for all the second derivatives of A.

10. Modelling Non-Spherical Cosmic Structures

So far we have examined sufficient conditions for the existence of spatial extrema of all covariant scalars of Szekeres models. We now concentrate on the specific case of spatial maxima and minima of the density that can be respectively associated with over-densities and density voids. In this section we discuss how to set up Szekeres models allowing for a non-trivial spatial distribution of such density extrema, providing as well a specific numerical example.

10.1. *Evolution equations*

Since the q-scalars and their exact fluctuations A_q, $\mathbf{D}^{(A)}$ fully determine the Szekeres covariant scalars $A = \rho, \mathcal{H}, \mathcal{K}$ and Σ, Ψ_2 through Eqs. (5)–(6), the dynamics of any Szekeres model can be completely described by evolution equations and constraints for these variables, such as the following system [Sussman, Delgado Gaspar & Hidalgo (2016)][4]:

$$\dot{\rho}_q = -3\rho_q \mathcal{H}_q, \tag{39}$$

$$\dot{\mathcal{H}}_q = -\mathcal{H}_q^2 - \frac{4\pi}{3}\rho_q + \frac{8\pi}{3}\Lambda, \tag{40}$$

$$\dot{\mathbf{D}}^{(\rho)} = -3(\mathcal{H}_q + \mathbf{D}^{(\mathcal{H})})\mathbf{D}^{(\rho)} - 3\rho_q \mathbf{D}^{(\mathcal{H})}, \tag{41}$$

$$\dot{\mathbf{D}}^{(\mathcal{H})} = -\frac{4\pi}{3}\mathbf{D}^{(\rho)} - \left(2\mathcal{H}_q + 3\mathbf{D}^{(\mathcal{H})}\right)\mathbf{D}^{(\mathcal{H})}, \tag{42}$$

$$\dot{a} = a\,\mathcal{H}_q, \tag{43}$$

$$\dot{\mathcal{G}} = 3\,\mathcal{G}\,\mathbf{D}^{(\mathcal{H})}, \qquad \mathcal{G} = \frac{\Gamma - \mathbf{W}}{1 - \mathbf{W}}, \tag{44}$$

where we have also included the evolution equations (43) and (44) of the metric functions a and Γ. The system (39)–(44) must comply with the time preserved algebraic constraints (9)–(11).

Any Szekeres model becomes uniquely specified by the initial conditions needed for the numerical integration of Eqs. (39)–(44), which are the following minimal set of six initial value functions

$$\rho_{q0},\ \Lambda,\ \mathcal{K}_{q0},\ X,\ Y,\ Z, \tag{45}$$

from which the remaining initial value functions $\mathcal{H}_{q0}$, $\mathbf{D}_0^{(\rho)}$, $\mathbf{D}_0^{(\mathcal{H})}$ can be found by means of Eqs. (7) and (9) evaluated at $t = t_0$ (considering that $a_0 = \mathcal{G}_0 = 1$).

10.2. *Numerical example*

Considering initial value functions A_{q0} and the dipole parameters X, Y, Z that comply with the regularity conditions (20) to avoid shell crossings and with the conditions for convergence to a suitable FLRW model as $r \to \infty$, we present below an example of a Szekeres model with a prescribed spatial distribution of over-densities and voids [Sussman, Delgado Gaspar & Hidalgo (2016)].

[4]This system was derived in [Sussman & Bolejko (2012)] for the relative fluctuations $\Delta^{(A)} = \mathbf{D}^{(A)}/A_q$.

10.2.1. *Initial value functions of the LT seed model*

In order to comply with initial conditions (23)–(24) we choose

$$\mu_{q0}(r) = \frac{1}{2} + m_0 \frac{\sin(\frac{\pi}{2}\chi)\cos(\frac{\pi}{2}\chi)}{1+(\frac{1}{2}\chi)^3}, \quad \kappa_{q0}(r) = k_0 \frac{\sin(\frac{\pi}{2}\chi)\cos(\frac{\pi}{2}\chi)}{1+(\frac{1}{2}\chi)^{\frac{7}{5}}}, \tag{46}$$

where $\mu_{q0} = (4\pi/3)\rho_{q0}/H_s^2$, $\kappa_{q0} = \mathcal{K}_{q0}/H_s^2$, $\chi = r/r_p$, with $r_p = 0.025$ Mpc, $m_0 = 0.01$, $k_0 = 0.00061$, while H_s is the Hubble length at $t = t_0$ (last scattering surface). These initial value functions are analogous to the radial profiles displayed by Fig. 1, admitting five local homogeneity spheres (besides $r = 0$) whose radial values are given further below. These initial conditions also comply with convergence to an asymptotic spatially flat ΛCDM background: in the limit $r \to \infty$ we have $\kappa_0 \to 0$ and $2\mu_{q0} \to 1$.

10.2.2. *The dipole parameters*

In order to obtain a different clear-cut precise angular direction at each Δ_*^i, it is convenient to choose X, Y, Z as piecewise functions. From Eqs. (13)–(14) we see that setting $Z = 0$ places the angular extrema in the equatorial plane $\theta_\pm = \pi/2$, while keeping only one of X or Y different from zero leads to a dipole orientation along the directions $\phi_\pm = 0, \pi$ (rectangular axis x) or $\phi_\pm = \pi/2, 3\pi/2$ (rectangular axis y). A simple but illustrative arrangement consists in placing the density spatial extrema in alternative directions along the rectangular axes x and y. Therefore, we choose $Z = 0$ together with the following forms for X and Y

$$X(r) = \begin{cases} X_1(r), & 0 \le r \leqslant r_*^1 \\ 0, & r_*^1 \le r \leqslant r_*^2, \\ X_2(r), & r_*^2 \le r \leqslant r_*^3, \\ 0, & r_*^3 \le r \leqslant r_*^4 \\ X_3(r), & r_*^4 \le r \leqslant r_*^5 \\ 0, & r > r_*^5, \end{cases}, \qquad Y(r) = \begin{cases} 0, & 0 \le r \leqslant r_*^1 \\ Y_2(r), & r_*^1 \le r \leqslant r_*^2, \\ 0, & r_*^2 \le r \leqslant r_*^3, \\ Y_4(r), & r_*^3 \le r \leqslant r_*^4 \\ 0, & r > r_*^4, \end{cases}, \tag{47}$$

where the local homogeneity spheres (besides the origin) are marked by $r_*^i = (2i-1)r_p/2$, with $i = 1, \ldots, 5$, while the functions X_1, X_j, Y_j (for $j = 2, 3, 4, 5$) are

$$X_1(r) = r \sin 2\pi\chi, \qquad X_j(r),\, Y_j(r) = \sin\left(\pi\chi - \frac{(2j-3)\pi}{2}\right).$$

Notice that the model becomes identical to the LT seed model for $r > r_*^5$, as the Szekeres dipole $\mathbf{W}$ vanishes identically for these radial ranges.

10.2.3. *Graphical depiction of the over-densities and density voids*

By integrating the system (39)–(44) for the initial conditions (46) and (47), we obtain the density contrast $\delta = (\rho - \bar{\rho})/\bar{\rho}$, where $\bar{\rho}$ is the density of the ΛCDM background. Figure 2 displays the level curves of δ along the equatorial plane $\theta = \pi/2$, with the left and right panels respectively depicting δ at the initial last scattering time t_0 and at present day cosmic time $t \sim 13.7 \times 10^9$ yrs. Besides the local density minimum at the origin (central density void), a total of five structures are clearly displayed in both panels: three over-densities (local density maxima in red/yellow shading) and two density voids (local density minima in blue shading), each located in the curve $\mathcal{B}_+(r)$ (solid line segments) in each one of the five intervals Δ^i_* in the previously prescribed directions specified by Eq. (47). The saddle points in the curves $\mathcal{B}_-(r)$ are not displayed. It is important to remark that Fig. 2 displays a coordinate representation of a Riemannian space with nonzero curvature, hence the apparent reflection symmetry between the curves $\mathcal{B}_\pm(r)$ (solid line segments) and between the location of different extrema is a coordinate effects, as these structures are at different proper radial distance $\ell = \int_0^r \sqrt{g_{rr}}\mathrm{d}r$ from the origin because of the angular dependence of g_{rr} in Eq. (1).

We have mentioned (see Sec. 7.2) that the locations of spatial extrema of the scalars (31) are not comoving (i.e. the values $r_{\pm e}$ change with t).

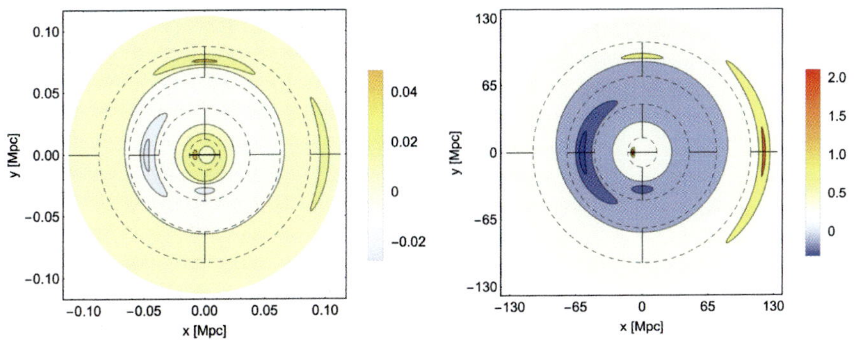

Fig. 2. **The density contrast.** The figure depicts the level curves of the density contrast in the equatorial plane $\theta = \pi/2$ at the initial time (last scattering surface) $t_0 \sim 3 \times 10^5$ years (left panel) and at present day cosmic time $t \sim 13.7 \times 10^9$ yrs (right panel). The radial coordinates r^i_* of the local homogeneity spheres are displayed as dashed circles, while the curves of angular extrema $\mathcal{B}_\pm(r)$ that follow from Eq. (47) are solid line segments in each interval Δ^i_* between the r^i_*. Over-densities appear yellow and red and density voids in shades of blue.

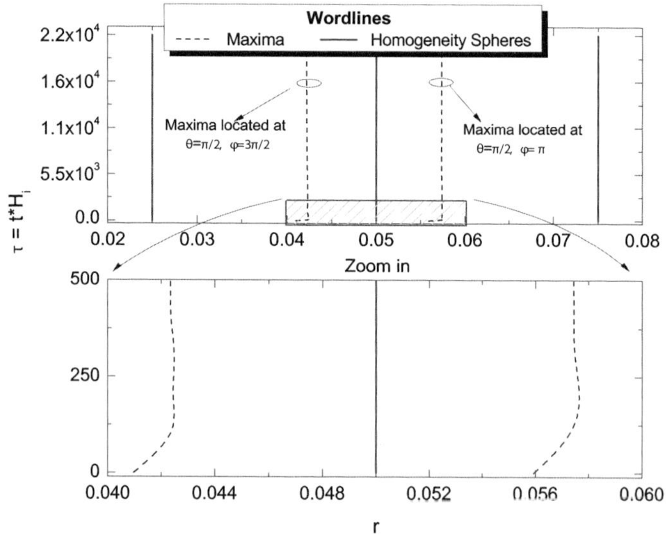

Fig. 3. **Worldlines of over-densities.** The figure depicts the worldlines of two over-densities evolving between local homogeneity spheres. This non-comoving evolution leads to non-trivial peculiar velocities of the over-densities with respect to a comoving frame that can be identified with the CMB.

This fact defines a trajectory of each over-density and density void inside the comoving intervals Δ_*^i. We display in Fig. 3 such trajectory for two over-densities in a configuration similar to that of Fig. 2. This non-comoving evolution of the over-densities and density voids allows us to compute peculiar velocities of these structures with respect to a comoving frame that can be associated with the CMB.

11. Conclusions

We have undertaken a comprehensive study of the sufficient conditions for the existence of spatial extrema (local maxima, minima and saddle points) for all t of the main covariant scalars of Szekeres models: the density ρ, the Hubble expansion $\mathcal{H}$, the spatial curvature $\mathcal{K}$ and the eigenvalues Σ, Ψ_2 of the shear and electric Weyl tensor, all of which are expressible in terms of the q-scalars and their fluctuations A_q, $\mathbf{D}^{(A)}$ from Eqs. (5)–(6).

By describing the models in spherical spatial coordinates, we have shown how to obtain these conditions by looking separately at the radial and angular location of the extrema, i.e. extrema of the radial profile of the

LT seed model and of the Szekeres dipole **W**. This leads to a well defined procedure to set up Szekeres models in which the spatial extrema of each one of these covariant scalars can be placed in roughly desired arbitrary locations in the spherical coordinate system (r, θ, ϕ) (see the implementation of this procedure in the numerical example of Sec. 10).

While the procedure we describe above is valid for the extrema of every covariant scalars, the spatial extrema of the density are specially relevant because they define spatial distributions of cosmic structure: over-densities can be identified with local density maxima and density voids with local density minima that (given the appropriate boundary conditions) can be immersed in a FLRW background. Hence, our results allow us to set up Szekeres models that provide a fully relativistic non-perturbative description of multiple spatial distributions of cosmological structure in the form of over-densities and density voids that can be defined for arbitrary scales. Since the evolution of these structures is not comoving with that of the FLRW background (see Fig. 3), these models define naturally non-trivial peculiar velocities of cosmic structure with respect to a comoving frame that can be associated with the CMB.

We have presented in Sec. 10 a simple but illustrative numerical example constructed along these lines, comprising of five intervals Δ_*^i separating five local homogeneity spheres, with the dipole parameters defined in order to place three over-densities and two density voids in perpendicular directions (see Fig. 2). However, more elaborated examples can be easily set up to describe evolving spatial distributions of an arbitrary number of over-densities and two density voids located in roughly desired angular and radial positions. The following is a list of potential applications for Szekeres models which describe this type of non-perturbative cosmic structures:

- Structure formation. The procedure we have described can be applied to collapsing regions immersed in ever expanding exteriors. Therefore, we can describe the evolution of a sort of multi-particle and non-spherical generalisation of the well known spherical collapse model.
- Growth suppression and distortion of the redshift space. By constructing models that provide a coarse-grained description of evolving observed structure (galactic clusters, superclusters and void regions), we can achieve a non-perturbative and fully relativistic approach to this problem.
- Since the locations of the spatial maxima and minima of the density are not comoving, we can define and study their peculiar velocities with respect to the frame associated with the FLRW background in which the

CMB is comoving. In particular, we can use this study of peculiar velocities to re-examine the kinematic Sunyaev–Zel'dovich effect by means of non-perturbative and fully relativistic methods.

Further possible applications are gravitational lensing, testing various aspects of cosmological observations and (possibly) to re-examine the possibility of setting up a large void multipolar structure to verify if observations can be fit without assuming the existence of dark energy or cosmological constant. We will undertake the study of these potential applications in future publications.

References

Biswas, T., Notari, A. and Valkenburg, W., *JCAP* **11**, 030 (2010).

Bolejko, K., *Phys. Rev.* **D75**, 043508 (2007).

Bolejko, K., *Gen. Rel. Gravit.* **41**, 1737 (2009).

Bolejko, K., Ahsan Nazer, M. and Wiltshire, D. L., *JCAP* **1606**, 035 (2016).

Bolejko, K. and Célérier, M. N., *Phys. Rev.* **D82**, 103510 (2010).

Bolejko, K., Célérier, M. N. and Krasiński, A., *Class. Quant. Grav.* **28**, 164002 (2011).

Bolejko, K., Krasiński, A., Hellaby, C. and Célérier, M. N., *Structures in the Universe by Exact Methods: Formation, Evolution, Interactions*, Cambridge University Press, (2009).

Bolejko, K. and Sussman, R. A., *Phys. Lett.* **B697**, 265, (2011).

Bondi, H., *Mon. Not. R. Astron. Soc.* **107**, 410 (1947); reprinted with historical introduction in: *Gen. Rel. Grav.* **11**, 1783 (1999).

Buckley, R. G. and Schlegel, E. M., *Phys. Rev.* **D87**, 023524 (2013).

Bull, P., Clifton, T. and Ferreira, P. G., *Phys. Rev.* **D85**, 024002 (2012).

Ellis, G. F. R., Maartens, R. and MacCallum, M. A. H., *Relativistic Cosmology*, Cambridge University Press (2012).

Goode, S. W. and Wainwright, J., *Phys Rev* **D26**, 3315 (1982).

Ishak, M., Peel, A. and Troxel, M. A., *Phys. Rev. Lett.* **111**, 251302 (2013).

Ishak, M., Richardson, J., Garred, D., Whittington, D., Nwankwo, A. and Sussman, R. A., *Phys. Rev.* **D78**, 123531 (2008).

Krasiński, A., *Inhomogeneous Cosmological Models*, Cambridge University Press (1997).

Krasiński, A. and Bolejko, K., *Phys. Rev.* **D83**, 083503 (2010).

Lemaître, G., *Ann. Soc. Sci. Brux.* **A53**, 51 (1933); English translation, with historical comments: *Gen. Rel. Grav.* **29**, 637 (1997).

Marra, V. and Notari, A., *Class. Quant. Grav.* **28**, 164004 (2011).

Nwankwo, A., Ishak, M. and Thompson, J., *JCAP* **1105**, 028 (2011).

Peel, A., Ishak, M. and Troxel, M. A., *Phys. Rev.* **D86**, 123508 (2012).

Plebański, J. and Krasiński, A., *An Introduction to General Relativity and Cosmology*, Cambridge University Press (2006).

Raccanelli, A., Zhao, G. B., Bacon, D. J. *et al.*, arXiv:1108.0930 (2008); Teyssier, R., Pires, S., Prunet, S. *et al.*, (2008), arXiv:0807.3651 (2008).
Redlich, M. *et al.*, *Astron. and Astroph.* **570**, A63 (2014).
Szekeres, P., *Commun Math Phys* **41**, 55 (1975a).
Szekeres, P., *Phys. Rev.* **D12**, 2941 (1975b).
Sussman, R. A., *Class. Quant. Grav.* **27**, 175001 (2010).
Sussman, R. A., *Class. Quant. Grav.* **30**, 235001 (2013a).
Sussman, R. A., *Class. Quant. Grav.* **30**, 065015 (2013b).
Sussman, R. A. and Bolejko, K., *Class. Quant. Grav.* **29**, 065018 (2012).
Sussman, R. A. and Delgado Gaspar, I., *Phys. Rev.* **D92**, 083533 (2015).
Sussman, R. A., Delgado Gaspar, I. and Hidalgo, J. C., *JCAP* **1603**, 012 (2016).
Sussman, R. A., Hidalgo, J. C., Dunsby, P. K. S. and German, G. *et al.*, *Phys. Rev.* **D91**, 063512 (2015).
Tolman, R. C., *Proc. Nat. Acad. Sci. USA* **20**, 169 (1934); reprinted, with historical comments in: *Gen. Rel. Grav.* **29**, 931 (1997).
Vrba, D. and Svitek, O., *Gen Rel Grav* **46**, 1808 (2014).
Walters, A. and Hellaby, C., *JCAP* **1212**, 001 (2012).
Zibin, J. P. and Moss, A., *Class. Quant. Grav* **28**, 164005 (2011a).
Zibin, J. P. and Moss, A., *Phys. Rev.* **D84**, 123508 (2011b).

Chapter 6

Cosmology after Einstein

Marc Lachièze-Rey

APC — Astroparticule et Cosmologie (UMR 7164)
Université Paris 7 Denis Diderot 10
rue Alice Domon et Lonie Duquet F-75205 Paris Cedex 13
mlr@apc.univ-paris7.fr

Cosmology is the discipline considering the natural phenomena in their totality; more specifically, of what is ordering them in their totality. It really became a physical science after Einstein and his theory of general relativity. Before him, the discipline had no substantial object. The frame for physical phenomena was reduced to Newtonian space and time, which do not present specific properties, and do not evolve, as it is the case for the global distribution of matter.

Contents

1. The Initial Einstein's Model (1917)

The starting point of the modern, relativistic cosmology may be considered as the 1917 paper of Einstein, in which he proposed the first cosmological model. Although entirely revolutionary (see below), it was still very far from our present cosmology. The development of the new discipline was permitted by subsequent observational results. It resulted from a synthesis between these results and the theory, which was fully accomplished by Georges Lemaître.

The theory of general relativity (hereafter GR) was achieved in 1915. Its main achievement was the replacement of (Newtonian) space and time by curved space-time, as a scene for the physical phenomena. Soon after, in 1917, Einstein initiated the modern cosmology by proposing the first relativistic cosmological model [Einstein (1917)]. Its characteristic is the assimilation of the notion of universe to the chronogeometrical structure (the shape) of space-time, together with its material content. In GR, this chronogeometry should be determined by the material (or, better, energetic) content, according to the Einstein's equation. In other words, space-time must be a (global) solution of his theory. The curvature tensor is identified with the gravitational field. This gave a substantial content to cosmology, whose first task then appeared to establish the (global) chronogeometry of space-time. In addition, Einstein constructed his model with three main prejudices.

The first prejudice, under the appellation of the *cosmological principle*, states that the universe presents the same appearance from all points of space and in all spatial directions. This very powerful prescription implies the Copernican principle (we do not occupy a special place in the universe) and forbids for instance the presence of a center or of a frontier to space-time. This constraints space-time to admit spatial sections with maximal symmetry, thus homogeneous and isotropic. There are only three kinds of maximally symmetric 3-dimensional Riemanian manifolds: Euclidean $\mathbb{R}^3$, spherical S^3 or hyperbolic H^3 with zero, positive or negative constant scalar curvature respectively, and labelled by values 0, 1 or -1 of a spatial curvature parameter k. Thus the spatial sections of space-time must be of one of the three types.

Independent of any dynamical equation, the cosmological principle implies that the metric of space-time can be written (in a suitable frame) under the FLRW form

$$g = dt^2 - R(t)^2 \, g_k,$$

where t is a parameter called *cosmic time* (see below), the *scale factor* $R(t)$ a function of it, and g_k the spatial metric of one of the three constant curvature spaces above. (This general form was not recognized by Einstein at that time.)

His second prejudice was that the universe remains static, i.e., maintains the same appearance at any moment. In other words, the world does not evolve and Einstein had no reason to think differently.

The third prejudice was that space (spatial sections) should be of finite extent. The main motivation was to avoid the necessity of prescribing limiting conditions at infinity for solving the equations of the theory, since that would imply that these solutions depend on these conditions and not only on the material content. This motivation was inspired by readings of the physicist and philosopher Ernst Mach [Eisenstaedt (1989)]: "From the equations of the general theory of relativity it can be deduced that this total reduction of inertia to reciprocal action between masses — as required by E. Mach, for example — is possible only if the universe is spatially finite." [Einstein (1921)].

The possibility of a space (a Riemanian manifold) of finite extension and without border was entirely new. It was only allowed by Riemanian geometry. It corresponds to the choice of S^3 for the spatial sections (see Sec. 5). With the staticity requirement, this completely fixes the chronogeometry of a 4-dimensional cylinder.

But Einstein realized that no distribution of usual matter alone, in the RHS of Einstein's equations, could allow such a solution. This led him to amend his theory by adding a new term: the cosmological constant Λ. Its repulsive gravitational effect compensates the gravitational attraction of the matter (of the stars) present in the universe, to allow a static model. With the correct value of this constant, its model was a perfect solution of his theory (see Sec. 4.2).

In the same year, the dutch astronomer Willem de Sitter proposed a different model.[1] This new solution of the theory was empty of any matter, at the great dissatisfaction of Einstein: "From my point of view, it would be entirely unsatisfactory if it were possible to think of the universe without matter. The field $g_{\mu\nu}$ must be determined by matter and without it, it cannot exist. This is the heart of what I understand as the postulate of the relativity of inertia." (quoted by [de Sitter (1917b)]). The solution, now called the *de Sitter space-time*, is a 4-dimensional hyperboloid (with maximal symmetry of space-time, not only of space). Its constant curvature

[1]One of his main motivations concerned *covariance*: the static Einstein model seemed to imply some kind of "absolute time" that de Sitter rejected as in contradiction with the spirit of general relativity [Merleau-Ponty (1965)].

identifies with the positive cosmological constant, the only source of gravitation. In 1924, the British physicist Arthur Eddington proposed that 36 redshifts, among the 41 spectral shifts discovered by Vesto Slipher, favor the de Sitter models.

2. Galaxies and the Expanding Universe

In the 19th century, the extension of the material word was identified with that of our own galaxy, the Milky Way. In particular, the nebulae were considered as gas clouds, or stellar systems, internal to it. On the other hand, the philosopher Immanuel Kant (among others) had suggested one century before that other *island-universes* may exist, out of our galaxy and very far from it.

At the beginning of the 20th century, the idea was regaining some popularity. The American astronomer Vesto Slipher performed spectroscopic observations of spiral nebulae at Lowell Observatory.[2] He measured their redshifts, interpreted as radial velocities via the Doppler–Fizeau effect. In 1914 he had accumulated 30 redshifts, and 45 in 1925 [Peacock (2013); Freeman (2013)]. He was very surprised by his own results which suggested very large velocities, many hundreds km/sec. He doubted that, with such large velocities, these objects could be confined inside our galaxy and suggested that they may be external to it, very far from us. Observations were not sufficient to bring a definite answer and in 1920, a *Great Debate* between Heber Curtis and Harlow Shapley remained inconclusive. After a long story, the situation was clarified in 1924 by Edwin Hubble when he identified some Cepheid stars in the nearby nebula M31 of Andromeda. After long observations, he was able to calibrate their luminosities and deduce the distance to M31, showing definitely that it was external to our galaxy, confirming the hypothesis of island-universe.

A second surprise came from the fact that (with very few exceptions), all nebulae showed redshifts, not blueshifts, i.e., *recession* velocities. In the first decades of the century Slipher suggested some kind of expansion but there was no theoretical context to interpret it [Peacock (2013); Freeman (2013)]. The redshift measurements were confirmed and extended by Hubble with his colleague Milton Humason at Mount Wilson Observatory (California). He was also able to measure the distances of the galaxies and

[2]One may find some interesting documentation on http://www.roe.ac.uk/~jap/slipher/.

it appeared that these measurements were expressing a systematic motion. This was announced in the famous 1929 Hubble's paper [Hubble (1929)], where he expressed what is presently known as the *Hubble law*: the recession velocity is proportional to the distance: $V = H_0 D$, with a proportionality parameter H_0, now called the *Hubble constant* that he estimated to be around 500 km/sec/Mpc. This phenomenological result found however no theoretical explanation at the time, as it appeared from the conclusions of the meeting at the Royal Astronomical Society in London in 1930, where Eddington expressed "the need for intermediate solutions" [Einstein (1921); de Sitter (1917b)].

The situation was clarified by the Belgian physicist Georges Lemaître, who had already published and theoretically explained the "Hubble law" before Hubble himself. After reading the conclusions of the meeting, he wrote to Arthur Eddington (his former adviser) to draw attention on the paper "Un univers homogène de masse constante et de rayon croissant, rendant compte de la vitesse radiale des nébuleuses extragalactiques" that he had previously (1927) published in the *Annales de la Société Scientifique de Bruxelles*.[3]

The Lemaître's paper presents a family of solutions of general relativity describing an expanding universe (the Einstein model appears as a limiting case). The "size" of the universe is described by a scale factor $R(t)$ which increases with the cosmic time t (see Sec. 6). He emphasized that his solutions implied redshifts for the galaxies and he derived the Hubble law two years before its publication by Hubble himself! Comparing with the 42 redshift measurement results available, he deduce a value of H_0 of 625 km/sec/Mpc.[4] This estimation, as well as that of Hubble was based on erroneous data. (After a long story, the Hubble constant is now estimated at 68 km/sec/Mpc.) In any case, Lemaître appears as the real "inventor" of the cosmic expansion, since Hubble attributed his redshift observations and his empirical law as an effect of the galactic motions.

The 1927 paper of Lemaître remained almost unnoticed. However he was red by Einstein who declared that, although the mathematics were correct, the physical content was "abominable". On the other hand, Eddington received the paper with enthusiasm. After the letter from Lemaître, he

[3]An english translation by Eddington, "A homogeneous universe of constant mass and increasing radius accounting for the radial velocities of extragalactic nebulae", was published later in MNRAS.

[4]This disappeared in the 1931 English translation; see [Luminet (2013)].

published a translation of his paper (see above) which led to the recognition of the cosmic expansion. Eddington also recognized the interest of the cosmological constant which was for him a necessary ingredient to the GR theory, providing "a fundamental length scale for the universe" [Merleau-Ponty (1965)]. He favored one member of the family of models, presently known as the *Eddington–Lemaître model.* It describes an expanding universe, which tends asymptotically to the Einstein model in the distant past. This is in accordance with his proof (in 1930) that the Einstein model is unstable and that the universe could not remain static.

In fact, solutions similar to those of Lemaître had been formally found before (in 1922) by the Russian mathematician Alexander Friedmann.[5] Friedmann was the first to propose global non-static solutions of general relativity: "the goal of this notice is the proof of the possibility of a universe whose spatial curvature is constant with respect to the three spatial coordinates and depend on time, e.g. on the fourth coordinate" [Friedmann (1922)]. In particular he showed that the cosmological principle allowed to reduce the tensorial Einstein equations to a pair of differential equations, presently known as the *Friedmann equations.* However, not aware of the redshifts observations, Friedmann failed to recognize the pertinence of his results for the real universe [Merleau-Ponty (1965)]. Hence the appellation *Friedmann–Lemaître models* (Lemaître was unaware of Friedmann's results in 1927). In any case, Lemaître was the first to establish a firm link between theory and observations, and to interpret correctly the galactic redshifts as an effect of the chronogeometry of space-time. (see more in [Ellis (1990)]).

In 1931, the community of astronomers and physicists, including Einstein, recognized the cosmic expansion. As it is well known, Einstein rejected the cosmological constant that he introduced before (see Sec. 4.2). In 1932, he published with de Sitter a new expanding model (*Einstein–de Sitter model*) without cosmological constant, with flat spatial sections. But this model was confronted to the "age problem": the (overestimated) value of H_0 given by Hubble led to an "age of the universe" about 2 Gyrs, much smaller than the current estimations of the age of our planet.

The same year, Lemaître [Lemaitre (1931)] presented a new solution, sometimes called an *hesitating universe*, where the influence of the

[5]Einstein received and read Friedmann's paper and rejected it firstly as erroneous. Quite soon he recognized that he himself was in error and admitted the correctness of the mathematical result, although he estimated "hardly possible to give [to it] a physical meaning". He admitted the cosmic expansion only in 1931.

cosmological constant creates a particular dynamics: a first rapidly expanding phase initiated by some kind of "explosive event" (which will be later called a Big Bang); then a stagnation period with almost no expansion; and then an accelerating expansion phase including the present period. This models does not suffer from the age problem. The GR equations imply a singular initial state of the expansion, where all distances tend to zero, and where the curvature and matter density tend to infinity: a beginning of space and time. This will become later the so called Big Bang models.

Also, Lemaître had a premonitory intuition, motivated by the recent developments in quantum physics, that the particular physical conditions in the primordial universe imply the occurrence of quantum effects. More precisely, he suggested that all the material content of the universe was at that time forming a unique quantum, which he called the *primeval atom*. During the subsequent evolution, this atom will disintegrate and finally become the matter distribution that we know: "the atom-world was broken into fragments, each fragment into still smaller pieces [...] The evolution of the world can be compared to a display of fireworks that has just ended: some few red wisps, ashes and smoke. Standing on a cooled cinder, we see the slow fading of the suns, and we try to recall the vanishing brilliance of the origin of the worlds." In fact, he even anticipated quantum gravity or cosmology: "In atomic processes, the notions of space and time are no more than statistical notions: they fade out when applied to individual phenomena involving but a small number of quanta. If the world has begun with a single quantum, the notions of space and time would altogether fail to have any sense at the beginning and would only begin to get some sensible meaning when the original quantum would have been divided in a sufficient number of quanta. If this suggestion is correct, the beginning of the world happened a little before the beginning of space and time. Such a beginning of the world is far enough from the present order of nature to be not at all repugnant."

But the idea precisely appeared "repugnant" (in his own terms) to Eddington. And Einstein estimated the model "too close to the Christian dogma of creation." Although Lemaître carefully avoided to mention a possible "creation" of the universe, he was very soon accused of *concordism*, the desire to harmonize the scientific model with the religious.[6] But Einstein

[6]Although Lemaître was a Catholic priest, he always distinguished his scientific work from its religious faith, see, e.g., [Lambert (2015)].

finally recognized the work of Lemaître. After a seminar of his in Pasadena in 1933, he declared "This is the most beautiful and satisfactory explanation of creation to which I have ever listened."

3. Toward the Big Bang Models

A variety of cosmological models have been subsequently proposed and discussed by different authors. In 1933, Lemaître proved that, under very general conditions, a singularity is unavoidable at the beginning of the cosmic expansion, in the framework of strict GR. This anticipated the famous singularity theorems by Hawking and Penrose in the 1960s (see, e.g., [Hawking and Ellis (1973)]). Lemaître also anticipated the assimilation of the cosmological constant to vacuum energy (see below) and suggested the presence of a cosmic background radiation.[7]

The new cosmological models did not encounter big interest. Cosmology appeared as a very abstract discipline, far from experiments or observations. Moreover Lemaître was suspected of concordism. The Einstein–de Sitter model became the most popular representative of this family, but the conflict with the age of the universe was a serious handicap. Very soon, the concurrent *steady state models* became more popular (see [Merleau-Ponty (1965)] for a detailed analysis of them). They had been initiated by the American Thomas Gold and the British Hermann Bondi [Bondi and Gold (1948)], and later improved by Fred Hoyle [Hoyle (1948)]. The authors extended the cosmological principle to a *perfect cosmological principle* which states that the appearance of the universe remains the same at any moment of time. Since the cosmic expansion dilutes the cosmic matter, this requires a compensating mechanism able to maintain a constant density: Hoyle postulated a *continuous creation of matter* (one proton mass per liter and per Gyr) under the form of a *creation field* (or C-field). This of course contradicted all the physical laws and Einstein qualified it as "romantic speculation." The resulting chronogeometry is identical to that of the de Sitter model. In its initial version, the model was not justified by any dynamical theory like GR, but Hoyle reconciled it with GR.

These models remained very popular, especially in the public, up to the 1950s, when they were excluded by observational results of Martin

[7]He also mention the possibility that the topology of the universe could be multi-connected, see Sec. 5 and [Lachièze-Rey and Luminet (1995)].

Ryle, who was counting the number of radio sources as a function of their brightness.[8]

3.1. *Nuclear physics*

In the 1940s the nuclear physicists realized that the primordial universe, according to Lemaître's views, provided the physical conditions for nuclear reactions. Questions about the origin of the elements had been already discussed for many decades. The apparent uniformity of the elementary abundances shown by observations was difficult to reconcile with a stellar origin. In addition the physical conditions inside stars did not appear adequate for their synthesis, so that it appeared reasonable to search for their origin in a process involving superdense conditions.

After preliminary studies by various authors [Alpher and Herman (1990)], the foundations of the *primordial cosmic nucleosynthesis* were established by Georges Gamow in the late 1940s. After unsuccessful trials to explain the origin of *all* elementary nuclei, he favored a non-equilibrium nucleosynthesis of the *light* nuclei from the dense matter (called *Ylem* by R. Alpher) in the primordial universe. The well-known "$\alpha\beta\gamma$-paper" [Alpher, Bethe and Gamow (1948)][9] was followed by a sequence of papers by Gamow and his coworkers R. A. Alpher and R. Herman, leading finally to the modern Standard Model in the early 1950s [Alpher and Herman (1990)]. From their calculations, they had deduced that a background of cold electromagnetic radiation should fill the universe. Its discovery in 1964 initiated the confidence in the Big Bang models.

3.2. *The cosmic microwave background (CMB)*

The first explicit mention of a relic radiation from the primordial universe was made in 1948 by Alpher and Herman.[10] They estimated originally a

---

[8]The relation between observed redshift and luminosity depends on the curvature of space-time. This is a typical case of *cosmological tests* which allow to decipher a part of the chronogeometry of space-time.

[9]In this 1st April letter to *Physical Review*, Gamow added the name of Bethe for humoristic reasons: "It seemed unfair to the Greek alphabet to have the article signed by Alpher and Gamow only, and so the name of Dr. Hans A. Bethe (in absentia) was inserted in preparing the manuscript for print." Later (in the July 1949 issue of *Reviews of Modern Physics* to celebrate Einstein's 70th birthday) he mentioned the "theory of the origin of atomic species recently developed by Alpher, Bethe, Gamow, and Delter."

[10]Although it had been suggested before by Lemaître.

temperature of 5 K, with subsequent various new estimates by themselves and by Gamow. Such predictions were renewed in the 1960s by Robert Dicke and collaborators, at Princeton University, in the frame of oscillating models where the universe follows a sequence of contractions–expansions separated by cosmic bounces where the scale factor reaches a minimum radius. And the team decided to build a radiometer for detecting that radiation.[11] But two radio astronomers Arno Penzias and Robert Wilson (working at the Bell Company laboratories in New Jersey) accidentally[12] detected it before the Dicke's radiometer was ready. The two teams published joined papers in the July 1965 issue of the *Astrophysical Journal.* Penzias and Wilson were awarded the Physics Nobel Prize in 1978. The detection was confirmed in 1965 by Dicke and his team (at the wavelength of 3.2 cm). Very soon, many other experiments checked the isotropy and the thermal character of the radiation. And it appeared clearly that the Big Bang models were the only possible explanations. This was their consecration. On the other hand, this relic radiation appears to be a very valuable fossil of the primordial universe.

In 1992, the COBE satellite confirmed in a spectacular way the black-body nature of the CMB radiation, and its isotropy up to a level of 10^{-5}. The team also detected its first anisotropies. This led to the 2006 Physics Nobel Prize to his principal investigators Georges Smoot and John Mather. This detection of anisotropies was highly expected since the first precise calculations by [Peebles and Yu (1970)]. They reflect the influence of the primordial fluctuations (inhomogeneities in the cosmic matter distribution) which should have been present in the primordial universe, in order to initiate the process of gravitational instability ultimately leading to the formation of stars and galaxies.

As it is well known, many subsequent observations of the CMB have been performed by experiments at various wavelengths and with various sky coverage. Mentioning only BOOMERanG (2000), WMAP (2003, 2010) and Planck (2013–2015). The analysis of their results provided a lot of information of cosmological interest. In particular, they measured the total cosmic

[11] An upper limit of 20 K had been previously found in 1946 by Dicke and collaborators; see [Peebles (2016)], also for other early observations.

[12] They were using an antenna originally built for communication with the telecommunication Echo satellite, that they were transforming to perform radio observations of the galaxy. They detected a "noise" of unknown origin at the wavelength of 7.35 cm. The characteristics of this signal (no seasonal variations, strong isotropy over the sky) led finally to its identification with the CMB.

density to be very near the critical value [Lachièze-Rey (1995)]. In 2001, the 2dFGRS galaxy redshift survey measured the contribution of *matter* to be near 25% of that value, supporting the existence of a positive cosmological constant or dark energy (see Sec. 4.2). These observations, combined to other astronomical data, have constrained the cosmic parameters with about 1% precision, from which result the so-called "concordance" cosmological models. Cosmology has become a precision observational science.

3.3. *Modern cosmology*

After the confirmation of the Big Bang models, the main task of cosmology became to identify the best member in their family to represent our universe. The second one was to understand the distribution of the galaxies, and more generally of the cosmic structures, as well as their formation process in an initially homogeneous space-time.[13] Lemaître was among the first to attack this questions.

In the 1970s, there were mainly two schools: the *hierarchical clustering* (bottom-up) scenario around Jim Peebles and the *pancake models* (top-down) developed by Zel'dovich and the Russian team [Einasto (2015)]. This was (and is still) accomplished thanks to different kinds of observations.

At the same period began the exploration of the large scale distribution of cosmic objects like galaxies and galaxy clusters (see [Einasto (2015)] for a recent report). Various statistical tools were introduced, starting from the correlation functions, to characterize them and to make the link with structure formation models [Peebles (1980)]. After the discovery of the Local Supercluster, and later of other superclusters, of large filaments and voids, it was progressively recognized that the large-scale three-dimensional distribution of luminous matter showed a complicated hierarchical network, the *cosmic web*. After a lot of observational and theoretical advancements, the complete characterization of this network, the realization of its spatial limits, its links with cosmology and with the scenarios for structure formation remain a subject of present investigations (which include new topics among which the use of *baryonic acoustic oscillations* and gravitational lensing observations).

---

[13]It is often forgotten that the question about the physical origin of the homogeneity of space (the horizon problem) has no meaning in the frame of the Big Bang models since this homogeneity is by definition assumed at the beginning; this question becomes relevant only in initially non-homogeneous models.

4. Dark Issues

4.1. *Dark matter*

The so called "dark matter problem" is indeed a dynamical problem. It was first identified by the Swiss astronomer Fritz Zwicky [Zwicky (1933)]. After measuring high velocities for galaxies inside the Coma cluster, he realized that a dynamical analysis based on the Newton law implied there is over 10 times more gravitating mass than the visible one. This conclusion was reinforced in the 1970s, after numerical simulations leading to theoretical arguments concerning the stability of rotating galactic discs. At the same period, observations of stars, and of the hydrogen gas at the 21 cm wavelength, in spiral galaxies indicated flat rotation curves suggesting similar conclusions [Oort (1940); Roberts (1966); Rubin and Ford (1970)]. They have been later confirmed by various types of observations, including X-ray observations of galaxy clusters and various cases of gravitational lensing. A natural *answer* to the problem, which became soon the most popular, is the presence of an invisible component of matter — dark matter — reaching about 0.3 in units of critical density.

Such a component should have strong impact on cosmology. It was realized in the 1980s that it is not possible to account for cosmic structure formation without it. Since that period, more and more results appeared to confirm its existence at the cosmological scale; in particular, in the last decades, the observed fluctuations of the CMB have confirmed the presence of *baryonic acoustic oscillations* (BAOs) in the primordial universe. Their amplitude and characteristics strongly depend on the energetic content and on the geometric properties of the universe. They imply, with uncertainties around 1%, a value 0.3 for matter, and 0.05 for baryonic matter (in units of critical density). The BAO's have also been recorded in the large scale distribution of galaxies.

On the other hand, in as early as the 1970s, the density of *bayonic* matter was estimated to be in the order of 0.05 (in critical units), in accordance with primordial Big Bang nucleosynthesis calculations. This implies that dark matter, if present, must be mainly non-baryonic, a conclusion largely reinforced later by arguments from galaxy formation and from CMB and large scale structure observations. Nothing however in the *Standard Model of particle physics* can do the job. However, hundreds of "possible candidates" beyond that model have been proposed, the most popular being the WIMPs (Weakly Interacting Massive Particle), including a hypothetical *neutralino* (the lightest supersymmetric partner in supersymmetric models). But the absence of any detection in the last decades, despite numerous experiments of various kinds, leaves only a narrow range of possibilities. In addition, the results of recent observations concerning the dynamics of galaxies appear very difficult to be reconciled with the dark matter hypothesis at these scales (see [Blanchet and Le Tiec (2009)] and references therein). This suggests the possibility that the solution to the dark matter problem may not be ... dark matter, at least at these scales. Other possibilities like *modified gravity* (a modification of our gravitation theory) are widely explored today (see, e.g., [Blanchet and Le Tiec (2009)], and the url http://www.college-de-france.fr/site/en-francoise-combes/course-2014-2015.htm).

4.2. *Cosmological constant or dark energy?*

It is notorious[14] that, after the discovery of cosmic expansion, Einstein declared that his past introduction of the cosmological constant was "the biggest blunder of his life." But it is possible in fact that his biggest blunder was indeed that declaration itself.

He originally introduced it in 1917 as a repulsive term in his equations, to allow for static solutions. The motivation disappeared after the discovery of the cosmic expansion, hence his declaration. On the other hand, many cosmologists like de Sitter, Eddington and particularly Lemaître remained strong defenders of this constant. In its usual formulation, Λ has the dimension of an inverse squared length, so that it may be seen as a fundamental length in nature, a macroscopic counterpart to the microscopic Planck length. Its status in the theory would be that of a fundamental constant, like that of the Newton's constant G. And like for G, the theory does not predict any natural value so that, similarly, its value should be deduced from the observations; in this case, from the cosmic acceleration.

During most of the 20th century, there have been two main motivations (beside theoretical ones) favoring $\Lambda \neq 0$: the age of the universe and galaxy formation. At the time of the discovery of expansion, the observational data suggested a value of the Hubble constant around 500 km/sec/Mpc. For the most popular cosmological models at the time, this led to an age of the universe in the order of 2 Gyr, incompatible with the estimated age

[14] As reported by Gamow; see however [Weinstein (2013)].

of our planet. This led one part of the community to favor the *steady state* models; the other part (around Lemaître) to invoke the cosmological constant which implied a greater age for a given value of H_0. After correction of the observational results in 1952, and the following reestimation of the Hubble constant (although with big uncertainty at the time), the age of the universe (without Λ) remained marginally incompatible with the ages of the oldest stars and a universe without Λ thus appeared problematic.

Lemaître was also the first to understand that galaxy formation requires a cosmological constant. The main reason is that it offers more time for the process of gravitational instability to be efficient. Initial attempts to explain galaxy formation were first accomplished in a universe of pure baryonic matter. Given the constraints from nucleosynthesis and the limits to initial inhomogeneities from CMB observations, it appeared that such models could not be efficient enough. The next generation (in the 1980s) was performed in a universe with critical density, dominated (about 95%) by *cold* dark matter (CDM).[15] Then, with the developments of numerical simulations, it was recognized that galaxy formation indeed required a cosmological constant and this was at the origin of the popularity of the so called ΛCDM models (possibly including some component of hot dark matter) at the end of the 20th century. Although ΛCDM became the standard cosmological model, very few discussions appeared about the nature, properties and influence of Λ.

The situation changed in the late 1990s when the observations of type Ia supernovae (the 2011 Nobel Prize in physics) showed an acceleration of the cosmic expansion, via the cosmic redshift–luminosity relation. This was later confirmed by CMB and large scale structure observations. It remained quite mysterious why this discovery was often presented as an unexpected surprise rather than the (exact) confirmation of the prediction from a positive cosmological constant Λ.

In any case, this settled the question of the nature of Λ. In its original introduction, Einstein considered it as a physical constant, with a status comparable, e.g., to that of G. In 1934 Lemaître [Lemaître (1934)]

[15]In order to be compatible with primordial nucleosynthesis calculations, most dark matter has to be non-baryonic. Non-baryonic dark matter has been classified into two main families, *hot* or *cold*, according to its kinematic behavior in the primordial universe.

remarked that the contribution of Λ to the cosmic dynamics could not be distinguished to that of a substance with equation of state $p = -\rho$, and he suggested that this characterizes the quantum vacuum: "The theory of relativity suggests that, when we identify gravitational mass and energy, we have to introduce a constant. Everything happens as though the energy in vacuo would be different from zero. In order that motion relative to vacuum may not be detected, we must associate a pressure $p = -\rho c^2$ to the density of energy $\rho\, c^2$ of vacuum." This was the premonition of what has been later (in 1967) discovered by the Russian physicist Yakov Zel'dovich, and subsequently called *dark energy*, defined as a substance with repulsive gravitational influence and thus accelerating the cosmic expansion. As it is well known, there have been a huge number of propositions for this mysterious contribution. This is not the place to discuss the merits of a genuine cosmological constant versus dark energy but let us simply recall some elementary facts (see also, e.g., [Bianchi and Rovelli (2010); Burgess (2013)]):

- A genuine cosmological constant predicts exactly all the characteristics of the cosmic acceleration as it is observed now ($w = -1$). If one assumes that cosmic acceleration is due to dark energy or to modified gravity, it would appear as an unexpected coincidence that these effects exactly mimic Λ.
- A genuine cosmological constant appears as the most *economic* explanation to the cosmic acceleration. It requires ... one constant only. All dark energy models require (in addition to a new physics) fine tunings under the form of constants or functions (e.g., masses or potentials of candidates).
- If one estimates that Λ is not part of the genuine general relativity theory, it appears as the simplest modification of it which already explains all cosmic data at the present precision. A motivation for more complicated modifications could be to explain *at the same time* the cosmic acceleration and the dark matter problem [Blanchet and Le Tiec (2009)].
- It is often claimed that the cosmological constant has an unexpectedly strong value, and some people refer to this as the (first) "cosmological constant problem." This is exactly the opposite since the theory does not predict any order of magnitude for Λ; only dark energy is faced with this very large discrepancy ("the worst prediction in the history of physics") that would be better called the "dark energy problem."

5. The Topology of Space-Time

The Einstein equations of general relativity take their meaning in a given differential manifold, thus with fixed topological and differential structures given *a priori*. On the other hand, a given metric cannot live, in general, in any differential manifold. For instance, a metric with constant positive spatial curvature implies a closed topology for the spatial sections. It results that information concerning the metric, which may come from the resolution of the Einstein equations or from other considerations, may constrain the topology. But even in this case, the topology of space-time (at least of its spatial part) remains partially unconstrained.

The standard cosmological models are based on the simplest assumption, that the topology of the spatial sections (of space-time) is "standard", i.e., simply connected. De Sitter [de Sitter (1917a,c)] was the first to examine the possibility of non-trivial (multi-connected) topology of space. He was followed by Einstein himself and Eddington [Eddington (1923)]: and then by Friedmann [Friedmann (1924)]: "The knowledge we have about the spatial curvature gives as yet no direct hint about the finiteness or infiniteness. To definitely decide about its finiteness one needs additional conditions. As a criterion for the distinctness of points we may take the principle that two points through which more than one geodesic can be drawn are not distinct. It is clear that this principle allows the possibility of 'ghosts': objects and their own images occurring at the same point. This formulation of the sameness and distinctness of points implies that a space of positive curvature is always finite. However it does not allow us to settle the question of the finiteness of a space of negative curvature"; and later Lemaître [Lemaitre (1929)].

This gave rise to the field of *cosmic topology* (see [Lachièze-Rey and Luminet (1995)]), based on the assumption that spatial sections are multi-connected. For instance, the simplest non-standard 3-dimensional Riemannian manifold with constant positive curvature, the projective plane $\mathbb{P}^3$, was considered originally by de Sitter and called by him "elliptical space". Although such models do not differ *locally* from their standard counterparts, they give rise to some observable effects which may allow us to recognize (or not) their validity.

First, a given source (galaxy) would give a collection of *ghost images*. Different authors searched (in vain) for such images and a general method of *cosmic crystallography* was constructed in 1996 [Lehoucq *et al.* (1996)]. Also, topology may give rise to various observable effects in the CMB, the two main being a modification of the anisotropy spectrum [Caillerie *et al.* (2007)] and the appearance of cosmic circles [Cornish *et al.* (1998)]. The present results constraint the various models at sizes sufficiently large so that topology could not affect our current observations [Luminet (2016)].

6. The Cosmic Time

Most often and especially in popularization texts, cosmic events are expressed in terms of *cosmic time.* Although there is no defined notion of *time* in general relativity, *cosmic time* is a well defined entity, at least for some solutions of the theory including the Big Bang cosmological models. Cosmic time is a particular case of a *time function.* It admits an intrinsic and covariant definition: for a given event (i.e., here and now), it is defined as the maximal value of the proper durations of all future directed histories admitting that event as a final event [Lachièze-Rey (2013)]. Thus cosmic time is defined only for those models where this maximum exists. This is of course the case for the Big Bang models. One interesting property of cosmic time is that it coincides with the proper durations measured by a particular class of comoving observers which includes, at some level of approximation, the terrestrial observers: they read cosmic time (coinciding with *universal time* at this level of approximation) on their clocks. In 1923, Hermann Weyl suggested to interpret cosmic time as the indication of a "cosmic clock" represented by an imaginary congruence of comoving observers, corresponding to the galaxies in a good approximation [Rugh and Zinkernagel (2010)].

However, cosmic time is far from being a measurable quantity. Observations of a given event (e.g., a supernova explosion) may provide various quantities, but not its date in cosmic time. The latter may only be reconstructed after having assumed a particular cosmological model and in any case has no objective physical relevance. Other time functions like, e.g., the cosmic scale factor are however measurable. In addition, despite its appellation, cosmic time does not share the usual properties of time. For instance, two events sharing a common value of cosmic time are not *simultaneous*,[16] even for a comoving observer of the kind above (although events sharing a common value of *conformal* time are) [Lachièze-Rey (2001)]. The proper duration of an history (a measurable quantity) does not coincide in

[16]Simultaneity has only a subjective meaning: for a given observer, it is defined from the Einstein synchronization procedure [Lachièze-Rey (2001)].

general with the lapse of cosmic time between its initial and final events [Lachièze-Rey (2014)]. For this reason, other time functions (e.g., conformal time, scale factor, ...) may be of better use and may appear closer to the usual notion of time. The existence and properties of cosmic time have been largely discussed in the first years of relativistic cosmology, in particular by Weyl and de Sitter [Rugh and Zinkernagel (2010)].

It may be interesting to mention the model proposed by the logician Kurt Gödel in 1949, at the occasion of Einstein's seventieth birthday [Gödel (1949)]. It corresponds to a solution of general relativity where not only *time*, but even (global) *time functions* do not exist. Gödel's motivation was to show explicitely the impossibility of the existence of time in general relativity. His solution is mostly known for the presence of *closed time curves* which violate the principle of causality. This initiated the scientific studies about "time travel" [Lachièze-Rey (2013)].

References

Alpher, R. A., Bethe, H. A. and Gamow, G., 1948, The origin of chemical elements. *Phys. Rev.* **73**(7), 803–804.

Alpher, R. A. and Herman, R., Early work on "Big Bang" cosmology and the cosmic blackbody radiation, in *Modern Cosmology in Retrospect*, Bertotti *et al.* ed., Cambridge University Press, pp. 97–113.

Bianchi, E. and Rovelli, C., 2010, Why all these prejudices against a constant?, http://arxiv.org/abs/1002.3966v1.

Blanchet, L. and Le, Tiec A., 2009, Dipolar dark matter and dark energy, *Phys. Rev.* **D80**, 023524.

Bondi, H. and Gold, T. 1948, The steady state theory of the expanding universe, *MNRAS*, **108**, 252–270.

Burgess, C.P., 2013, The Cosmological constant problem : Why it's hard to get dark energy from micro-physics, http://arxiv.org/abs/1309.4133v1.

Caillerie, S., Lachièze-Rey, M., Luminet, J-P., Lehoucq, R., Riazuelo, A., 2007, A new analysis of Poincaré dodecahedral space model, *A&A* **476**(2), 691–696, http://arxiv.org/abs/0705.0217v2.

Cornish, N. J., Spergel, D. N., Starkman, G. D., 1998, Circles in the sky: finding topology with the microwave background radiation, *Class. Quant. Grav.*, 15, 26572670.

de Sitter, W., 1917a, *MNRAS*, 78, 3.

de Sitter, W., 1917b, On the relativity of inertia. Remarks concerning Einstein's latest hypothesis, *Proc. Acad. Sci. Amsterdam* **L9**, 1217–1225.

de Sitter, W., 1917c, *Proc. Acad. Sci. Amsterdam*, **20**, 229.

Eddington, A. S., 1923, *The Mathematical Theory of Relativity*, Cambridge University Press, Chap. 5.

Einasto, J., 2015, Yakov Zeldovich and the cosmic web paradigm, in *The Zeldovich Universe: Genesis and Growth of the Cosmic Web*, Proceedings IAU Symposium No. 308, 2015.

Einstein, A., 1917, Cosmologist betrachtungen zur allgemeinen relativitätstheorie, *Preussische Akademier der Wissenschaften*, Sitzungsberichte, 142–152.

Einstein, A., 1921, Geometry und erfahrung, in *Sitzungsberichte der Preussische Akademie der Wissenschaften*, **1**, 123–130.

Eisenstaedt, J., 1989, Cosmology: a space for thought on general relativity, in *Proceedings of the Seminar on the Foundations of Big Bang Cosmology*, Barcelona, Spain, 1987. F. Walter Meyerstein, ed. Singapore: World Scientific, **27L**, 295.

Ellis, G. F. R., 1990, The transition to the expanding universe, in *Modern Cosmology in Retrospect*, Bertotti *et al.* ed., Cambridge University Press, 97–113.

Freeman, K., 2013, Slipher and the nature of the nebulae, arXiv:1301.7509v1.

Friedmann, A., 1922, *On the curvature of space, A. Z. Phys.* **10**, 377–386.

Friedmann, A., 1924, On the possibility of a world with constant negative spatial curvature, *Z. Phys.* **21**, 326.

Gödel, K., 1949, An example of a new type of cosmological solution of Einstein's field equations of gravitation, *Rev. Mod. Phys.* **21**, 447–450.

Hawking, S. W. and Ellis, G. F. R., 1973, *The Large Scale Structure of Spacetime*, Cambridge University Press.

Hoyle, F., 1948, A new model for the expanding universe, *MNRAS*, **108**, 372–382.

Hubble, E., 1929, A relation between distance and radial velocity among extragalactic nebulae, *Proceedings of the National Academy of Sciences*, **15**(3).

Lachièze-Rey, M., 1995, *Cosmology: A First Course*, Cambridge University Press.

Lachièze-Rey, M., 2001, Space and observers in cosmology, *A&A*, **376**, 17–27, http://arxiv.org/abs/gr-qc/0107010.

Lachièze-Rey, M., 2013, *Voyager dans le Temps: La Physique Moderne et la Temporalité*, Seuil Sciences Ouvertes.

Lachièze-Rey, M., 2014, In search of relativistic time, *Studies in History and Philosophy of Modern Physics*, **46**, Part A, 38–47.

Lachièze-Rey, M. and Luminet, J.–P., 1995, Cosmic topology, *Phys. Rep.*, **254**, 135–214.

Lambert, D., 2015, *The Atom of the Universe: The Life and Work of Georges Lemaître*, Copernicus Center Press.

Lehoucq, R., Lachièze-Rey M., and Luminet, J.-P., 1996, Cosmic crystallography, *A&A*, **313**, 339–346.

Lemaître, G., 1929, *Rev. Quest. Sci.* 189–216.

Lemaître, G., 1931, The expanding universe, *MNRAS*, **91**, 490–501.

Lemaître, G., 1934, Evolution of the expanding universe, *Proceedings of the National Academy of Science*, USA.

Luminet, J. P., 2013, Editorial note to "Homogeneous Universe of Constant Mass and Increasing Radius accounting for the Radial Velocity of Extra-Galactic Nebulae" by Georges Lemaître(1927), arXiv1305.6470v1.

Luminet, J.-P., 2016, The status of cosmic topology after planck data, http://arxiv.org/abs/1601.03884.
Merleau-Ponty, J., *Cosmologies du XXe Siècle. Étude Épistémologique et Historique des Théories de la Cosmologie Contemporaine*, Gallimard, Paris.
Oort, J. H., 1940, *ApJ*, **91**, 273.
Peacock, J. A., 2013, Slipher, galaxies, and cosmological velocity fields, in *Origins of the Expanding Universe: 1912–1932*, ASP Conference Series, Vol. 471, Astronomical Society of the Pacific, http://arxiv.org/abs/1301.7286v1.
Peebles, P. J. E., 1980, *The Large Structure of the Universe*, Princeton University Press.
Peebles, P. J. E., 2016, Robert Dicke and the naissance of experimental gravity physics, http://arxiv.org/abs/1603.06474v1.
Peebles, P. J. E. and Yu, J. T., 1970, Primeval adiabatic perturbation in an expanding universe, *ApJ* **162**, 815–836.
Roberts, M. S., 1966, *ApJ*, **144**, 639.
Rubin, V. C. and Ford, W. K. J., 1970, *ApJ*, **159**, 379.
Rugh, S. E. and Zinkernagel, H., 2010, Weyl's principle, cosmic time and quantum fundamentalism, http://arxiv.org/abs/1006.5848.
Weinstein, G., 2013, George Gamow and Albert Einstein: Did Einstein say the cosmological constant was the "Biggest Blunder" he ever made in his life?, http://arxiv.org/abs/1310.1033.
Zwicky, F., 1933, Die rotverschiebung von extragalaktischen nebeln, *Helv. Phys. Acta*, **6**, 110.

Chapter 7

Highlights of Standard Model Results from ATLAS and CMS

Cristina Biino

Istituto Nazionale di Fisica Nucleare
Torino, Italy
The ATLAS and CMS collaboration
European Organisation for Nuclear Research (CERN)
Geneva, Switzerland
cristina.biino@to.infn.it

In this contribution we describe in detail the technical characteristics of the major experimental facilities at CERN, the latest data involving highlights of Standard Model results from ATLAS and CMS — in particular the recent discovery of the Higgs boson — and the operational capability of this laboratory in exploring the frontiers of physics in the description of the properties of the tiniest components in the first moments of the Universe.

Contents

1. The Large Hadron Collider at CERN

The LHC is the world's largest and most powerful particle accelerator collider, one of the largest and truly global scientific projects ever. It has been designed to provide pp collisions at a centre-of-mass energy of 13 TeV and a nominal intensity luminosity $L = 10^{34}\,\mathrm{cm}^{-2}\,\mathrm{s}^{-1}$. Pb–Pb collisions and asymmetric p–Pb collisions are also available.

The main motivation is the study of the nature of the electroweak symmetry breaking and the physics phenomena at the TeV energy scale.

The LHC accelerator [Evans *et al.* (2008)] and [Bruning *et al.* (2004)], is using the tunnel built in the eighties for LEP (the Large Electron–Positron collider), the previous CERN accelerator, and still the largest electron–positron accelerator ever built. The size of a collider accelerator is a function of the radius of the machine and the strength of the dipole magnetic field that keeps particles in their circular orbits. At LHC the strong magnetic field is produced by dipole superconducting electromagnets, delivering a magnetic field of 8.3 T, at the limit of modern technology.

The proton–proton collider ring is 27 km long and the tunnel is situated about 100 m underground in the Geneva area.

The protons are coming from the Super Proton Synchrotron (SPS) in (2808) bunches of about 1.15×10^{11} particles, and are kept in the circular orbit by 1,232 superconductive dipole magnets and 392 quadrupole magnets which focus the beams.

The electromagnets are built from coils of special Nb-Ti electric cable and operated in a superconducting state, efficiently conducting electricity without resistance or loss of energy. This requires cooling the magnets to −271.3°C. For this reason, the accelerator ring is connected to a cryogenic distribution system of liquid helium.

A collider has the advantage over an accelerator, where a beam collides with a stationary target, that the available collision energy is the sum of the energies of the two beams whereas in the second case it is proportional to the square root of the energy of the particle hitting the target. Moreover the advantage of circular accelerators over linear accelerators (linacs) is that the ring topology allows continuous acceleration using a limited number of radio-frequency cavities, as the particle can transit repeatedly through them.

At LHC the hadrons form two beams travelling in opposite directions, in separate beam pipes kept at high vacuum, which collide head-on at four points where the two rings of the machine intersect. Just prior to

Table 1. LHC collider main parameters.

Diameter and circumference	8.5 and 27 Km
Proton beam energy	7 TeV (world record energy)
Luminosity	$10^{34}\text{cm}^{-2}\text{s}^{-1}$
Number of bunches and spacing	2808, 25 ns
Number of collisions per second	600 million
Number of turns per second	11,245
Machine current and beam stored energy	0.5 A, 362 MJ
Operating temperature	1.9 K
Number of magnets	9,593
Peak magnetic dipole field	8.33 T
Number of RF cavities	8 per beam
Magnet stored energy	8,800 MJ
Power consumption	about 120 MW
Cost	6.0×10^9 CHF

collision, special magnets are used to "squeeze" the particles closer together to increase the chances of collisions. At the interaction points, where the experiments are located, collisions happen every 25 ns, corresponding to a bunch crossing frequency of 40 MHz.

The main parameters of the machine are given in Table 1.

2. The LHC Detectors

The LHC experiments are installed in underground caverns built at the four collision points of the LHC beams. Two are general purpose and very big detectors, named ATLAS and CMS, designed to cover the widest possible range of physics at the TeV scale, from the search for the Higgs boson to supersymmetry (SUSY), extra dimension and dark matter. The other two are specialized detectors: ALICE, to study quark–gluon plasma and LHCb, designed to study the asymmetry between matter and antimatter in the interactions of B-particles.

At design luminosity of $10^{34}\,\text{cm}^{-2}\,\text{s}^{-1}$ the 7 TeV proton beams provide on average 20 collisions per crossing, producing hundreds of particles. In order to sustain such a rate the detectors require very high granularity and radiation resistance. Maintaining detector performance in the high radiation areas immediately surrounding the proton beams is a significant engineering challenge.

2.1. *CMS (Compact Muon Solenoid) detector*

The overall size of the detector is set by the muon tracking system that in turn makes use of the steel return yoke of the CMS magnet.

The central feature of the CMS apparatus is the big superconducting solenoid of 6 m internal diameter delivering a uniform magnetic field of 3.8 T. Immersed in the magnetic field are the inner tracker, the crystal electromagnetic calorimeter (ECAL) and the brass-scintillator hadron calorimeter (HCAL). The tracker consists of three layers of silicon pixel detectors (66M channels) followed by 10 layers of silicon micro-strips (10M channels) measuring charged particles within the pseudo-rapidity range $|\eta| < 2.5$. It provides a transverse momentum resolution of about 0.7% for 1 GeV/c charged particles, and a 15 μm vertex position accuracy. ECAL, made of 75,848 lead-tungstate scintillating crystals, has an energy resolution of about 0.5% for high energy electromagnetic showers, with a low mass Higgs decay to two photons as a bench-mark channel. Muons are measured in different gas-ionization detectors (Drift Tubes, Cathode Strip and Resistive Plate Chambers). The momentum resolution, combined with tracker, is 1% at low p_T and 5% at 1 TeV/c. The overall CMS detector is compact and hermetic, segmented in wheels and disks. It was completely assembled on surface and lowered in the collision hall cavern slice by slice.

The CMS experiment was completely assembled in the fall of 2008 after more than a decade of design, construction and installation.

During the following two years, cosmic ray data were collected on a regular basis. These data enabled CMS to align the detector components, both spatially and temporally. In addition, the CMS calorimetry has been crosschecked with test beam data, thus providing an initial energy calibration. The CMS magnet has been field mapped. The trigger and data acquisition systems have been run at full speed. Monte Carlo simulation of the CMS detector at a detailed geometric level has been tuned using the first collected data.

A detailed description of the CMS experiment can be found elsewhere, [CMS [1] (1994)], and the overall layout is shown in Fig. 1. To appreciate the performance of the detector the di-muon mass distribution collected with various di-muon triggers during the first 1.1 fb^{-1} of data taking in 2011 is shown in Fig. 2.

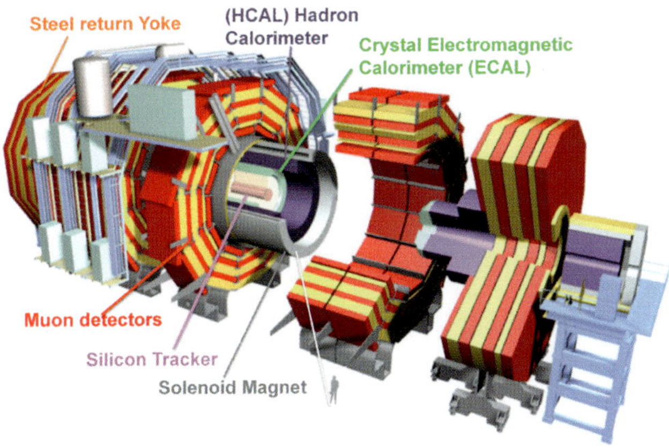

Fig. 1. A perspective view of the CMS detector.

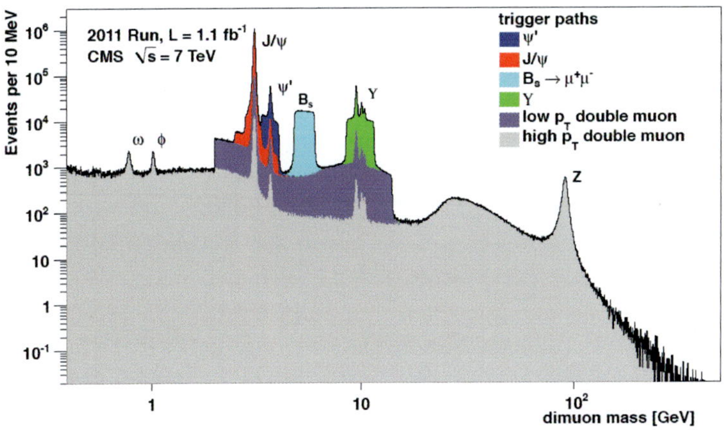

Fig. 2. CMS dimuon mass spectrum collected with various dimuon triggers, from the first 1.1 fb^{-1} of data taking. The coloured paths correspond to dimuon triggers with low p_T thresholds collected in a narrow mass window, while the light gray continuous distribution represents events collected with a dimuon trigger with high p_T thresholds. The dark gray band was a "quarkonium" dimuon trigger collected during the first 220 pb^{-1} only.

2.2. *ATLAS (Toroidal LHC Apparatus) detector*

The ATLAS detector consists of a series of ever-larger concentric cylinders around the interaction point where the proton beams from the LHC

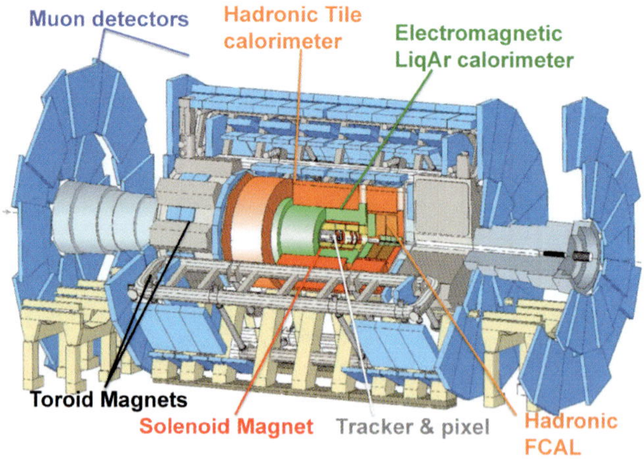

Fig. 3. A perspective view of the ATLAS detector.

collide. It can be divided into four major parts: the Inner Detector, the Calorimeters, the Muon Spectrometer and the Magnet Systems. A detailed description of the ATLAS experiment can be found in [ATLAS [1] (1999)] and the overall layout is shown in Fig. 3.

The multiple layers of detectors are complementary: the Inner Detector (ID) tracks particles precisely, the calorimeters measure the energy of easily stopped particles, and the muon system makes additional measurements of highly penetrating muons. The two magnet systems bend charged particles in the Inner Detector and the Muon Spectrometer, allowing their momenta to be measured. The ID consists of a silicon pixel detector and a silicon micro-strip detector, both covering a pseudo-rapidity range $|\eta| < 2.5$, followed by a transition radiation straw-tube tracker (TRT) covering $|\eta| < 2$. The ID is surrounded by a thin superconducting solenoid providing a 2 T axial magnetic field. A highly segmented lead/liquid argon (LAr) sampling electromagnetic calorimeter measures the energy and the position of the electromagnetic showers. An iron/scintillator tile calorimeter measures hadronic showers in the central region. The Muon Spectrometer (MS) surrounds the calorimeters and is designed to detect muons in the pseudorapidity range up to $|\eta| = 2.7$. The MS consists of one barrel and two end-cap parts. A system of three large superconducting air-core toroid magnets, each with eight coil, provides magnetic fields with a bending integral of about 2.5 T $\times$ m in the barrel and up to 6 T $\times$ m in the end-caps. Drift-tube chambers in both the barrel and the end-cap regions and cathode strip chambers are used as precision chambers, whereas resistive plate chambers in the barrel and thin gap chambers in the end-caps are used as trigger chambers. The chambers are arranged in three layers, so high-p_T particles traverse at least three stations with a lever arm of several meters.

2.3. *LHCb (Large Hadron Collider Beauty Experiment) detector*

The aim of the LHCb experiment is to record the decay of particles containing b and anti-b quarks, collectively known as *B mesons*. The experiments 4,500 tonne detector is specifically designed to filter out these particles and the products of their decay.

The LHCb detector is a single-arm forward spectrometer covering the pseudo-rapidity range $2 < |\eta| < 5$. B mesons formed by the colliding proton beams (and the particles they decay into) stay close to the line of the beam pipe, and this is reflected in the design of the detector. The other LHC experiments surround the entire collision point with layers of sub-detectors, like an onion, but the LHCb detector stretches for 20 metres along the beam pipe, with its sub-detectors stacked behind each other like books on a shelf.

A detailed description of the LHCb experiment can be found in [LHCb [1] (1998)] and the overall layout is shown in Fig. 4.

The detector includes a high-precision tracking system (VErtex LOcator or VELO) consisting of a silicon-strip vertex detector surrounding the pp interaction region, a large area silicon strip detector located upstream of a dipole magnet with a bending power of about 4 T m, and 3 stations of silicon strip detectors and straw drift tubes placed downstream.

The VELO picks out B mesons from the multitude of other particles produced, a tricky task as their short lives is spent close to the beam. To find them, the VELO's detector elements must be positioned at a distance of just 5 mm from the point where protons collide. To prevent damage from the LHC proton beams, the sensitive detector elements are held away while the beams are being injected, but once safe, the silicon elements are moved mechanically towards the beam. The VELO measures the distance between the point where protons collide (and where B particles are created) and the point where the B particles decay, producing other particles that VELO can detect. The B particles are therefore never measured directly, their presence is inferred from the separation between these two positions.

The combined tracking system has a momentum resolution of 0.6% at 100 GeV/c, and a decay time resolution of 50 fs. The resolution of

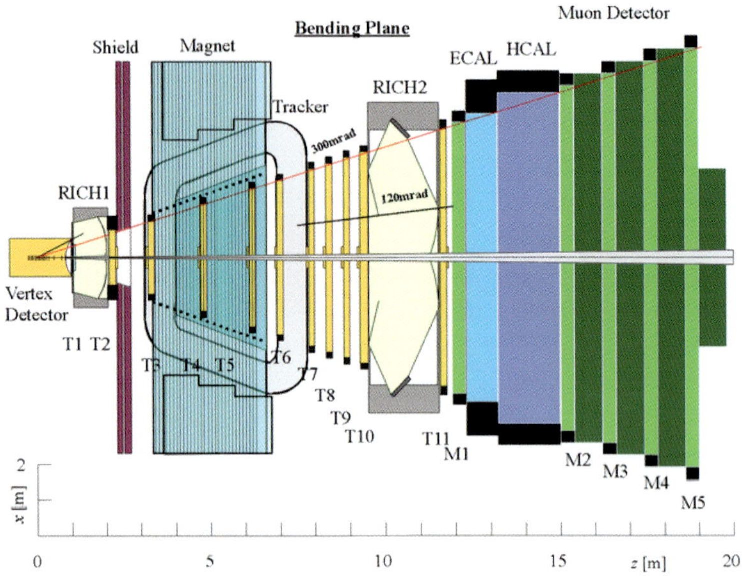

Fig. 4. A longitudinal view of the LHCb detector.

the impact parameter, defined as the transfer distance of closest approach between the track and a primary interaction, is about 20 μm for tracks with large transverse momentum. The transverse component is measured in the plane normal to the beam axis.

Charged hadrons resulting from the decay of B mesons (pions, kaons and protons) are identified by using two ring-imaging Cherenkov (RICH) detectors, measuring the emission of Cherenkov radiation. This phenomenon, often compared to the sonic boom produced by an aircraft breaking the sound barrier, occurs when a charged particle passes through a certain medium (in this case a dense gas) faster than light does. As it travels, the particle emits a cone of light, which the RICH detectors reflect onto an array of sensors using mirrors. The shape of the cone of light depends on the particles velocity, enabling the detector to determine its speed. This information is then combined with a record of the particle trajectory (collected using the tracking system and a magnetic field) to calculate its mass, charge, and therefore its identity.

Photon, electron and hadron candidates are identified and their energy measured by a calorimeter system consisting of scintillating pad and preshower detectors, an electromagnetic calorimeter, and a hadronic calorimeter. Muons are identified in the outer tracker, a system made up of alternating layers of iron and thousands of gas-filled straw tubes.

3. LHC Run 1 Data Taking and Operating Conditions

The LHC has operated in 2011 and 2012 mainly at centre-of-mass energies, $\sqrt{s}$, of 7 and 8 TeV. The fantastic increase in the performance is shown in Fig. 5 and Fig. 6 [LHC Lumi Group (2013)].

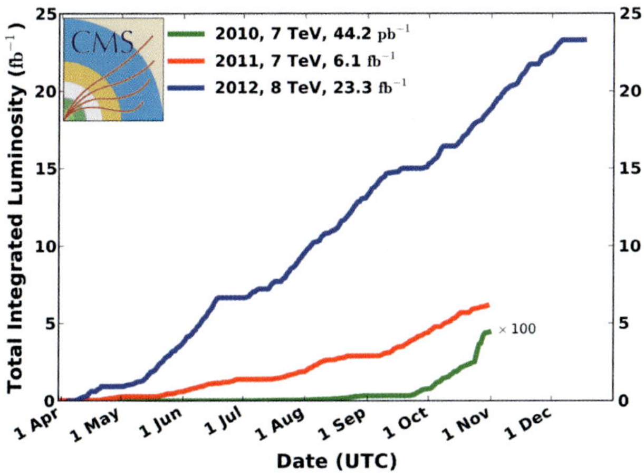

Fig. 5. LHC integrated luminosity as a function of the day, during 2010, 2011 and 2012 pp data taking, as measured by the CMS detector [LHC Lumi Group (2013)].

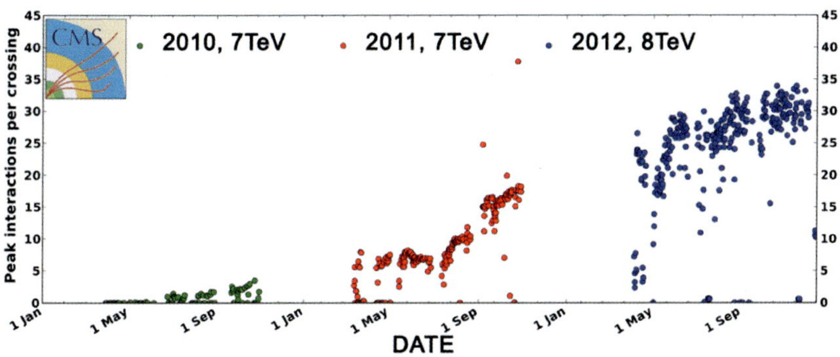

Fig. 6. LHC peak interaction per bunch crossing as a function of the day, during 2010, 2011 and 2012 pp data taking, as measured by the CMS detector [LHC Lumi Group (2013)].

The total luminosity delivered to ATLAS and CMS is $\simeq$ 30 fb^{-1} ($\simeq$ 7 fb^{-1} at 7 TeV and $\simeq$ 22 fb^{-1} at 8 TeV). The peak instantaneous luminosity of 7.7×10^{33} cm^{-2} s^{-1}, reached in November 2012, was very close to the design luminosity but at half the design energy and at 50 ns, twice the design beam crossing time. Detectors performed equally well with 95% of all sub-systems channels operational and recording 95% of the collisions. As a result of the excellent collection efficiency and high quality of the data, the results are based upon $\simeq$90% of all beam crossings. The pile-up is defined as the multiple interaction in the same bunch crossing. Both collaborations did an excellent job in extracting precision results in a very harsh pile-up environment, with up to 38 simultaneous pp collisions, and an average of 9 (21) interactions per bunch crossing in 2011 (2012).

The integrated luminosity recorded by LHCb was 38 pb^{-1} in 2010, 1.11 fb^{-1} in 2011 and 2.08 fb^{-1} in 2012.

LHCb took the majority of the data at a luminosity of 3.5×10^{32} cm^{-2} s^{-1}. This was 1.75 times more than the design luminosity as shown in Fig. 7. In 2011 a luminosity levelling procedure was introduced at the LHCb interaction point. By adjusting the transverse overlap of the beams at the LHCb, the instantaneous luminosity could be kept stable to within about 5% during a fill, as illustrated in Fig. 8.

4. Physics Objects

The physics objects reconstructed in particle physics analyses are: electrons, photons, muons, jets and missing transverse energy (MET). Electrons are detected using information coming from a tracking system as well as energy deposited in an electromagnetic calorimeter. Photons are reconstructed using energy depositions in the electromagnetic calorimeter, which are not associated to tracks in the tracking system. Muons are detected using dedicated spectrometers located at the outside of the detectors.

In the LHC detectors the only established stable particles that cannot be detected directly are the neutrinos; their presence is inferred by measuring a momentum imbalance among detected particles. For this to work, the detectors must be "hermetic", meaning they must detect all non-neutrinos produced, with no blind spots.

Jets are formed from hadrons detected in the electromagnetic and hadronic calorimeters and are clustered together within a given area. MET represents the imbalance in energy in the transverse plane derived from the reconstructed objects of the event. It is an important quantity for dark

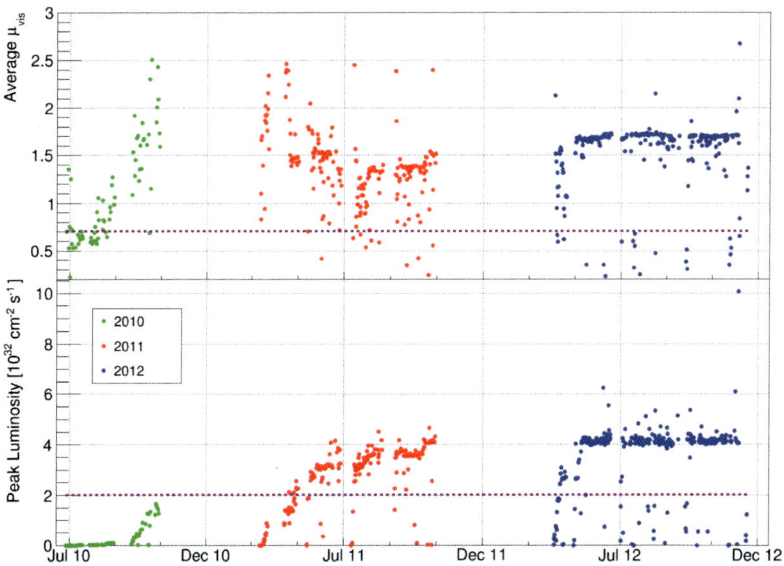

Fig. 7. Pile-up (top) and instantaneous luminosity (bottom) at the LHCb interaction point in the period 2010–2012. The dotted lines show the design values [LHCb [2] (2015)].

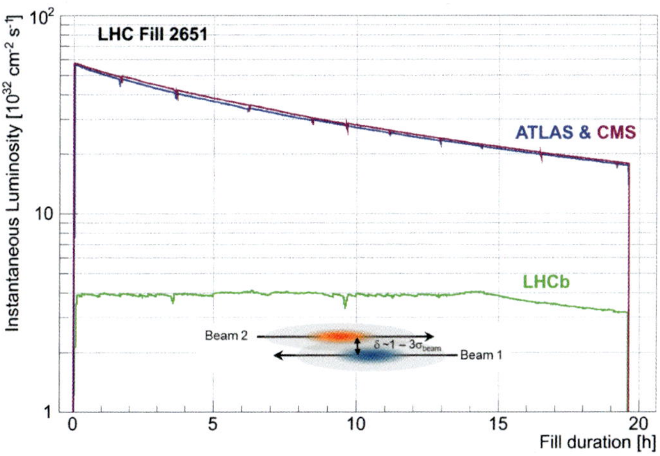

Fig. 8. Development of the instantaneous luminosity for ATLAS, CMS and LHCb during LHC fill 2651. For LHCb, the luminosity is kept stable by adjusting the transversal beam overlap. The difference in luminosity towards the end of the fill between ATLAS, CMS and LHCb is due to the difference in the final focusing at the collision points, commonly referred as the β^* [LHCb [2] (2015)].

matter searches at the LHC, since any observed excess, above SM expectations, of events with large MET indicate the possible presence of dark matter candidates, which cannot be directly detected by the detectors.

5. The Discovery of the Higgs Boson

The SM is the most successful theory to describe the elementary particles and their interactions. Its key prediction is the electroweak symmetry breaking (EWSB) through a complex scalar field. There are three force carriers in electroweak (EW) theory, the photon, the W and Z bosons, but the photon is massless while W and Z are quite massive. In the SM, this symmetry breaking is achieved through the Higgs mechanism and an immediate consequence of this mechanism is the existence of a physical particle, the Higgs boson, the only SM particle not yet observed before the start of the pp collisions LHC program. The Higgs scalar field is crucial for the validity of the SM: the gauge bosons W and Z obtain their masses via interaction with this field. Moreover the SM predicts that all the elementary fermions obtain their mass interacting with the Higgs scalar field, through the so-called Yukawa mechanism [Higgs (1964)]. The Higgs mass is the only free parameter in the theory (from perturbative unitary consideration $m_H < 1$ TeV) and the Higgs is expected to have vacuum quantum numbers $J^P = 0^+$ [Cornwall *et al.* (1973)].

One of the primary goals of the physics program of LHC was naturally the exploration of the EWSB mechanism and the study of the mass generation of the elementary particles. What we knew about the Higgs boson in 2011, after collecting $\simeq$5 fb^{-1} of data at LHC, was that we could NOT exclude, at 95% CL, a narrow mass range above 117 GeV.

5.1. *The Higgs boson production and decay*

At a hadron collider there are four major Higgs production modes (through two different couplings), as shown in Fig. 9: gluon-gluon fusion ($\sigma \simeq 10\,$pb); weak vector boson fusion ($\simeq$ 1 pb); associate production with a W or Z or with a top-quark pair. The light quarks and gluons in the protons have small direct coupling to the Higgs boson (H), therefore it is necessary to first produce massive particles, which have a larger coupling to the H. Figure 10 shows the expected production cross section as a function of the H mass [Dittmaier *et al.* (2011)].

Given the Higgs mass, many decay channels are open, as shown in Fig. 11. The H prefers to decay to the heaviest kinematically allowed pair

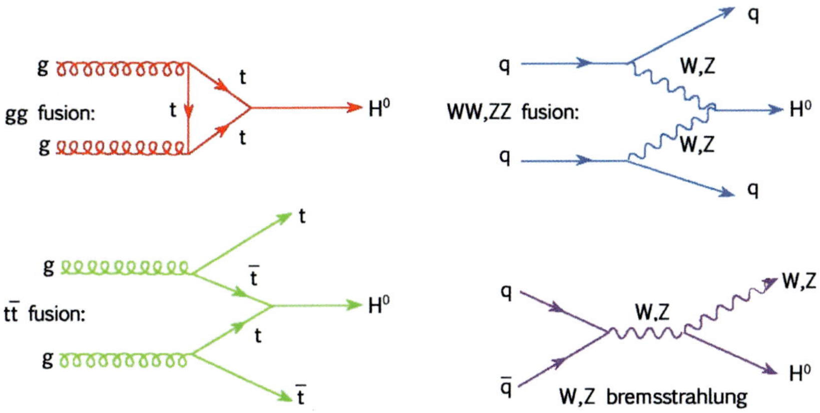

Fig. 9. Feynman diagrams for the leading SM Higgs boson production modes: gg fusion, vector boson fusion, associate production with a top-quark pair, associate production with a W/Z boson.

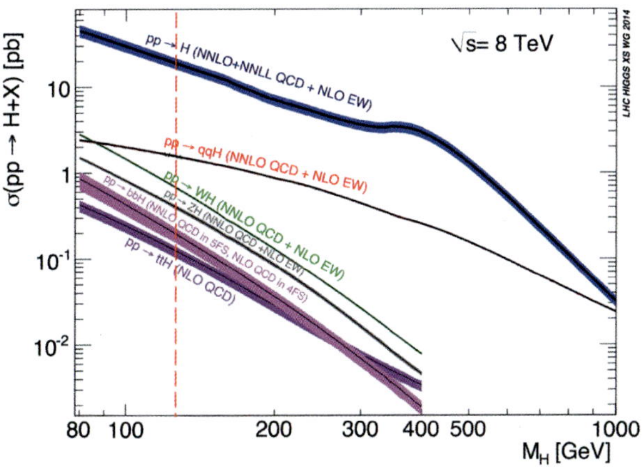

Fig. 10. SM Higgs boson production cross section as a function of the Higgs mass, for pp collisions at $\sqrt{s} = 8$ TeV [Dittmaier *et al.* (2011)].

of particles. Besides the branching fraction (BR), the signal-to-background ratio is also important for the observation of a decay channel. At the LHC, the five most sensitive Higgs boson decay channels are the modes $\gamma\gamma$, WW, ZZ, $\tau\tau$ and $b\bar{b}$. H $\rightarrow b\bar{b}$ has the largest branching ratio (BR) but suffers from large background processes. The WH and ZH associated production channels are used, with W or Z decaying leptonically. H $\rightarrow$ ZZ and H $\rightarrow \gamma\gamma$

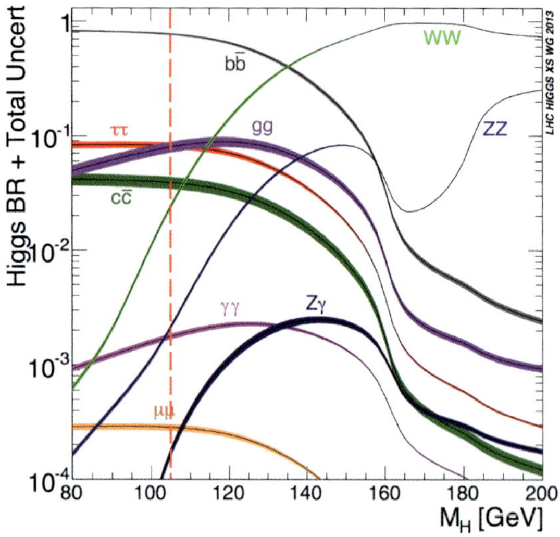

Fig. 11. SM Higgs boson Branching Fraction as a function of the Higgs mass. The red line corresponds to $m_H = 125$ GeV [Dittmaier *et al.* (2011)].

are the cleanest channels for signal observation. They have small BR but also small backgrounds. H $\rightarrow$ WW and H $\rightarrow \tau\tau$ have large BR but suffer from large backgrounds. Of all the channels the H $\rightarrow \gamma\gamma$ and H $\rightarrow$ ZZ $\rightarrow$ $4l$ are the ones providing full reconstruction of the H mass, a narrow mass peak with a typical experimental resolution of 1.6–2 GeV over a smooth background. In July 2012 the ATLAS and CMS collaborations reported, independently, the observation of a new particle compatible with the SM Higgs boson, based on the observation of the *golden* decay channels $\gamma\gamma$ and ZZ, and also WW. Subsequent measurements of the properties of this particle are all consistent with the SM Higgs interpretation.

5.2. $H \rightarrow \gamma\gamma$

The H $\rightarrow \gamma\gamma$ channel proceeds via loops of massive states (mainly top quarks and vector bosons) and provides a clean final state topology allowing the reconstruction of the mass of the H with high precision. The search for a narrow resonance in the $\gamma\gamma$ invariant mass distribution provides a clear evidence. The final state consists of two isolated photons originating from the primary vertex. The signal peak lays on top of a large but smooth background distribution. The signal to background ratio is 1:20.

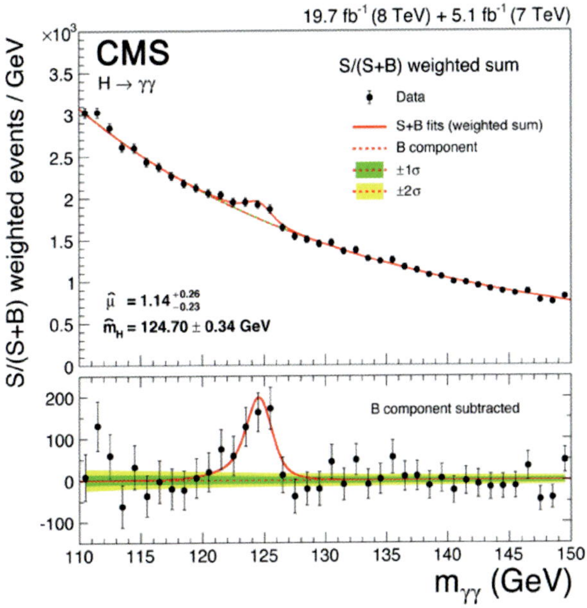

Fig. 12. The CMS $m_{\gamma\gamma}$ distribution as weighted sum of all event categories. S and B are the numbers of signal and background events in a small mass window around m_H for each event category. The lower plot shows the residual data after subtracting the fitted background component [CMS [2] (2014)].

The mass is measured by maximizing a likelihood with m_H as the parameter of interest. A clear signal is observed, see Fig. 12 and Fig. 13, with mass $m_H = 124.7 \pm 0.4$ GeV for CMS [CMS [2] (2014)], $m_H = 126.0 \pm 0.5$ GeV for ATLAS [ATLAS [3] (2014)]. The local significance is 5.7 and 5.2σ respectively. The signal strength defined relative to SM prediction is $\mu = \sigma/\sigma_{SM} \times$ BR/BR$_{SM} = 1.14^{+0.26}_{-0.24}$ for CMS and $\mu = 1.17 \pm 0.27$ for ATLAS. All results are in agreement with the SM expectation $\mu = 1$.

5.3. $H \to ZZ^* \to 4l$, for $l = e$ or μ

Events are selected by searching for two pairs of same flavour, opposite charge, isolated leptons, compatible with a ZZ* system, in which Z* indicates a boson off mass-shell ($m < m_Z$). The rate is very low but the final state is very clean. Figures 14 and 15 show the distribution of the four-lepton reconstructed mass for the sum of the 4e, 4μ, and 2e2μ channels for the low-mass region. Points with error bars represent the data, shaded

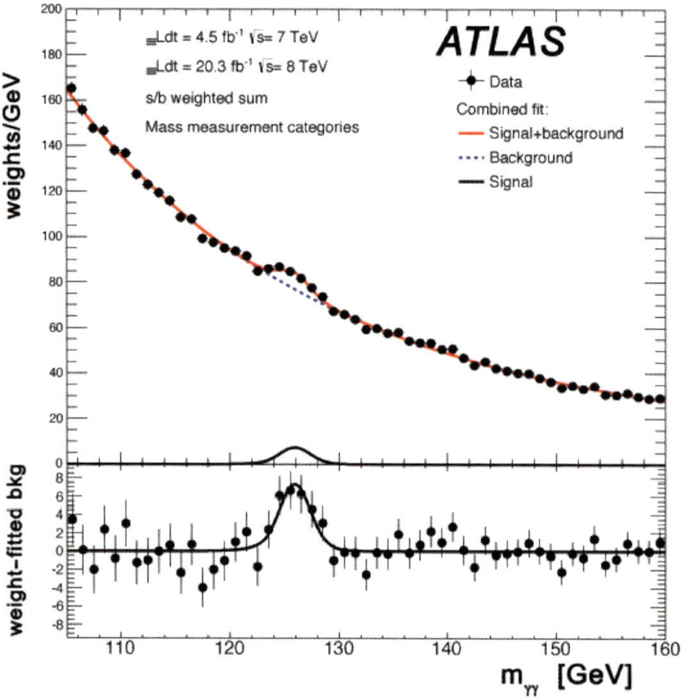

Fig. 13. The ATLAS $m_{\gamma\gamma}$ spectrum as weighted sum of all event categories. The red curve shows the fitted signal plus background model for $m_H = 125.4$ GeV. The dotted blue line shows the background component of the fit. The lower plot shows the residual data after subtracting the fitted background component [ATLAS [3] (2014)].

histograms represent the backgrounds, and the unshaded histogram the signal expectation for a mass hypothesis of $m_H \simeq 126$ GeV. Signal and ZZ* background are normalized to the SM expectation, Z+X background to the estimation from data. The Higgs peak is clearly visible with mass $m_H = 124.5 \pm 0.5$ GeV for ATLAS [ATLAS [3] (2014)] and $m_H = 125.6$ $\pm$ 0.4 GeV for CMS [CMS [3] (2014)]. This channel has the best signal to background ratio of 2:1.

Both ATLAS and CMS results are in good agreement with SM expectations. The local significance is 8.2 and 6.7σ respectively. The signal strength is $\mu = 1.7 \pm 0.4$ for ATLAS and $\mu = 0.9 \pm 0.3$ for CMS. The data strongly prefer the spin parity consistent with the vacuum one [ATLAS [2] (2013)] [CMS [4] (2014)], $J^{PC} = 0^{++}$, the value predicted for the SM Higgs. Each

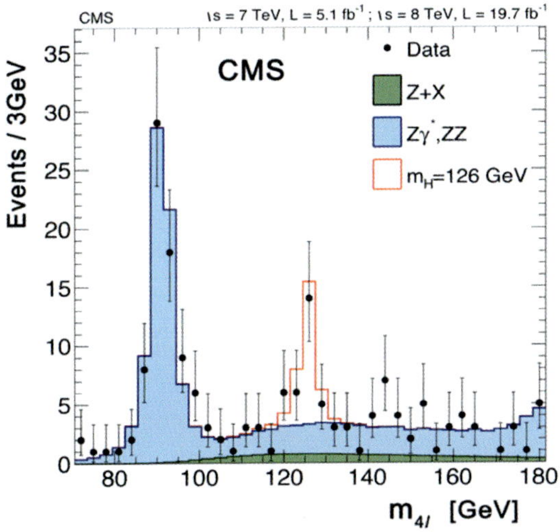

Fig. 14. CMS: the invariant mass distribution of the four-lepton for the selected candidates compared to the expected signal and background contributions, for m_H = 126 GeV hypothesis [CMS [3] (2014)].

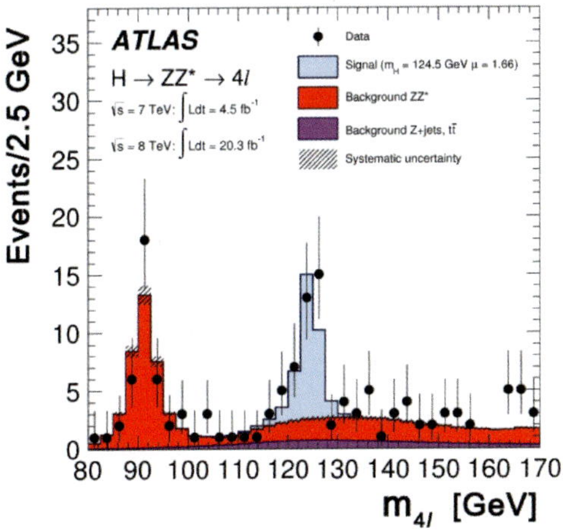

Fig. 15. ATLAS: distribution of the four-lepton invariant mass, for the selected candidates (filled circles) compared to the expected signal and background contributions (histograms), for a mass hypothesis m_H = 124.5 GeV, and normalized to the inclusive filled signal strength corresponding to the Higgs measurement [ATLAS [3] (2014)].

experiment ruled out the pseudoscalar and tensor hypotheses at 2–3σ level. The vector and pseudovector hypotheses are ruled out by the observation of the new boson in the $\gamma\gamma$ decay mode, as a consequence of the Landau–Yang [Landau (1948)] [Yang (1950)] theorem.

Measurement of the signal strength μ in the bosonic and fermionic final states lead to the same conclusion: the new boson is the predicted SM Higgs boson [CMS [8] (2015)].

In the SM the mass of the Higgs boson is not predicted. Its measurement is required for precise calculations of electroweak observables including the production and decay properties of the Higgs boson itself. To measure the mass we combine only the two high resolution channels from [ATLAS [3] (2014); CMS [2] (2014)] and [CMS [3] (2014)] and we obtain $m_H = 125.09 \pm 0.21_{\text{stat}} \pm 0.11_{\text{syst}}$. The Higgs mass is already measured to a remarkable precision, a better relative precision than the mass of the top, or any other quark.

Also the measurement of the Higgs mass is interesting in itself. The stability of the SM potential is a crucial issue for models of inflation that employ the Higgs boson. The vacuum stability depends strongly on m_H and on m_{top}. Our current results are in the meta-stability zone (with life time larger then the age of the universe): $m_{top} = 173$ GeV, $m_H = 125$ GeV but close to borders [Degrassi *et al.* (2012)], see Fig. 16. It is therefore very important to improve on the precision of the top-quark mass measurement.

The SM expectation for the Higgs boson width at $m_H = 125$ GeV is $\Gamma_H^{SM} \simeq 4$ MeV. The detectors have energy resolution in the GeV region, therefore direct measurements are dominated by instrumental resolution

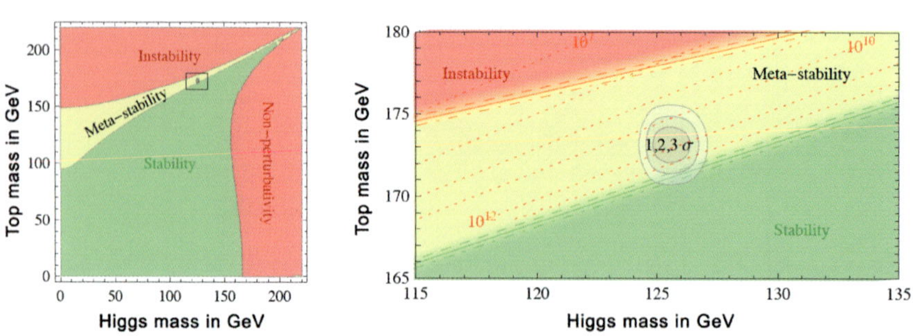

Fig. 16. Regions of absolute stability, meta-stability and instability of the SM vacuum in the m_{top} vs. m_H plane (left) and zoom in the region of preferred experimental range of m_H and m_{top} (right); area contours denote 1, 2 and 3 statistical σ.

which is about 3 orders of magnitude higher than the width predicted by the SM. However it is possible to constrain the Higgs boson width using its off-shell production and the decay into ZZ away from the resonance peak (high mass off peak region beyond $2m_Z$) [Campbell *et al.* (2014)] [Caola *et al.* (2014)] [Kauer *et al.* (2013)]. A measurement of the relative off-shell to on-shell production in the H→ ZZ channel provide direct information on Γ_H^{SM}. The resulting upper limit on the Higgs boson width obtained by CMS and ATLAS are $\Gamma_H < 22$ MeV [CMS [5] (2014)] and < 22.7 MeV [ATLAS [4] (2015)] respectively, at 95% CL.

6. Cross Section Measurements

Over the past decades the SM of particle physics has provided a remarkably accurate description of experimental results. In parallel with the development of the Higgs Boson analysis, there have been considerable progresses in understanding the SM backgrounds. The larger pp collision centre-of-mass energy available at the LHC opens up the phase space at high p_T for multiple jets recoiling against W/Z bosons. The W/Z + n jets production are major backgrounds to Higgs boson analysis in multi lepton final states when one or more jets are mis-identified as leptons in the detector. These processes have been precisely measured by ATLAS and CMS up to n $\geq$ 4 jets and compared with SM theoretical expectations from full calculations available at NLO. A compilation of ATLAS and CMS results are shown in Figs. 17, 18 and 19.

The SM processes are holding up remarkably well to scrutiny up to $\sqrt{s}$ = 8 TeV over several orders of magnitude in production cross sections, but we are just at the beginning of the "TeV" scale exploration.

7. Conclusions

We have exploited the Run 1 data: the majority of searches and analysis have been published but the LHC has so far produced only 1% of its ultimate data sample. A lot is already known about the Higgs boson: the decays to $\gamma\gamma$, ZZ and WW [ATLAS [5] (2015)] [CMS [6] (2014)] are established beyond doubt and its decays to τ leptons is also observed [ATLAS [6] (2015)] [CMS [7] (2014)]. The mass is known to the per mill level and its main couplings are known to about 20% level. LHC Run 2 data taking will allow an improved determination of the Higgs boson properties.

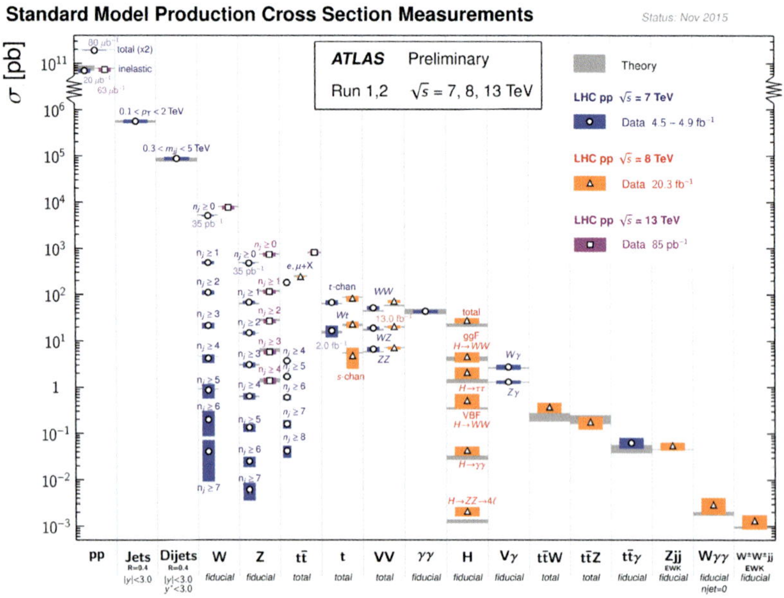

Fig. 17. Summary of several Standard Model total and fiducial production cross section measurements, corrected for leptonic branching fractions, compared to the corresponding theoretical expectations. All theoretical expectations were calculated at NLO or higher. The W and Z vector-boson inclusive cross sections were measured with 35 pb^{-1} of integrated luminosity from the 2010 dataset. All other measurements were performed using the 2011 dataset, the 2012 dataset, or the 2015 partial dataset. The luminosity used for each measurement is indicated by the color of the data point [ATLAS & CMS [1] (2015)].

It is not possible to cover all the hundred results from Run 1. Here are just a few examples of exciting SM results that have not been discussed in this chapter:

- First observation of the $B_s^0(\mu\mu)$ decay consistent with the SM
- First observation of direct CP violation in B decays
- Top mass measurement
- Single top production
- Observation of new heavy excited beauty bound states
- QCD PDF's measurements.

LHC is a very successful particle accelerator giving in its first 3 years a spectacular performance. The detectors brought the first major discovery and also a new program of searches and precision measurements.

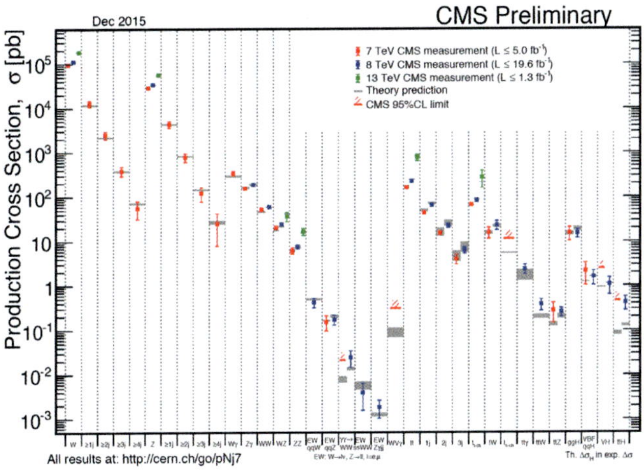

Fig. 18. CMS summary of SM cross section measurements with fiducial cross section measurements and limits, compared to theory. Total experimental uncertainties are shown as error bars. Vertical extent of the theory predictions indicates the theoretical uncertainty. Higgs results include theoretical uncertainties in the experimental error bars [ATLAS & CMS [1] (2015)].

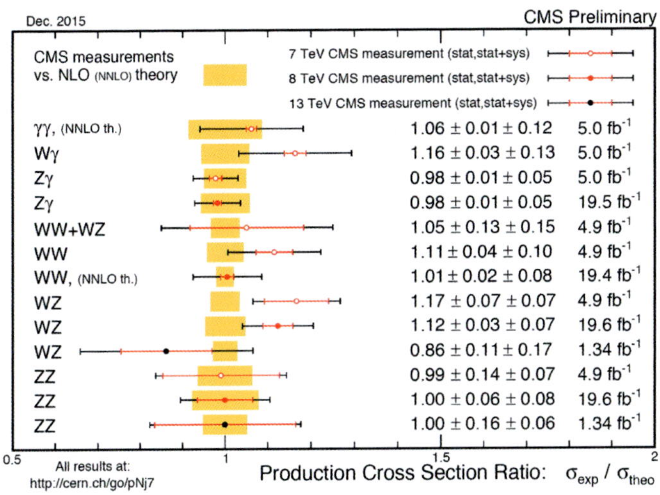

Fig. 19. CMS summary of ratios of experimental and theoretical cross section measurements [ATLAS & CMS [1] (2015)].

In 2015 LHC Run 2 data taking begun, operating at full energy of 13 TeV, 25 ns bunch crossing time and at high luminosity, increasing the power of searches by about a factor ten. A new era is starting, the era of LHC as the world Higgs factory.

LHC is a young machine and will be running for another 20 years at high intensity and with an exciting program ahead.

References

ATLAS Collaboration [1] (1999). CERN-LHCC-99-14 (1999). *JINST* **3**, S08003 (2008), ISBN: 978-92-9083-336-9.

ATLAS Collaboration [2] (2013). *Phys. Lett.* **B 726**, 120 (2013).

ATLAS Collaboration [3] (2014). *Phys. Rev.* **D90**, 052004 (2014) and arXiv: 1406.3827 [hep-ph] (2014).

ATLAS Collaboration [4] (2015). *Eur. Phys. J.* **C75**, 335 (2015) and arXiv:1503.01060 [hep-ph] (2015).

ATLAS Collaboration [5] (2015). *Phys. Rev.* **D92**, 012006 (2015) and arXiv:1412.2641 [hep-ph] (2014).

ATLAS Collaboration [6] (2015). *JHEP* **04**, 117 (2015) and arxiv:1501.04943 [hep-ph] (2015).

ATLAS & CMS Collaboration [1] (2015).
twiki.cern.ch/twiki/bin/view/CMSPublic/PhysicsResultsCombined.
twiki.cern.ch/twiki/bin/view/AtlasPublic/StandardModelPublicResults.

Bruning, O. E. *et al.* CERN-2004-003-V-1 (2004).

Campbell, J. M., Ellis, R. K., Williams, C., *JHEP* **04**, 060 (2014).

Caola, F., Melnikov, K., *Phys. Rev.* **D88**, 054024 (2013).

CMS Collaboration [1] (1994). CERN-LHCC-94-38 (1994), ISBN: 978-92-9083-338-3. *JINST* **3**, S08004 (2008).

CMS Collaboration [2] (2014). *EPJC* **74**, 3076 (2014).

CMS Collaboration [3] (2014). *Phys. Rev.* **D89**, 092007 (2014).

CMS Collaboration [4] (2014). *Phys. Rev.* **D92**, 012004 (2014) and arXiv:1411.3441 [hep-ph] (2014).

CMS Collaboration [5] (2014). *Phys. Lett.* **B736**, 64 (2014).

CMS Collaboration [6] (2014). *JHEP* **01**, 096 (2014).

CMS Collaboration [7] (2014). *JHEP* **05**, 104 (2014) and arXiv:1401.5041 [hep-ph] (2014).

CMS Collaboration [8] (2015). *Eur. Phys. J.* **C75**, 212 (2015); arXiv:1412.8662 [hep-ph] (2015).

Cornwall, J. M., Levin, D. N., and Tiktopoulos, G., *Phys. Rev. Lett.* **30**, 1268 (1973). Cornwall, J. M., Levin, D. N. and Tiktopoulos. G., *Phys. Rev.* **D10**, 1145 (1974) and *erratum Phys. Rev.* **D11**, 972 (1975). Llewellyn Smith, C. H., *Phys. Lett.* **B46**, 233 (1973). Lee, B. W., Quigg, C. and Thacker, H. B., *Phys. Rev.* **D16**, 1519 (1977).

Degrassi, G. *et al.*, *JHEP* **1208**, 098 (2012); arXiv:1205.6497 [hep-ph] (2012); CERN-PH-TH/2012-134 (2012).

Dittmaier, S. *et al.*, arXiv:1101.0593 [hep-ph] (2011). CERN-2011-002 (2011). arXiv:1307.1347 [hep-ph] (2013).

Evans, L. and Bryant, P., *JINST* **3**, S08001 (2008). ISBN: 978-92-9083-336-9.

Higgs, P. W., *Phys. Lett.* **64**, 132 (1964). *Phys. Rev. Lett.* **13**, 508 (1964). Englert, F., and Brout, R., *Phys. Rev. Lett.* **13**, 321 (1964). Guralnik, G. S., Hagen, C. R. and Kibble, T. W. B., *Phys. Rev. Lett.* **13**, 585 (1964). Higgs, P. W., *Phys. Rev.* **145**, 1156 (1966). Kibble, T. W. B., *Phys. Rev.* **155**, 1554 (1967).

Kauer, N., Passarino, G., *JHEP* **08**, 116 (2013). Kaur, N., *Mod. Phys. Lett.* **A28**, 1330015 (2013).

Landau, L. D., *Dokl. Akad. Nauk. Ser. Fiz.* **60**, 207 (1948).

LHCb Collaboration [1] (1998). CERN-LHCC-98-04 (1998). CERN-LHCC-2003-030 (2003). *JINST* **3**, S08005 (2008). ISBN: 978-92-9083-338-3.

LHCb Collaboration [2] (2015). *Int. J. Mod. Phys.* **A30**, 1530022 (2015) and arXiv: 1412.6352 [hep-ph].

LHC Lumi Group (2013). https://twiki.cern.ch/twiki/bin/view/CMSPublic/ History: r105.

Yang, C. N., *Phys. Rev.* **77**, 242 (1950).

Chapter 8

Beyond the Standard Model Searches at ATLAS and CMS

Géraldine Conti

The ATLAS collaboration
European Organisation for Nuclear Research (CERN)
Geneva, Switzerland
geraldine.conti@cern.ch

We describe in this contribution the results of searches for deviations from the Standard Model expectations performed with the LHC Run 1 data at ATLAS and CMS, referred to as "Beyond the Standard Model" (BSM) searches, which have been carried out in various areas, including BSM Higgs, supersymmetry, exotic physics and searches for the graviton, dark matter and thermal black holes.

Contents

1. Introduction

Despite the successful tests that the Standard Model (SM) has passed over the last decades of high-energy particle physics experiments, it remains incomplete based on experimental observations. For instance, it does not include masses of neutrinos; it does not explain the matter–antimatter asymmetry nor provide any dark matter candidate. In addition, the SM contains unsatisfactory features, such as the masses of the fundamental particles whose values are not understood, or the tremendous difference between the Higgs boson mass and the Planck mass — also called *hierarchy problem*. The SM must therefore be extended and its features explained.

The LHC Run 1 dataset described in Sec. 3 of Chapter 7 was used to discover a SM Higgs boson at a mass of 125 GeV. In parallel, an impressive number of searches for deviations from the SM expectations have been carried out in various physics areas, including BSM Higgs, supersymmetry and exotic physics. They are described in this chapter. The physics objects used (electrons, photons, muons, taus, jets and missing transverse energy) are described in Sec. 4 of Chapter 7.

2. BSM Higgs Searches

In the Higgs sector, searches for BSM physics are performed in two ways: finding decays beyond the SM of the discovered Higgs boson, or finding new Higgs bosons. They are discussed in Secs. 2.1 and 2.2, respectively. As no significant excess of events above SM expectation was observed with the Run 1 dataset, upper limits at 95% confidence level (CL) on the production cross section times decay branching fraction ($\sigma \times$BR) have been quoted as a function of mass.

2.1. *Exotic Higgs decays*

These decays are either prevented in the SM by conservation laws or they include new particles that are not part of the SM.

2.1.1. *Lepton flavor violating Higgs decays*

To test the presence of additional sources of lepton flavor violation (LFV),[1] the $H \to \tau\mu, \tau e, \mu e$ decays are searched for. The 95% CL upper limit on BR($H \to \tau\mu$) is $< 1.51\%$ [ATLAS & CMS [1] (2015)]. LFV is also tested

[1]In the SM, the only observation to date of LFV comes from neutrino oscillations.

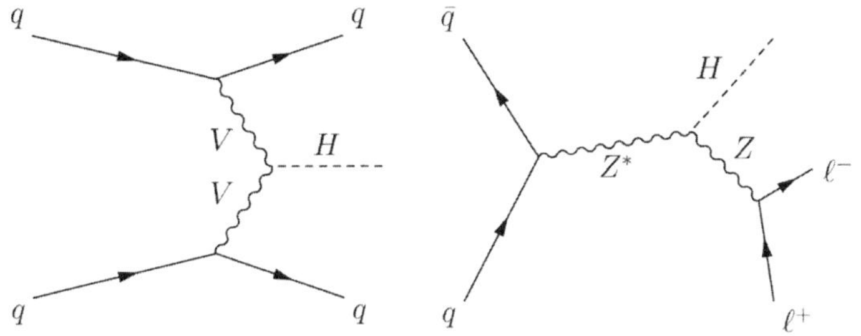

Fig. 1. Feynman diagrams corresponding to VBF Higgs production (left) and Higgs production associated with a Z boson (right) [ATLAS & CMS [3] (2014/2015)].

by searching for new heavy resonances decaying to $e\mu, e\tau, \mu\tau$ or for $Z \rightarrow e\mu$ decays [ATLAS [2] (2014/2015)].

2.1.2. *Higgs decays to invisible*

If dark matter is made of weakly interacting massive particles (WIMPs), called χ^0, it is possible that the Higgs boson couples to them, whence the importance to search for the $H \rightarrow \chi^0\bar{\chi}^0$ decay. To select these events, a large missing transverse energy (MET) is required, which is due to the two escaping χ^0. In addition, two topologies are considered, as shown in Fig. 1: events where the Higgs boson is produced via vector boson fusion (VBF) (left) or in association with a $Z/W^{\pm}$ boson (right).

VBF events are selected by requiring the presence of two jets with a large pseudorapidity gap and a large invariant dijet mass. For the Higgs boson production in association with a $Z/W^{\pm}$ boson, events with two leptons with an invariant mass consistent with the Z boson mass or at least two jets coming from the $Z/W^{\pm}$ hadronic decay are selected. The 95% CL upper limit on BR($H \rightarrow \chi^0\bar{\chi}^0$) is <28% from the VBF analysis. Combining the VBF and ZH production results, it is <25% for a 125 GeV Higgs boson [ATLAS & CMS [3] (2014/2015)].

Higgs boson decays to undetectable particles and photons are also studied in the context of supersymmetry (SUSY)[2] [CMS [4] (2015)].

2.2. *New Higgs bosons*

The presence of other Higgs bosons is motivated by various models, like the Electroweak Singlet (EWS) [Hill *et al.* (1987)], the two Higgs Doublet

[2]More details about supersymmetry are given in Sec. 3.

Model (2HDM) [Lee (1973)] or the Georgi–Machacek Higgs Triplet Model [Georgi & Machacek (1985)]. The 2HDM also appears in SUSY, which predicts the existence of two neutral CP-even Higgs bosons h and H, a CP-odd Higgs A and two charged Higgs bosons $H^{\pm}$. The discovered Higgs boson is compatible with 2HDM in the so-called "alignment" limit. In this limit, one of the two CP-even scalars, h or H, is identified with the 125 GeV Higgs boson and the condition $\cos(\beta-\alpha)\sim 0$ or $\sin(\beta-\alpha)\sim 0$ is satisfied, where $\tan(\beta)$ and α are, respectively, the ratio of the vacuum expectation values and the mixing angle of the two CP-even Higgs bosons.

2.2.1. *Charged Higgs bosons*

Light charged Higgs bosons ($m_{H^{\pm}} < m_t$) are searched for in $t\bar{t} \to (H^{\pm}b)(W^{\mp}b)$ events depicted in Fig. 2. Two $H^{\pm}$ decays are considered: $H^{\pm} \to c\bar{s}$ and $H^{\pm} \to \tau^{\pm}\nu$. The final states are characterised by the presence of b-tagged jets and a $W^{\mp}$ boson, and additionally of two jets for the $c\bar{s}$ decay or a τ and missing transverse energy in the $\tau^{\pm}\nu$ decay. Assuming $\mathrm{BR}(H^{\pm} \to c\bar{s}) = 100\%$, the 95% CL upper limit on $\mathrm{BR}(t \to H^{\pm}b)$ is 5% to 1% for $m_{H^{\pm}} = [90, 150]$ GeV [ATLAS & CMS [5] (2013/2015)]. The 95% CL upper limit on $\mathrm{BR}(t \to H^{\pm}b) \times \mathrm{BR}(H^{\pm} \to \tau^{\pm}\nu)$ ranges between 0.23% and 1.3% for $m_{H^{\pm}} = [80, 160]$ GeV [ATLAS [6] (2015)]. In the context of the Minimal Supersymmetric Standard Model (MSSM),[3] 95% CL limits are set on $\tan(\beta)$. Nearly all values of $\tan(\beta)$ above one are excluded for $m_{H^{\pm}} = [80, 160]$ GeV, as shown in Fig. 3 [CMS [7] (2015)].

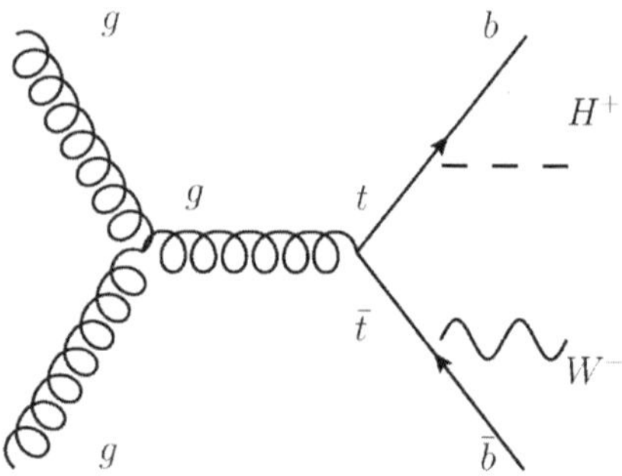

Fig. 2. Feynman diagram of $t\bar{t} \to (H^{\pm}b)(W^{\mp}b)$ decays [ATLAS [6] (2015)].

[3]More details about MSSM are given in Sec. 3.

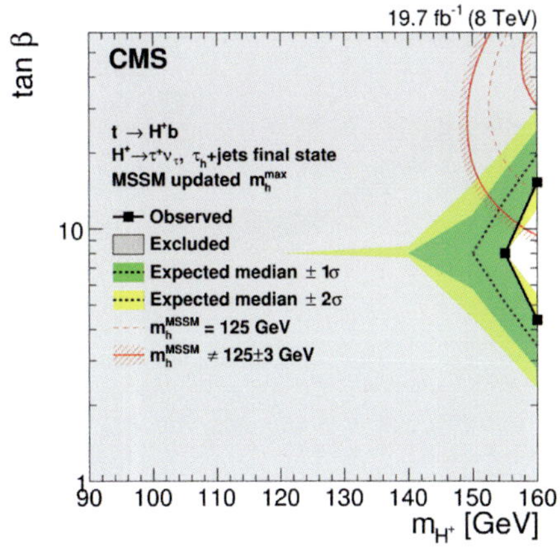

Fig. 3. The 95% CL exclusion limits on $\tan\beta$ as a function of m_{H+} in the MSSM model [CMS [7] (2015)].

Heavy charged Higgs bosons are searched for in $H^{\pm} \to \tau^{\pm}\nu$ and $H^{+} \to tb$ decays if $m_{H^{\pm}} > m_t$. For $m_{H^{\pm}} = [180, 1000]$ GeV, the 95% CL upper limits on $\sigma(pp \to \bar{t}H^{+} + X) \times \mathrm{BR}(H^{+} \to \tau^{+}\nu)$ ranges between 0.76 pb and 4.5 fb. In the context of the MSSM, a region of parameter space with high $\tan(\beta)$ is excluded for the mass range $m_{H^{\pm}} = [200, 250]$ GeV [ATLAS [6] (2015); CMS [7] (2015)].

The VBF production of a charged Higgs boson is not allowed at tree-level in the 2HDM, but it is in the Higgs Triplet Model. In order to test it, the VBF $H^{\pm} \to ZW^{\pm}$ decays are searched for. The 95% CL exclusion limit for $H^{\pm}$ VBF production cross section times $\mathrm{BR}(H^{\pm} \to ZW^{\pm})$ is shown in Fig. 4 [ATLAS [8] (2015)].

2.2.2. *Heavy CP-even Higgs Boson*

A heavy CP-even Higgs boson is searched for in various decays channels: $H \to VV(V = Z, W^{\pm})$ and $H \to \gamma\gamma$ [ATLAS & CMS [9] (2015/2016); ATLAS & CMS [10] (2014/2015)]. A heavy Higgs boson with SM-like couplings and decays is excluded in the range 145 GeV $< m_H <$ 1000 GeV, as shown in Fig. 5. It is also searched for in the $H \to hh$ channel, with the subsequent decays $hh \to b\bar{b}(b\bar{b}, \gamma\gamma)$, $\mu\mu\tau\tau$, multileptons and leptons+γ [ATLAS & CMS [11] (2015); ATLAS [12] (2015); CMS [13] (2014)]. 95% CL upper limits on the production cross section times $\mathrm{BR}(H \to hh)$ are shown in Fig. 6 for different decay modes.

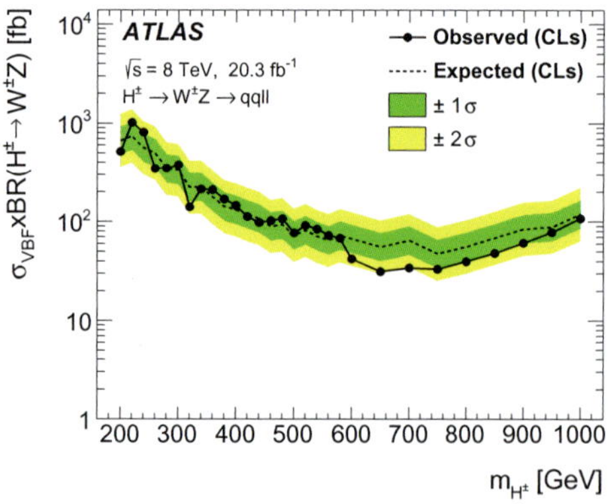

Fig. 4. The 95% CL exclusion limits for VBF production cross section of a $H^{\pm}$ times $\mathrm{BR}(H^{\pm} \to ZW^{\pm})$ [ATLAS [8] (2015)].

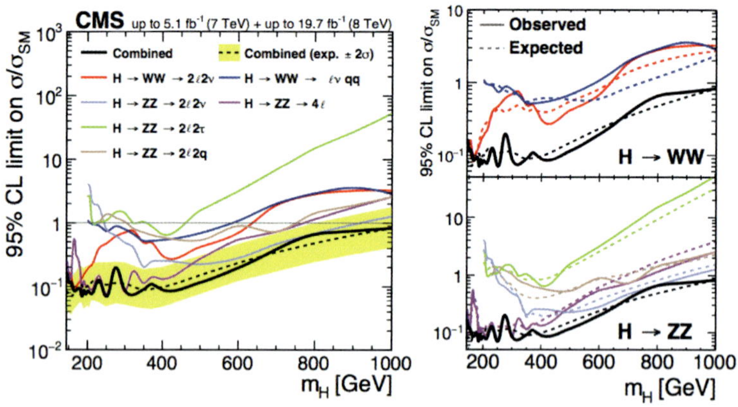

Fig. 5. 95% CL upper limits for the contributing final states and their combination in the search for a heavy CP-even neutral Higgs boson H [ATLAS & CMS [9] (2015/2016)].

2.2.3. *CP-odd Higgs boson*

A CP-odd Higgs boson A is searched for in the $A \to \tau^+\tau^-$ and $A \to Zh$ decay channels. The cross section limits are interpreted in a 2HDM as a function of $\tan(\beta)$ and m_A. An example is given in Fig. 7. A good complementarity of the exclusion regions is observed between the $A \to Zh$ and $A \to \tau^+\tau^-$ results [ATLAS & CMS [11] (2015); ATLAS & CMS [14] (2014/2015)].

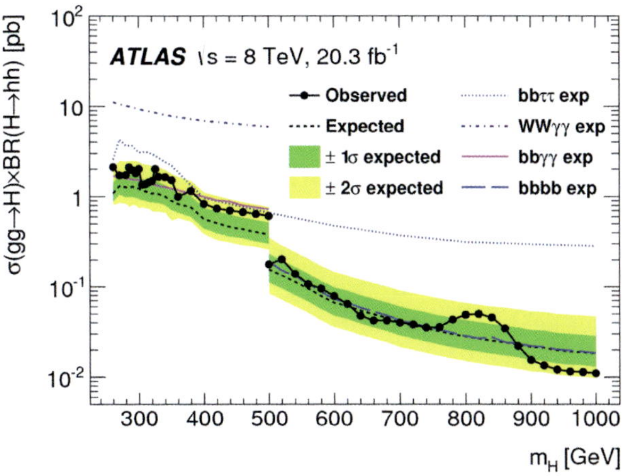

Fig. 6. 95% CL upper limits on the production cross section times BR($H \rightarrow hh$) [ATLAS & CMS [11] (2015)].

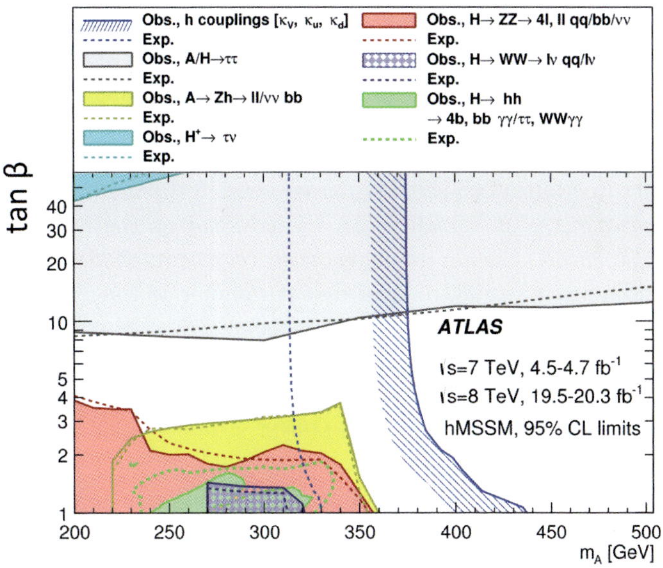

Fig. 7. 2HDM limits as a function of tan(β) and m_A [ATLAS & CMS [11] (2015)].

3. Supersymmetry Searches

Supersymmetry (SUSY) is considered as one of the most attractive BSM frameworks [Martin (1997)] and references therein. It provides an elegant solution to the hierarchy problem and a candidate for dark matter. It also suggests the high-energy unification of the weak, strong and electromagnetic couplings. For each SM particle, a superpartner of different spin (boson $\leftrightarrow$ fermion) is predicted in SUSY. The superpartners of the W bosons of the $SU(2)_L$ gauge fields are called "winos." The "bino" is the superpartner of the U(1) gauge field. The gluino is the superpartner of the gluon and the gravitino of the graviton. They all form together the "gauginos", superpartners of gauge fields. The superpartners of the Higgs field degrees of freedom are called "higgsinos." The binos, winos and higgsinos mix together to form linear combinations (or mass eigenstates) called "electroweakinos", four neutral ones $\chi^0_{1,2,3,4}$ (called "neutralinos") and two charged ones $\chi^{\pm}_{1,2}$ (called "charginos"). The superpartners of the quarks are called "squarks" and of the leptons "sleptons." The Higgs sector contains five physical Higgs bosons that have similar properties to those of the 2HDM described in Sec. 2.2.

The "Minimal Supersymmetric Standard Model" (MSSM) [Fayet (1976)] considers only the minimum number of new particles and interactions consistent with phenomenology. It has more than a hundred free parameters in addition to the SM parameters. To further reduce the number of parameters, universality can be assumed at the "Grand Unification" (GUT) scale.[4] Such a model is called "constrained MSSM" (cMSSM) [Chamseddine *et al.* (1982)]. The free parameter is $\tan\beta$ (introduced in Sec. 2.2) and the sign of the higgsino mass term μ is not fixed. Due to its simplicity, the cMSSM model was tested extensively at LHC Run 1. Combining the direct searches with astrophysics results and measurements of the Higgs boson properties, the cMSSM is today ruled out at 90% CL [Bechtle *et al.* (2015)].

In the MSSM, the baryon number B and lepton number L are no longer automatically conserved. Hence, a symmetry is added to restore it, called "R-parity" [Barger & Ma (1995)]. When R-parity is conserved, the lightest SUSY particle (LSP) cannot decay. The number of parameters of

[4] At the GUT scale, this implies that all the scalar particles have the same mass m_0. All the gauginos have the same mass $m_{1/2}$. All the trilinear couplings A_0 of the particles are the same.

the MSSM can be reduced to 19 under the following assumptions: no new sources of CP-violation, no Flavor Changing Neutral Currents (FCNC), first and second generation universality. R-parity is assumed to be conserved and the LSP is the lightest neutralino.[5] This model is called "phenomenological MSSM" (pMSSM) [Djouadi *et al.* (2007)].

The SUSY particle masses are free parameters in the theory. In addition, the hierarchy between these masses is not determined. Based on naturalness and limited fine-tuning of the theory considerations, mass spectrum scenarios can be derived. In the "natural SUSY" scenario [Barbieri *et al.* (1988)], higgsinos, gluinos and the third-generation squarks are constrained to be quite light, because of the corrections they bring to the Higgs boson mass. The light stop (superpartner of the top quark) is used to solve the hierarchy problem. Another scenario, called "Split SUSY" [Arkani-Hamed *et al.* (2005)], became more popular towards the end of Run 1 due to the absence of a SUSY signal and the discovery of a Higgs boson with a 125 GeV mass. In it, only fermions (gluinos and electroweakinos) are within LHC reach due to their moderate masses. The scalar sparticle masses (including the stop) can be arbitrarily high, hence fine tuning in the theory is accepted.

As no SUSY particle was found in the LHC Run 1 data, the results are used to set 95% CL exclusion limits. They are calculated in two frameworks: The first one is the pMSSM [ATLAS & CMS [15] (2015)]. The second is called the "Simplified Model" approach [Alwall *et al.* (2009)]. For a channel of interest, the branching fraction of a SUSY mother particle decaying to SM particles and the LSP is assumed to be 100%. The exclusion limits are set as a function of the SUSY mother particle mass and the LSP mass.

3.1. *Searches for strongly-produced SUSY particles*

Following the natural SUSY expectations, the strategy for Run 1 SUSY searches was to look with high priority for strongly-produced gluinos and third-generation squarks, since these particles were expected to be produced copiously at the LHC. The 95% CL exclusion limits for searches for gluino pair production with the subsequent $\tilde{g} \to t\bar{t}\tilde{\chi}_1^0$ gluino decay are shown in Fig. 8 as a function of the gluino mass and the LSP mass. The various coloured curves come from analyses considering different final states in terms of number of leptons, lepton charges and number of b-jets, which come from the top quark decays ($t \to W(\ell\nu; b\bar{b})b$). For a massless LSP,

[5] If the LSP is the gravitino, then the number of free parameters is 20.

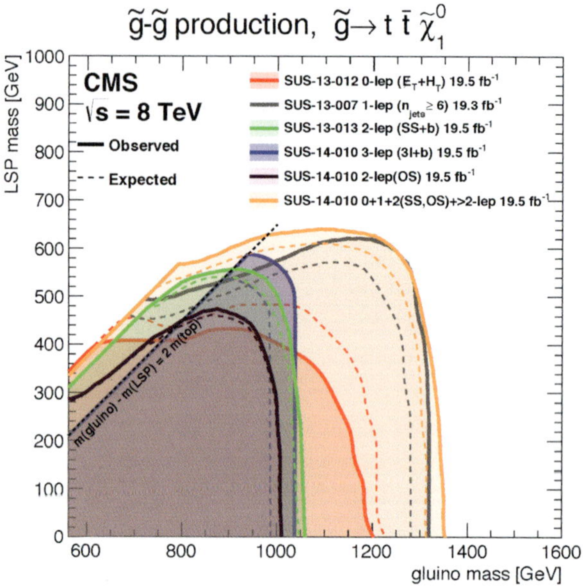

Fig. 8. 95% CL exclusion limits for strong production of gluino pairs with the gluinos decaying via the $t\bar{t}\tilde{\chi}_1^0$ decay [ATLAS & CMS [16] (2013/2014/2015)].

the best lower limit on gluino mass is >1300 GeV [ATLAS & CMS [16] (2013/2014/2015)].

The 95% CL exclusion limits for searches for stop pair production with the subsequent $\tilde{t} \to t^{(*)}\tilde{\chi}_1^0/c\tilde{\chi}_1^0$ decay are shown in Fig. 9. Different analyses have been designed to target three scenarios: $m_{\tilde{t}} > m_t$, $m_W < m_{\tilde{t}} < m_t$ and $m_W > m_{\tilde{t}}$. When the mass difference between the stop and the LSP is close to the mass of the top quark or the $W^{\pm}$, the kinematics of the SUSY signal becomes similar to the SM background and it is difficult to disentangle them, hence the weak limits in Fig. 9 close to the ($\mathrm{m}_{\tilde{t}}$-$\mathrm{m}_{\tilde{\chi}_1^0} = m_{W/t}$) diagonals. For a massless LSP, the lower limit on the stop mass is >750 GeV [ATLAS & CMS [16] (2013/2014/2015)].

3.2. *Searches for electroweakly-produced SUSY particles*

In the absence of strongly-produced SUSY particles, the electroweak pair production of charginos and neutralinos was thought to possibly be the dominant process at the LHC. The 95% CL exclusion limits for searches for $\tilde{\chi}_2^0\tilde{\chi}_1^{\pm}$ and $\tilde{\chi}_1^{\pm}\tilde{\chi}_1^{\mp}$ pair production are shown in Fig. 10 as a function of

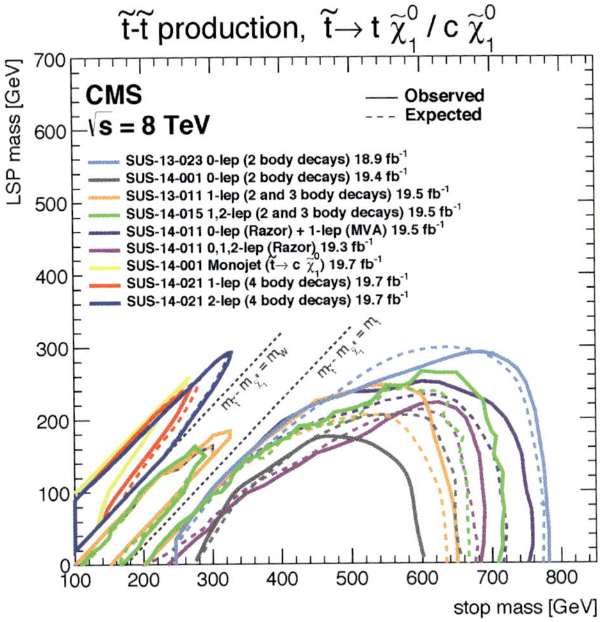

Fig. 9. 95% CL exclusion limits for strong production of stop pairs with the stops decaying via the $\tilde{t} \to t\tilde{\chi}_1^0/c\tilde{\chi}_1^0$ decay [ATLAS & CMS [16] (2013/2014/2015)].

the mass of the $\tilde{\chi}_2^0/\tilde{\chi}_1^{\pm}$ (assumed to be equal[6]) and the LSP mass. The decay channels include either $Z/W^{\pm}$ bosons or a SUSY Higgs boson h consistent with the SM Higgs boson. For the decay via $W^{\pm}Z$, and assuming a massless $\tilde{\chi}_1^0$, the lower limit on the $\tilde{\chi}_2^0/\tilde{\chi}_1^{\pm}$ mass is >425 GeV. For a $\tilde{\chi}_2^0/\tilde{\chi}_1^{\pm}$ mass of 100 GeV, the lower limit on the $\tilde{\chi}_1^0$ mass is >70 GeV [ATLAS & CMS [17] (2014/2015)].

In the pMSSM, the electroweak sector is described by four parameters: tan(β), μ, and M_1 and M_2 that are respectively the bino and wino mass parameters.[7] An example of 95% CL limits is given in Fig. 11. The value of the $\tan\beta$ parameter is fixed to 10. The exclusion limits are presented as a function of the M_2 and μ parameters for given M_1 values (here

[6]In the model used, $\tilde{\chi}_1^0$ is supposed to be bino-like, and $\tilde{\chi}_2^0/\tilde{\chi}_1^{\pm}$ wino-like.

[7]The other parameters are related to the gluino, squark and slepton masses, third generation trilinear couplings and the mass of the CP-odd Higgs boson A. Usually, very large slepton masses are assumed to study the electroweak sector.

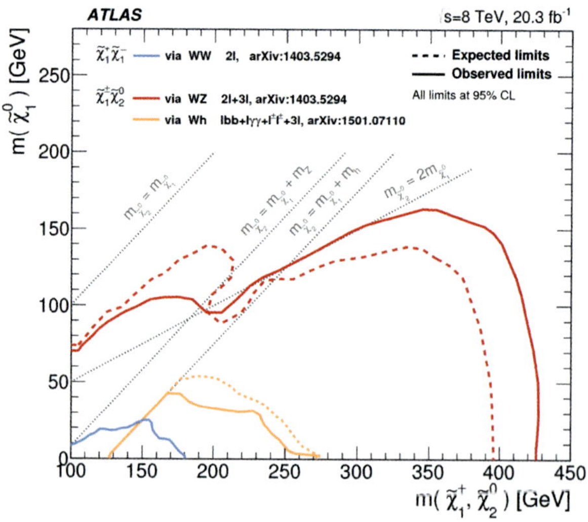

Fig. 10. 95% CL exclusion limits for electroweak production of charginos and neutralinos pairs decaying into various final states [ATLAS & CMS [17] (2014/2015)].

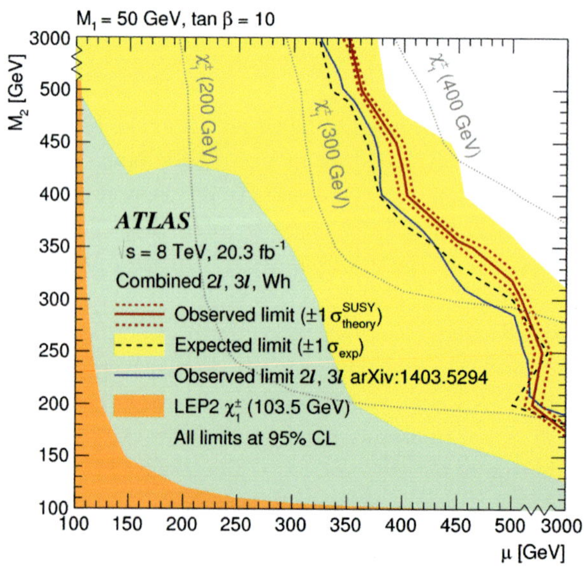

Fig. 11. 95% CL exclusion regions in the (μ, M_2) mass plane of the pMSSM with $\mu = 10$ and $M_1 = 50$ GeV [ATLAS & CMS [17] (2014/2015)]. Limits coming from LEP2 are also displayed [LEP2 (2001)].

$M_1 = 50$ GeV). A large part of the plane is excluded (green area) using the Run 1 data [ATLAS & CMS [17] (2014/2015)].

3.3. *Searches for long-lived SUSY particles*

In the absence of both strongly- and electroweakly-produced SUSY particles, it was thought that the SUSY signal may have been missed because the new particles live for some time before decaying.[8] For instance, stongly-interacting gluinos can have a long lifetime and would form colourless hadronic bond states (called "gluino R-hadrons") in the Split SUSY scenario. They would give rise to experimental signatures like decay vertices that are displaced compared to the production vertex, or activity in the calorimeter occurring in between *pp* collisions. They can also leave an anomalous large specific energy loss in the pixel detector.

Various searches were performed, which are nicely complementary to each other, spanning gluino $c\tau$ distances from $\sim 10^{-3}$ m to $\sim 10^{4}$ m as shown in Fig. 12. Interestingly, the strongest 95% CL limit to date on gluino mass comes from an analysis with displaced secondary vertices, at >1550 GeV, assuming a ~0.2 ns lifetime, because the background is predicted to be very low [ATLAS & CMS [18] (2013/2014/2015)].

4. Exotic Physics Searches

Apart from SUSY, many other BSM models have been extensively tested during LHC Run 1. Three types of effects are searched for in kinematic distributions: resonances in invariant mass spectra, excess in the tails of distributions, or higher event rates than expected from SM. Also, long-lived particles have been searched for [ATLAS & CMS [18] (2013/2014/2015)], as was already discussed in the context of SUSY searches in Sec. 3.3.

4.1. *Resonance searches*

New resonances are searched for in invariant mass spectra of di-objects: di-jets, di-leptons and di-bosons[9] or in a lepton+MET final states. These new resonances include heavy gauge bosons, Randall–Sundrum gravitons, excited fermions, vector-like quarks or scalar leptoquarks. These searches

[8]The analyses designed so far were supposing that the SUSY particles decay promptly.

[9]Searches for di-Higgs resonances are reported in Sec. 2.2.

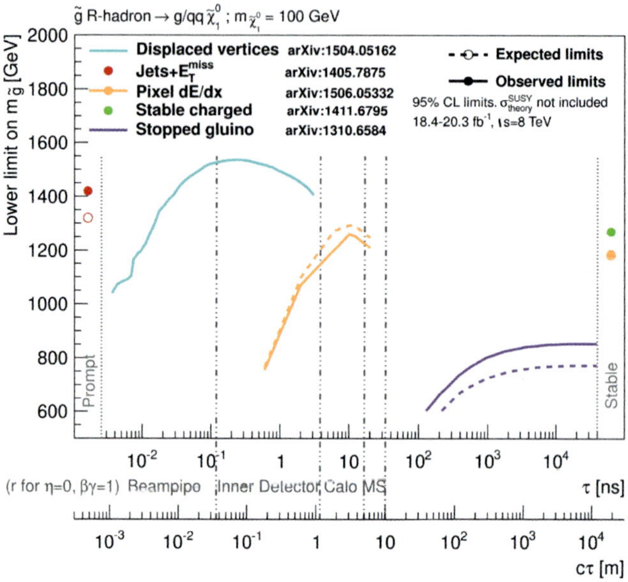

Fig. 12. 95% CL limits as a function of the mass and lifetime of a gluino R-hadron, which decays into a gluon or light quarks and a neutralino with a mass of 100 GeV. Areas below the curves are excluded. The dots on the right represents results for which the particle is assumed to be prompt or escaping the detector [ATLAS & CMS [18] (2013/2014/2015)].

are also sensitive to quantum black holes caracterized by resonance-like signatures in the di-object mass spectra.

Heavy gauge bosons W'/Z' are predicted in models like extended gauge model (EGM) and sequential SM (SSM) [Altarelli *et al.* (1990)]. The SSM model is based on the idea that, since there are sequential leptons and quarks, there may also be sequential gauge bosons. The new resonances decay in two vector bosons (Z/W), in $Z/W\gamma$, in Z/WH or directly into two jets or leptons. The 95% CL lower limit on the mass of a SSM W' boson is >3.2 TeV in the $\ell\nu$ final state and of a EGM W' boson >1.59 TeV in di-boson final states. The 95% CL lower limit on the mass of a SSM Z' boson is >2.9 TeV in di-lepton final states [ATLAS & CMS [19] (2014/2015/2016)].

In the Randall-Sundrum RS1 model [Randall & Sundrum (1999)], the hierarchy problem is overcome by adding one warped spatial dimension to space-time. It predicts the existence of Kaluza-Klein (KK) excitations of a spin-2 boson, the graviton. The model is characterized by two parameters, the mass M_1 of the first graviton mode G_{KK} and the coupling strength $\tilde{k} = k/\bar{M}_{Pl}$, where k is the curvature of the warped space and $\bar{M}_{Pl}$ the reduced Planck mass. RS1 gravitons are searched for in the $G_{KK} \to \ell^+\ell^-, \gamma\gamma$ decays channels. For $\tilde{k} = 0.1$, the 95% CL lower limit on the RS1 graviton mass is >2.7 TeV [ATLAS & CMS [19] (2014/2015/2016)].

Excited quark states (q*) are predicted in compositeness models [Baur *et al.* (1990)], in which the quarks and leptons are made of smaller constituents. Such models would help explain the three generation of fermions as well as the elementary particle mass spectrum. Excited quarks are expected to decay to $q^* \to q + X$ with $X = q, \gamma, Z, W^{\pm}$, hence they are searched for in di-jet, γ+jets and Wt final states. The decay rates depend on the excited-quark mass m_{q^*} and the compositeness scale Λ. The 95% CL lower limit on the excited quark mass is $m_{q^*} > 4.06$ TeV, where the compositeness scale is set to $\Lambda = m_{q^*}$. Single and pair production of excited leptons are searched for in an inclusive multilepton final state, considering the $\ell^* \to \ell + \gamma$ or $\nu^* \to \ell W$ decays. For excited leptons, the 95% CL lower mass limits are >3 TeV for excited electrons and muons, >2.5 TeV for excited taus, and >1.6 TeV for every excited-neutrino flavour, assuming $\Lambda = m_{\ell^*}$ [ATLAS & CMS [20] (2014/2015)].

Vector-like quarks (VLQ) are predicted in models like Little Higgs [Arkani-Hamed *et al.* (2002)] and Composite Higgs [Kaplan *et al.* (1984)] that try to address the hierarchy problem. VLQ couple preferentially to the third-generation quarks and can have flavour-changing neutral current decays.[10] Pair production of vector-like up-type (T) and down-type (B) quarks are therefore searched for in the $T\bar{T} \to W^{\pm}b/Zt + X$ or $T\bar{T}(B\bar{B}) \to Ht(Hb) + X$ decays. Final states are characterised by an isolated lepton, large MET and multiple jets. The 95% CL lower limits on the T and B vector-like quark masses range between 710 GeV and 950 GeV for all possible values of the branching ratios into the three decay modes, as shown in Figs. 13 and 14 [ATLAS & CMS [21] (2014/2015)].

Scalar leptoquarks (LQ) appear in many BSM theories. They could provide an explanation for the observed similarities between the quark and lepton sectors of the SM. They are colour-triplet bosons with fractional electric charge and carry non-zero values for both baryon and lepton

[10] An up-type quark T with charge $+2/3$ can decay not only to a W boson and a b-quark, but also to a Higgs or Z boson and a top quark ($T \to Wb, Zt, Ht$). Similarly for a down-type quark B with charge $-1/3$.

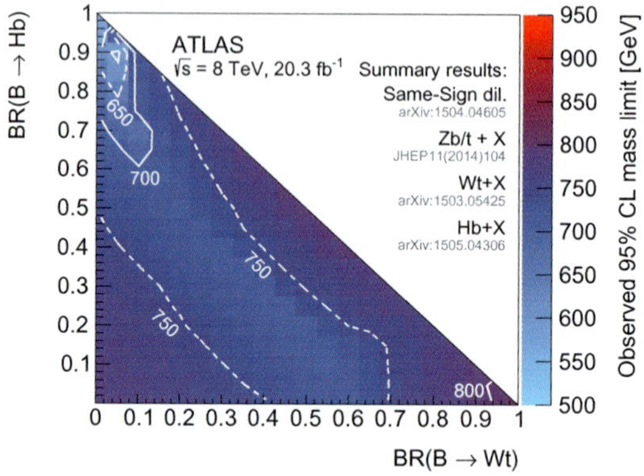

Fig. 13. The 95% CL observed limits on the mass of the B quark in the plane of BR($B \to Hb$) versus BR($B \to Wt$) [ATLAS & CMS [21] (2014/2015)].

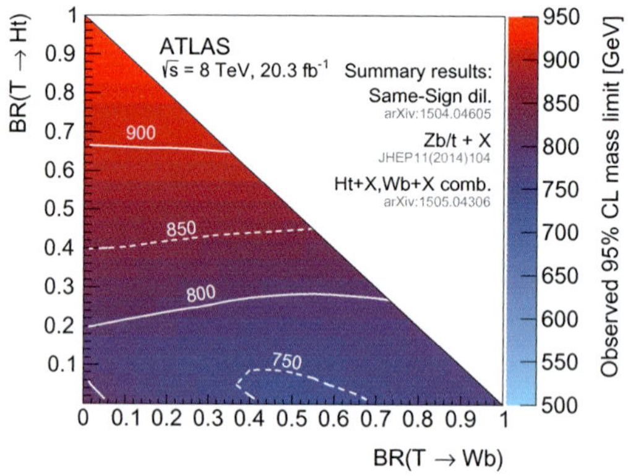

Fig. 14. The 95% CL observed limits on the mass of the T quark in the plane of BR($T \to Ht$) versus BR($T \to Wb$) [ATLAS & CMS [21] (2014/2015)].

numbers [Schrempp (1985)]. They are either scalar or vector bosons and are expected to decay directly to lepton-quark pairs. They can be pair-produced or singly-produced via the decay of a quark, $q \to (\mathrm{LQ})\ell$. Searches are performed using events with leptons and jets. 95% CL lower limits on the first generation LQ mass is $m_{\mathrm{LQ1}} > 1050$ GeV and on the second generation $m_{\mathrm{LQ2}} > 1070$ GeV. Third generation masses between 210 GeV $< m_{LQ3} <$ 640 GeV are excluded [ATLAS & CMS [22] (2014/2015/2016)].

Black holes are predicted to be produced at a collider in models with extra dimensions. Semi-classical black holes can be formed if the available energy is well above the higher-dimensional Planck scale [Dimopoulos & Landsberg (2001)]. They will then lose mass and angular momentum through Hawking radiation [Hawking (1975)]. Quantum black holes (QBH) lack a well-defined temperature or significant entropy. For this reason, this type of black holes produced at a mass scale just above the (higher-dimensional) Planck scale cannot decay thermally.[11] Instead, they decay into a limited number of particles in the final state. Production of QBHs are predicted in the Arkani-Hamed–Dimopoulus–Dvali (ADD) model [Arkani-Hamed *et al.* (1998)] and the Randall-Sundrum model [Randall & Sundrum (1999)]. They are characterised by two parameters, the number of extra dimensions n and the Planck scale m_D. QBHs have been searched for in di-jet and di-lepton final states. In the di-jet channel, the 95% CL lower limits on QBH mass range from >5 TeV to >6.3 TeV depending on n and m_D and the model, as shown in Fig. 15. In the di-lepton channel, the 95% CL lower limit on QBH mass is >3.65 TeV for the ADD model and >2.24 TeV for the RS model [ATLAS & CMS [20] (2014/2015); ATLAS & CMS [23] (2014/2015)].

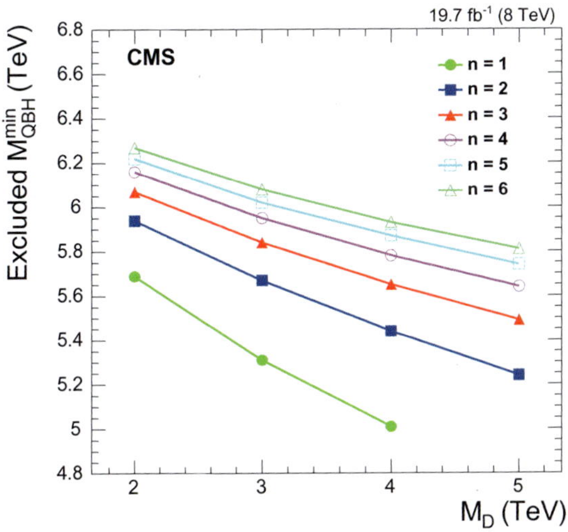

Fig. 15. 95% CL observed lower limits on QBH mass as a function of the Planck scale M_D and the number of extra dimensions n [ATLAS & CMS [20] (2014/2015)].

[11]This is why they are also called "non-thermal" black holes.

4.2. *Non-resonance searches*

Non-resonance searches can manifest themselves as excesses in event rate compared to SM prediction, or as shape differences in kinematic distributions. This type of search is sensitive to contact interaction models (CI) [Eichten *et al.* (1984)] and extra dimensions as predicted by the ADD model. In CI, the possible deviations compared to the SM are due to quark and lepton substructures. In ADD, they are due to virtual graviton-mediated processes. The CI model is characterised by the the energy scale parameter for the contact interaction Λ. The ADD model is parameterized by the scale characterizing the onset of quantum gravity[12] and the number of additional spatial dimensions n. The 95% CL limits obtained on the CI Λ scale and the ADD string scale using di-jet events are shown in Fig. 16, for different compositeness models and different parameterizations and number of extra-dimensions, respectively [ATLAS & CMS [23] (2014/2015)].

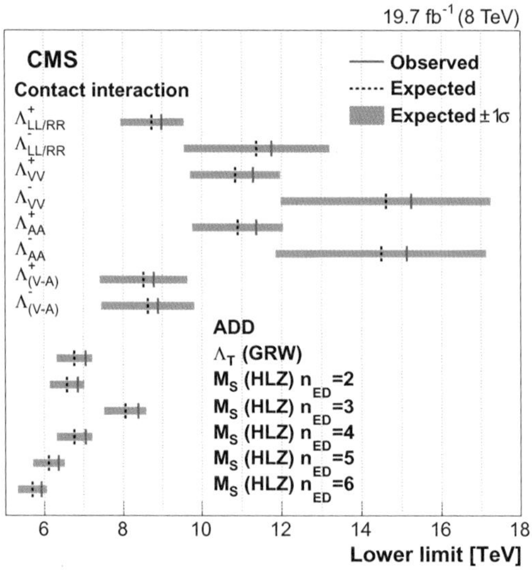

Fig. 16. 95% CL lower limits for the CI scales Λ for different compositeness models, for the ADD model scale with GRW parameterization Λ_T and for the ADD model scale with HLZ parameterization M_S [ATLAS & CMS [23] (2014/2015)].

[12]This scale is called Λ_T in the GRW parameterization and M_S in the HLZ parameterization.

At the LHC, black holes that decay through the emission of Hawking radiation are expected to emit a lot of SM particles of all types. Hence, a search for an excess of events with multiple high transverse momentum objects is performed. This signature is also characteristic of string ball states. Lower limits on black holes and string ball masses range from 4.6 to 6.2 TeV [ATLAS & CMS [24] (2013/2014/2015)].

In the SM, neutrinos could be their own anti-particles, called "Majorana" fermions. If this is the case, then the low mass scale of the neutrinos could be explained by a seesaw mechanism [Minkowski (1977)], which predicts right-handed (RH) heavy Majorana neutrino states N. This simplest extension of the SM including RH neutrinos is called "Minimal Type-I seesaw mechanism" (mTISM). The model is parameterized by the mass M_N of the RH neutrinos and the mixing parameters between the SM and Majorana neutrinos V_{iN} ($i = e, \mu, \tau$). The $W^+ \to N\ell^\pm \to (W^\mp\ell^\pm)\ell^\pm$ decay is searched for, as shown in Fig. 17. The characteristic signature is the presence of two leptons of same charge, two jets coming from the hadronic decay of the W and no MET. The 95% CL limit on the square of the mixing parameter $|V_{\mu N}|^2$ is shown in Fig. 18. For $\mathrm{m}_N = 90$ GeV, the limit is $|V_{\mu N}|^2 < 0.00470$ [ATLAS & CMS [25] (2014/2015)].

4.3. *Dark matter searches*

The direct search for dark matter (DM) at colliders makes use of initial-state radiation (ISR) activity, by requiring the presence of an object recoiling against the MET created by the weakly interacting massive particles escaping detection. This object can be an ISR jet, a γ, a $Z/W^\pm$ boson or a top quark. DM production in association with a pair of top quarks has also been studied. The strategy is to look for an excess of events in the tails of the MET distribution [ATLAS & CMS [26] (2014/2015)].

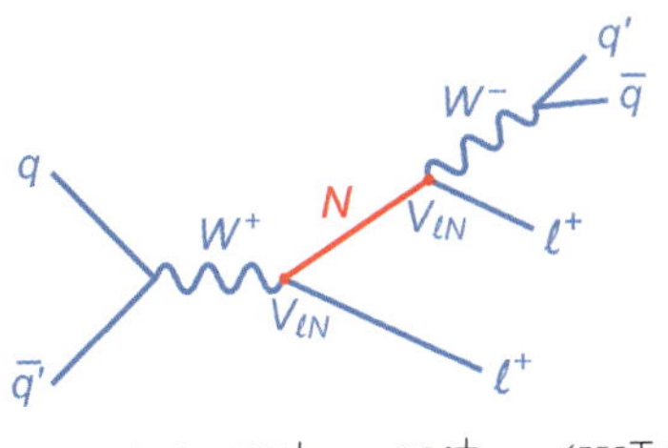

Fig. 17. Feynman diagram of the $W^+ \to N\ell^\pm \to (W^\mp\ell^\pm)\ell^\pm$ decay [ATLAS & CMS [25] (2014/2015)].

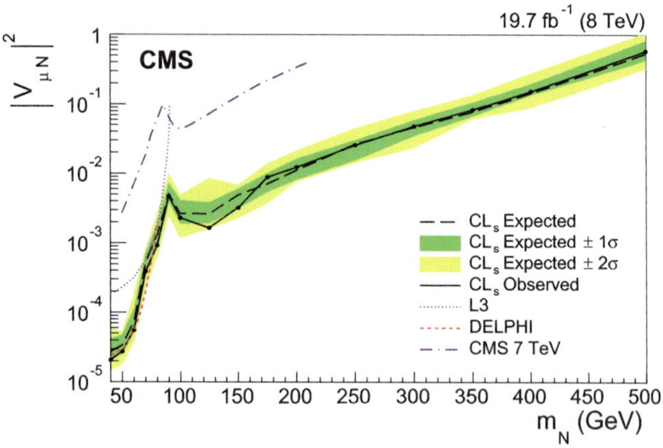

Fig. 18. Observed exclusion region in $|V_{\mu N}|^2$ as a function of the N mass. The region above the exclusion curve is ruled out [ATLAS & CMS [25] (2014/2015)].

The results are interpreted in two theoretical frameworks: an Effective Field Theory (EFT) [Goodmann *et al.* (2010)] and the Simplified Model approach (see Sec. 3). In EFT, the interaction of WIMPs with SM particles are described as a contact interaction. The free parameters are the suppression scale Λ and the DM mass. Depending on the nature of the mediator (scalar, pseudo-scalar, axial-vector or vector),[13] the DM couplings will be different. For the various possibilities, the limits on Λ are then converted into limits on DM-nucleon scattering cross sections, which allows for a comparison of the LHC results with the (in)direct DM detection results. In the Simplified Model approach, scans overs the mediator mass and the DM mass are performed for the different mediator natures.

In Fig. 19, the 90% CL limits on the spin-independent WIMP-nucleon scattering cross section are shown as a function of the DM mass M_χ for different scalar and vector operators describing the WIMP-SM particle interactions. The LHC results allow to access low-mass WIMP scenarios, which are nicely complementary to the results coming from (in)direct DM detection. In the spin-dependent scenario, the LHC limits are the best ones as

[13]The WIMP scattering on a nucleus can be spin-independent or spin-dependent. If the mediator is a scalar or a vector, this corresponds to the spin-independent case. If it is pseudo-scalar or axial-vector, this corresponds to the spin-dependent case.

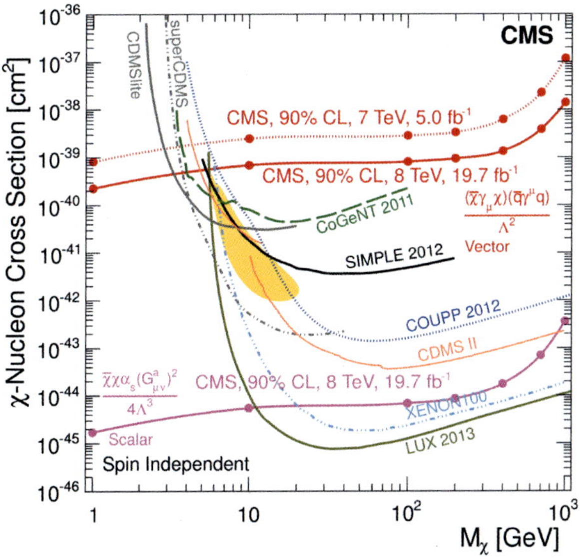

Fig. 19. 90% CL upper limits on WIMP-nucleon scattering cross section versus DM mass m_χ for the vector and scalar operators corresponding to the spin-independent case [ATLAS & CMS [26] (2014/2015)] and comparison with previously published results.

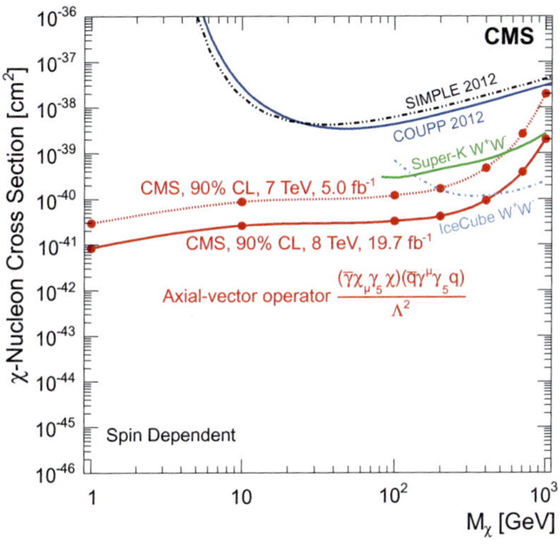

Fig. 20. 90% CL upper limits on WIMP-nucleon scattering cross section versus DM mass m_χ for the axial-vector operator corresponding to the spin-dependent case [ATLAS & CMS [26] (2014/2015)] and comparison with previously published results.

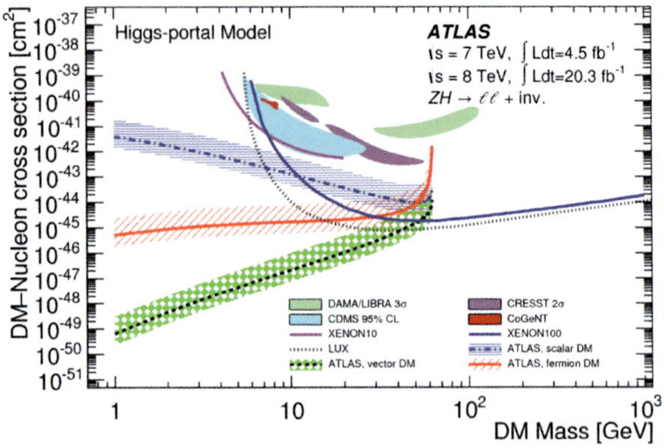

Fig. 21. 90% CL limits on WIMP-nucleon scattering cross section as a function of the DM mass for different natures of the DM candidate (fermion, scalar, vector) in the Higgs portal model and comparison with direct DM detection [ATLAS & CMS [3] (2014/2015)].

of today over almost the whole range of WIMP masses, as shown in Fig. 20 [ATLAS & CMS [26] (2014/2015)].

The $ZH \to \ell\ell + \chi^0\bar{\chi}^0$ analysis introduced in Sec. 2.1 allows to set limits on the Higgs portal model. In it, the Higgs boson acts as a mediator between a new physics sector and the SM particles. Results are shown in Fig. 21 for different natures of the DM candidate (fermion, scalar, vector) [ATLAS & CMS [3] (2014/2015)]. For vector WIMPs, the LHC limits are the best for WIMP masses below half the Higgs boson mass.

References

Altarelli, G., Mele, B., Ruiz-Altaba, M., *Z. Phys.* **C47**, 676 (1990).

Alwall, J., Schuster, P., Toro N., *Phys. Rev.* **D79**, 075020 (2009). Alves D. *et al.* *J. Phys. G: Nucl. Part. Phys.* **39**, 105005 (2012).

Arkani-Hamed, N., Dimopoulos, S., Dvali, G., *Phys. Lett.* **B429**, 263 (2002). Arkani-Hamed, N., Dimopoulos, S., Dvali, G., *Phys. Rev.* **D59**, 086004 (1999).

Arkani-Hamed, N. *et al.* *JHEP* **0207**, 034 (2002). Schmaltz, M., Tucker-Smith, D., *Ann. Rev. Nucl. Part. Sci.* **55**, 229 (2005).

Arkani-Hamed, N., Dimopoulos S., *JHEP* **06**, 073 (2005). Arkani-Hamed, N. *et al.* *Nucl. Phys.* **B709**, 3 (2005).

ATLAS & CMS Collaboration [1] (2015)., arXiv:1508.03372v1 [hep-ex] 2015, submitted to *JHEP*, *Phys. Lett.* **B749**, 337 (2015).

ATLAS & CMS Collaboration [2] (2014/2015). *PRL* **115**, 031801 (2015). *Phys. Rev.* **D90**, 072010 (2014).

ATLAS & CMS Collaboration [3] (2014/2015). *JHEP* **01**, 172 (2016). *Eur. Phys. J.* **C75**, 337 (2015). *Phys. Rev. Lett.* **112**, 201802 (2014). *JHEP* **11**, 206 (2015). *Eur. Phys. J.* **C74**, 2980 (2014).

ATLAS & CMS Collaboration [5] (2013/2015). *Eur. Phys. J.* **C73**, 2465 (2013). *JHEP* **12**, 178 (2015).

ATLAS Collaboration [6] (2015). *JHEP* **03**, 088 (2015).

ATLAS Collaboration [8] (2015). *Phys. Rev. Lett.* **114**, 231801 (2015).

ATLAS & CMS Collaboration [9] (2015/2016). *JHEP* **01**, 032 (2016). arXiv:1507.05930v1 [hep-ex], submitted to *EPJC*, *JHEP* **10**, 144 (2015).

ATLAS & CMS Collaboration [10] (2014/2015). *PRL* **113**, 171801 (2014). *Phys. Lett.* **B750**, 494 (2015).

ATLAS & CMS Collaboration [11] (2015). *Phys. Rev.* **D92**, 092004 (2015). Auxiliary figures in: https://atlas.web.cern.ch/Atlas/GROUPS/PHYSICS/PAPERS/HIGG-2013-33/, *Phys. Lett.* **B749**, 560 (2015).

ATLAS Collaboration [12] (2015). *Phys. Rev.* **D92**, 052002 (2015).

ATLAS & CMS Collaboration [14] (2014/2015). *JHEP* **11**, 056 (2014). *Phys. Lett.* **B744**, 163 (2015). *Phys. Lett.* **B748**, 221 (2015). arXiv:1510.01181 [hep-ex], submitted to *Phys. Lett. B*.

ATLAS & CMS Collaboration [15] (2015). *JHEP* **10**, 134 (2015). *EPJ* **C73**, 325 (2015).

ATLAS & CMS Collaboration [16] (2013/2014/2015). *JHEP* **10**, 054 (2015). *Eur. Phys. J.* **C75**, 510 (2015). *JHEP* **06**, 116 (2015). *Eur. Phys. J.* **C73**, 2677 (2013). *JHEP* **06**, 055 (2014). *Phys. Lett.* **B733**, 328 (2014). *JHEP* **01**, 163 (2014). *Phys. Lett.* **B745**, 5 (2015).

ATLAS & CMS Collaboration [17] (2014/2015). *Phys. Rev.* **D93**, 052002 (2016). *JHEP* **11**, 189 (2015). *Eur. Phys. J.* **C74**, 3036 (2014). *Phys. Rev.* **D90**, 092007 (2014).

ATLAS & CMS Collaboration [18] (2013/2014/2015). *Phys. Rev.* **D92**, 072004 (2015). *JHEP* **09**, 176 (2014). *Eur. Phys. J.* **C75**, 407 (2015). *JHEP* **01**, 068 (2015). *Phys. Rev.* **D88**, 112003 (2013). *Eur. Phys. J.* **C75**, 362 (2015). *Physics Letters* **B743**, 15 (2015). *JHEP* **11**, 088 (2014). *Phys. Rev.* **D92**, 012010 (2015). *Phys. Rev. Lett.* **114**, 061801 (2015). *J. High Energy Phys.* **01**, 096 (2015). *Eur. Phys. J.* **C75**, 151 (2015). *Phys. Rev.* **D91**, 012007 (2015). *Phys. Rev.* **D91**, 052012. (2015) .

ATLAS & CMS Collaboration [19] (2014/2015/2016). *JHEP* **09**, 037 (2014). *Eur. Phys. J.* **C75**, 69 (2015). *Phys. Rev.* **D90**, 052005 (2014). *Eur. Phys. J.* **C75**, 209 (2015). *Phys. Lett.* **B738**, 428 (2014). *JHEP* **12**, 55 (2015). *JHEP* **07**, 157 (2015). arXiv:1406.4456 [hep-ex]. submitted to *Phys. Lett. B.*, *Eur. Phys. J.* **C75**, 263 (2015). *JHEP* **08**, 148 (2015). *Phys. Rev.* **D92**, 032004 (2015). *JHEP* **04**, 025 (2015). *JHEP* **08**, 173 (2014). *Phys. Lett.* **B740**, 83 (2015). *Phys. Rev.* **D91**, 092005 (2015). *JHEP* **02**, 145 (2016). *Phys. Rev.* **D93**, 012001 (2016). arXiv:1508.04308 [hep-ex], submitted to PRL., *Phys. Lett.* **B748**, 255 (2015). *JHEP* **04**, 025 (2015). *JHEP* **08**, 174 (2014). *JHEP* **08**, 173 (2014).

ATLAS & CMS Collaboration [20] (2014/2015). *Phys. Rev.* **D91**, 052007 (2015). *Phys. Lett.* **B728C**, 562 (2014). *JHEP* **08**, 138 (2015). arXiv:1510.02664 [hep-ex], submitted to JHEP., *Phys. Rev.* **D91**, 052009 (2015). *Phys. Lett.* **B738**, 274 (2014). *JHEP* **08**, 173 (2014).

ATLAS & CMS Collaboration [21] (2014/2015). *JHEP* **08**, 105 (2015). *Phys. Rev.* **D91**, 112011 (2015). *JHEP* **11**, 104 (2014). *JHEP* **10**, 150 (2015). *Phys. Rev.* **D93**, 112009 (2016). *Phys. Rev.* **D93**, 012003 (2016). *JHEP* **06**, 080 (2015). *PLB* **729**, 149 (2014). *PRL* **112**, 171801 (2014).

ATLAS & CMS Collaboration[22] (2014/2015/2016). *Eur. Phys. J.* **C76** (1), 1 (2016). *JHEP* **07**, 042 (2015). *Phys. Rev.* **D93**, 032005 (2016). *Phys. Rev.* **D93**, 032004 (2016). *Phys. Lett.* **B739**, 229 (2014).

ATLAS & CMS Collaboration [23] (2014/2015). *Eur. Phys. J.* **C74**, 3134 (2014). *Phys. Rev. Lett.* **114**, 221802 (2015). *JHEP* **04**, 025 (2015). *Phys. Lett.* **B746**, 79 (2015).

ATLAS & CMS Collaboration [24] (2013/2014/2015). *JHEP* **08**, 103 (2014). *JHEP* **07**, 032 (2015). *Phys. Rev.* **D88**, 072001 (2013). *JHEP* **07**, 178 (2013).

ATLAS & CMS Collaboration [25] (2014/2015). *JHEP* **07**, 162 (2015). *PLB*, **748**, 144 (2015). *Eur. Phys. J.* **C74**, 3149 (2014).

ATLAS & CMS Collaboration [26] (2014/2015). *Eur. Phys. J.* **C75**, 299 (2015). *Phys. Rev.* **D90**, 012004 (2015). *Phys. Rev. Lett.* **112**, 041802 (2014). *JHEP* **11**, 118 (2014). *Phys. Rev.* **D91**, 012008 (2015). *Phys. Rev. Lett.* **115**, 131801 (2015). *Phys. Rev.* **D93**, 072007 (2016). *Eur. Phys. J.* **C75**, 79 (2015). *Eur. Phys. J.* **C75**, 235 (2015). *Phys. Lett.* **B755**, 102 (2016). *Phys. Rev. Lett.* **114**, 101801 (2015). *Phys. Rev.* **D91**, 092005 (2015). *JHEP* **06**, 121 (2015).

Barbieri, R., Giudice, G. F., *Nucl. Phys.* **B306**, 63 (1988). de Carlos, B., Casas, J. A., *Phys. Lett.* **B309**, 320 (1993).

Barger, V., Ma, E., *Phys. Rev.* **D51**, 1332 (1995).

Bechtle P. *et al.* arXiv:1508.05951v1 [hep-ph] (2015), to be submitted to *EPJC*.

Baur, U., Spira, M., Zerwas, M., *Phys. Rev.* **D42**, 815 (1990). Cakir, O., Mehdiyev, R., *Phys. Rev.* **D60**, 034004 (1999). Bhattacharya, S. *et al. Phys. Rev.* **D80**, 015014 (2009).

Chamseddine, Ali H., Arnowitt, R., Pran N., *Phys. Rev. Lett.* **49**, 970 (1982). Barbieri, R., Ferrara, S., Savoy Carlos, A, *Phys. Lett.* **B119**, 343 (1982). Ibanez, L. E, *Phys. Lett.* **B118**, 73 (1982). Hall, L. J., Lykken, J. D., Weinberg, S., *Phys. Rev.* **D27**, 2359 (1983). Ohta, N., *Prog. Theor. Phys.* **70**, 542 (1983). Kane, G. L. *et al. Phys. Rev.* **D49**, 6173 (1994).

CMS Collaboration [4] (2015). *Phys. Lett.* **B753**, 363 (2016).

CMS Collaboration [7] (2015). *JHEP* **11**, 018 (2015).

CMS Collaboration [13] (2015). *Phys. Rev.* **D90**, 112013 (2014).

Dimopoulos, D., Landsberg, G., *Phys. Rev. Lett.* **87**, 161602 (2001). Giddings, S. B., Thomas, S, *Phys. Rev.* **D65**, 056010 (2002).

Djouadi, A., Kneur, J. -L., Moultaka, G., *Comput. Phys. Commun.* **176**, 426 (2007). Berger C. F. *et al. JHEP* **0902**, 023 (2009). Cahill-Rowley, M. W. *et al. Eur. Phys. J.* **C72**, 2156 (2012).

Eichten, E. *et al. Rev. Mod. Phys.* **56**, 579 (1984).

Fayet, P., *Phys. Lett.* **B64**, 159 (1976). Fayet, P., *Phys. Lett.* **B69**, 489 (1977). Farrar, G., Fayet, P., *Phys. Lett.* **B76**, 575 (1978). Fayet, P, *Phys. Lett.* **B84**, 416 (1979). Dimopoulos, S., Georgi, H., *Nucl. Phys.* **B193**, 150 (1981).

Georgi, H., Machacek, M., *Nucl. Phys.* **B262**, 463 (1985). Cheung, K., Ghosh, D. K., *JHEP* **0211**, 048 (2002).

Goodman, J. *et al. Phys. Rev.* **D82**, 116010 (2010).

Hawking, S. W., *Comm. in Math. Phys.* **43** (3), 199 (1975).

Hill, A., van der Bij, J., *Phys. Rev.* **D36**, 3463 (1987). Veltman, M., Yndurain, F., *Nucl. Phys.* **B325**, 1 (1989). Binoth, T., van der Bij, J., *Z. Phys.* **C75**, 17 (1997). Schabinger, R., Wells, J. D., *Phys. Rev.* **D72**, 093007 (2005). Patt, B., Wilczek, F, arXiv:hep-ph/0605188 (2006). Pruna, G. M., Robens, T., *Phys. Rev.* **D88**, 115012 (2013). Lopez-Val, D., Robens, T., *Phys. Rev.* **D90**, 114018 (2014).

Kaplan, D. B., Georgi, H., Dimopoulos, S., *Phys. Lett.* **B136**, 187 (1984). Agashe, K., Contino, R., Pomarol, A, *Nucl. Phys.* **B719**, 165 (2005).

Lee, T., *Phys. Rev.* **D8**, 1226 (1973).

LEP2 Collaboration (2001). LEPSUSYWG//01-03.1.

Martin, P. S., arXiv:hep-ph/9709356 (1997).

Minkowski, P., *Phys. Lett.* **B67**, 421 (1997). Mohapatra, R. N., Senjanovic, G., *Phys. Rev. Lett.* **44**, 912 (1980). Mohapatra, R. N., Senjanovic, G., *Phys. Rev.* **D23**, 165 (1981). Davidson, A., Wali, K. C., *Phys. Rev. Lett.* **60**, 1813 (1988). Schechter, J., Valle, J. W. F., *Phys. Rev.* **D22**, 2227 (1980).

Randall, L., Sundrum R., *Phys. Rev. Lett.* **83**, 3370 (1999).

Schrempp, B., Schrempp, F., *Phys. Lett.* **B153**, 101 (1985).

Chapter 9

Results from LHCb

Katharina Müller

Physik Institut University of Zurich
Winterthurerstrasse 190, CH-8057 Zurich, Switzerland
LHCb Collaboration
European Organization for Nuclear Research (CERN)
CH-1211 Geneva 23, Switzerland
kmueller@physik.uzh.ch

In this contribution we describe a wide range of selected physics results from the LHCb experiment, demonstrating its unique role, both as a heavy flavour experiment and as a general purpose detector in the forward region.

Contents

1. Introduction

The LHCb experiment [LHCb (2008)] is a forward spectrometer fully instrumented in the pseudorapidity range $2 < \eta < 5$. It is designed for precision tests of the Standard Model (SM) and optimised for indirect searches for physics beyond the SM (BSM) through precision measurements of CP violating phases and rare heavy-quark decays. These measurements are complementary to direct searches by the CMS and ATLAS experiments. Processes that are strongly suppressed in the SM, such as flavour-changing neutral current (FCNC) $b \to s$ transitions, are particularly interesting since BSM contributions may enter via loop processes at the same level as SM contributions. Thanks to its unique kinematic coverage, not accessible by other experiments, LHCb also has an ambitious program of SM measurements. In addition, LHCb also participated in data taking of proton–ion collisions.

The LHCb detector has an excellent vertex and momentum resolution needed to separate primary and secondary vertices, provides good invariant mass resolution, and has a highly efficient and flexible trigger, able to trigger on particles with low transverse momentum, p_{T}. Two RICH detectors allow particle identification over a wide momentum range.

A few recent LHCb results which show the broad physics potential of the experiment will be presented in the following. The analyses are based on data collected between 2010 and 2013, corresponding to an integrated luminosity of 1 fb^{-1} at a centre-of-mass energy of 7 TeV and 2 fb^{-1} at 8 TeV. Furthermore results are reported of a proton-ion run corresponding to 1.6 nb^{-1} of proton-lead collisions at a centre-of-mass energy per proton-nucleon pair of $\sqrt{s_{NN}} = 5$ TeV.

2. Selected Results from LHCb

2.1. *Angular analysis of $B \to K^{(*)}\mu^+\mu^-$ decays*

The rare decay $B^0 \to K^{*0}\mu^+\mu^-$ proceeds via a b- to s-quark flavour-changing neutral current transition that is very sensitive to BSM. This process is forbidden in the SM at tree level and suppressed by the GIM mechanism, therefore new, heavy particles can enter in competing processes and can significantly change the branching fraction and the angular distribution of the final state particles. Angular observables are particularly interesting, since they allow to probe several BSM models with a wide range of effective couplings with quarks and leptons. The kinematics can be described by the square of the invariant mass of the di-muon pair, q^2, and three helicity angles. Several observables can be constructed with small uncertainties due to the form factors [Descotes-Genon *et al.* (2013)]. The analysis performed on the 1 fb^{-1} data sample showed a local 3.7σ discrepancy with respect to SM predictions in one observable, P_5', while all other measurements were found to be in agreement with the SM, with a mild tension in the forward-backward asymmetry of the di-muon system A_{FB} [LHCb [3] (2013)].

In the analysis of the full 3 fb^{-1} dataset the full angular distribution is used and a complete set of CP-averaged observables F_L (the longitudinal polarisation fraction of the K^{*0}) , A_{FB} and S_{3-9} is extracted simultaneously [LHCb [16] (2015)]. This allows the correlations between the measured quantities to be computed. The observable S_5, which is strongly related to the previously observed P_5', is in poor agreement with the SM prediction in the region $4.0 < q^2 < 8.0$ GeV^2/c^4 as shown in Fig. 1. In this q^2 range the P_5' measurements is only compatible with the SM prediction at a level of 3.7σ. Again a mild tension of A_{FB} with respect to SM predictions is observed. Several explanations for the deviation and the consistency of all the measurements of $b \to s$ transitions have been brought up [Descotes-Genon *et al.* (2015)] such as the existence of an additional boson, Z', with a mass above 10 TeV [Gauld (2013)]. It has also been pointed out that the discrepancy might be explained by QCD effects ignored in the SM predictions [Lyon & Zwicky (2014)].

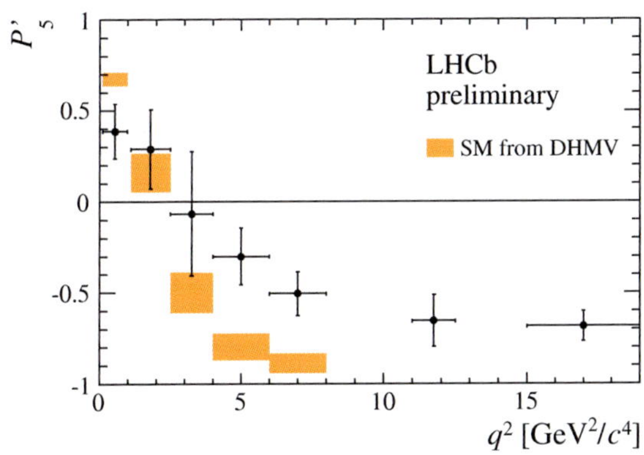

Fig. 1. The observable P_5' in bins of the di-muon mass squared q^2 (Fig. 6 from [LHCb [16] (2015)]) compared to the SM predictions from [Descotes-Genon *et al.* (2012)].

2.2. *Lepton universality*

Lepton universality (LU) is an accidental symmetry of the SM, which implies the couplings between the gauge bosons and the three families of leptons to be equal. No definitive observation of a deviation has yet been made. However, many BSM models contain additional interactions that violate this principle.

LU has mainly been tested measuring the coupling of the SM gauge bosons to the different leptons, however FCNC transitions where new particles can have a visible contribution are ideal to test LU. One of such measurements is the ratio of the branching fractions of the $B^+ \to K^+\mu^+\mu^-$ and $B^+ \to K^+e^+e^-$ decays, where theory uncertainties in the branching fractions largely cancel. Using electrons is difficult since there is a lower trigger efficiency and a poorer mass resolution due to bremsstrahlung. Figure 2 shows the K^+e^+e mass distribution for $B^+ \to K^+e^+e^-$ candidates, which is safely far from the radiative tail of the $J/\psi \to e^+e$ decay. The value of the ratio of branching fractions for the dilepton invariant mass squared range $1 < q^2 < 6\ \mathrm{GeV}^2/c^4$ is measured to be $0.745^{+0.074}_{-0.090}(\mathrm{stat}) \pm 0.036(\mathrm{syst})$ [LHCb [7] (2014)]. This value is the most precise measurement of this quantity to date and is compatible with the SM prediction of unity within 2.6 standard deviations. The result may be interpreted as a possible indication of a new vector particle that would couple more strongly to muons and interfere destructively with the SM vector current.

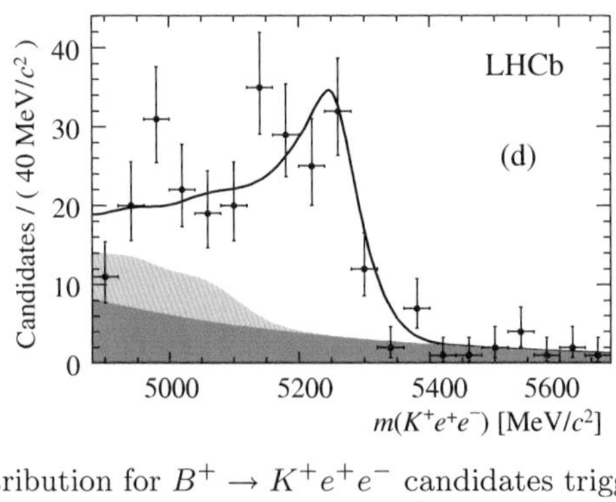

Fig. 2. Mass distribution for $B^+ \to K^+e^+e^-$ candidates triggered by one of the electrons at the hardware trigger (Fig. 2 from [LHCb [7] (2014)]). The total fit model is shown in black, the combinatorial background component by the dark shaded region and the background from partially reconstructed b-hadron decays by the light shaded region.

LU is also tested in the branching fraction ratio $\mathcal{R}(D^*) = \mathcal{B}(\overline{B}^0 \to D^{*+}\tau^-\overline{\nu}_\tau)/\mathcal{B}(\overline{B}^0 \to D^{*+}\mu^-\overline{\nu}_\mu)$. The tau lepton is identified in the decay mode $\tau^- \to \mu^-\overline{\nu}_\mu\nu_\tau$. The semitauonic decay is sensitive to contributions from BSM particles that preferentially couple to the third generation of fermions, in particular additional charged Higgs bosons.

A multidimensional fit to kinematic distributions of the candidate $\overline{B}^0$ is used to statistically disentangle the signal, normalisation component and the residual backgrounds. The ratio is measured to be $\mathcal{R}(D^*) = 0.336 \pm 0.027(\text{stat}) \pm 0.030(\text{syst})$ [LHCb [17] (2015)] and is the first measurement of this quantity at a hadron collider. It is 2.1 standard deviations larger than the value expected from LU in the SM. The measurement is in good agreement with results from BaBar [Lees *et al.* (2012)] and Belle [Bozek *et al.* (2010)]; together they are at the level of 4 sigmas larger than the SM expectation.

2.3. $B_s^0(B^0) \to \mu^+\mu^-$ *in combination with CMS*

The decays $B_s^0 \to \mu^+\mu^-$ and $B^0 \to \mu^+\mu^-$ are very suppressed in the SM, in addition to be FCNC processes they also have a helicity suppression, since one of the leptons would be emitted with the wrong helicity. These decays are particularly sensitive to BSM models with an extended Higgs sector, such as supersymmetry or two Higgs doublet models as their SM branching fractions are very small ($\mathcal{O}(10^{-9})$) and therefore BSM contributions might be of the same order of magnitude as contributions from the SM. For example, in Minimal Supersymmetric Standard Models, the branching fraction increases with the sixth power of the ratio of the two Higgs vacuum expectation values present in this model.

The first evidence for the decay $B_s^0 \to \mu^+\mu$ was presented by the LHCb collaboration in 2012 [LHCb [1] (2012)]. Both, CMS and LHCb later published results from all data collected at centre-of-mass energies of 7 and 8 TeV. The measurements had comparable precision and were in good agreement, although none of the individual results had sufficient precision to constitute the first definitive observation of the $B_s^0 \to \mu^+\mu^-$ decay. The combined measurement resulted in the first observation of the decay $B_s^0 \to \mu^+\mu^-$ and the first evidence for $B^0 \to \mu^+\mu^-$ [CMS & LHCb (2014)] with measured branching fractions of $\mathcal{B}(B_s^0 \to \mu^+\mu^-) = (2.8^{+0.7}_{-0.6}) \times 10^{-9}$ and $\mathcal{B}(B^0 \to \mu^+\mu^-) = (3.9^{+1.6}_{-1.4}) \times 10^{-10}$ in agreement with the SM prediction at the 1.2σ and 2.2σ level, respectively [Bobeth (2013)]. This measurement sets strong constraints to supersymmetry models in particular and to

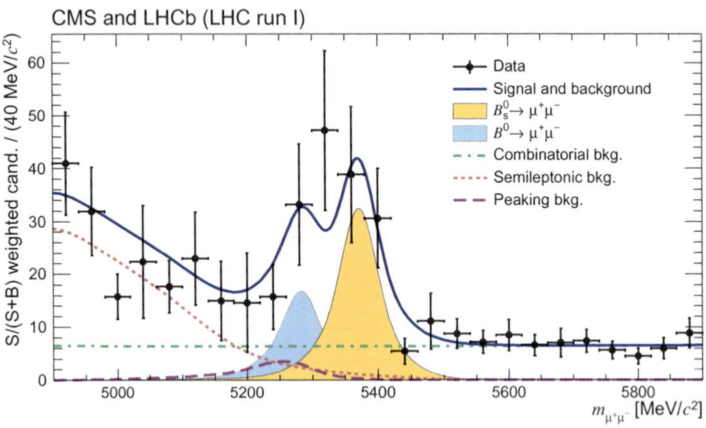

Fig. 3. Distribution of the di-muon invariant mass for $B_s^0(B^0) \to \mu^+\mu^-$ candidates (Fig. 2 from [CMS & LHCb (2014)]).

models with large effective scales in general. The di-muon invariant mass distribution of the candidates from CMS and LHCb is shown in Fig. 3.

2.4. *Photon polarisation in $b \to s\gamma$ transitions*

The radiative flavour-changing neutral current decay, $B^+ \to K^+\pi^-\pi^+\gamma$, is used to study the photon polarisation. The SM predicts that the photon emitted from the electroweak penguin loop in $b \to s\gamma$ transitions is predominantly left-handed, since the recoiling s quark that couples to a W boson is left-handed. Several BSM models, compatible with all current measurements, predict that the photon acquires a significant right-handed component, in particular, due to the exchange of heavy fermions [Atwood *et al.* (1997)].

A measurement of the photon polarisation can be performed by studying the distribution of the angle between the photon and the plane formed by the three hadrons. An up-down asymmetry, A_{ud}, of the photon direction with respect to this plane can be defined which is proportional to the photon polarisation.

The LHCb experiment has published a first measurement of A_{ud} using the full dataset with 13876 ± 153 signal candidates [LHCb [8] (2014)]. The measurement is done separately in four regions of the $K\pi\pi$ invariant mass, designed to separate different resonant contributions as indicated in Fig. 4. Combining the four regions, evidence for non-zero photon polarisation is

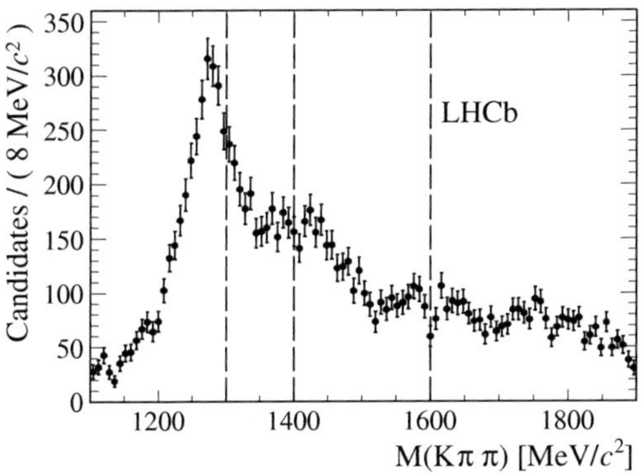

Fig. 4. Background subtracted $M_{K\pi\pi}$ distribution of the $B \to K\pi\pi\gamma$ candidates (Fig. 2 from [LHCb [8] (2014)]).

observed at the level of 5.2σ. This is the first observation of photon polarisation in radiative b-hadron decays. Though clear evidence for polarisation is seen, the structure of the $K\pi\pi$ mass spectrum is rather complex and more work on the experimental and theory side is needed to convert this observation into a measurement of the photon polarisation.

2.5. *Lepton flavour and baryon number violation in τ decays*

Lepton flavour and baryon number are conserved in the SM. Decays such as $\tau^- \to \mu^-\mu^+\mu^-$ occur in the SM via neutrino mixing $\nu_\tau \to \nu_\mu$ and are suppressed by a factor of $\Delta m_\nu^2/M_W^2$ where Δm_ν is the mass difference of the two muons and M_W the mass of the W boson. Hence, the branching fractions for such decays are smaller than $\mathcal{O}(10^{-40})$ in the SM, whereas many BSM models predict branching fractions of up to 10^{-8}.

LHCb searched for the decay $\tau^- \to \mu^-\mu^+\mu^-$ [LHCb [9] (2014)] using the full 3 fb^{-1} data sample. No evidence for any signal is found. The limit at 90% confidence level obtained by LHCb is $\mathcal{B}(\tau^- \to \mu^-\mu^+\mu^-) < 4.6 \times 10^{-8}$. In combination with results from the B-factories, this limit improves the constraints placed on the parameters of a broad class of BSM models [Amhis *et al.* (2014)].

2.6. *Measurement of the γ angle*

The angle γ of the unitarity triangle is the weak phase between the elements of the CKM matrix V_{cb} and V_{ub}. It has the weakest experimental constrains and therefore its measurement is an important test of the CKM consistency. The best sensitivity is achieved through a combination of measurements that determine γ along with several other hadronic parameters. The time integrated analyses of $B \to DK$, $B \to D\pi$ and $B \to DK^*$ decays are used, with the D decaying into hh, $K\pi\pi\pi$ or $K_s hh$ final states, here h can be a charged pion or a kaon. In these decays γ arises from the interference of $b \to u$ and $b \to c$ transitions. The effect of $D^0\overline{D}^0$ mixing is taken into account. In addition, the time dependent results from the tree level decay $B_s \to D^{\mp}K^{\pm}$ are included. Here, the sensitivity comes through CP violation in the interference of mixing and decay amplitudes. Combining all the results the value of γ is measured to be $(73^{+9}_{-10})^\circ$ at 68% CL. [LHCb [10] (2014)]. Most of the analyses included are based on 1 fb^{-1} of data and are being updated to 3 fb^{-1}. LHCb expects to achieve a precision of 7°, with the analyses of the 3 fb^{-1}. The precision of the γ measurement at LHCb is expected to improve significantly with the data of the next LHC run.

2.7. *Quantum numbers of* $X(3872)$ *and* $Z(4430)^-$

The $X(3872)$ state was discovered in $B^{+,0} \to X(3872)K^{+,0}$, $X(3872) \to \pi^+\pi^- J/\psi$ decays by the Belle experiment [Choi *et al.* (2003)] and subsequently confirmed by other experiments. However, the nature of this state remains unclear. The $X(3872)$ state is narrow, has a mass very close to the $D^0\overline{D}^0$ threshold and decays to $\rho^0 J/\psi$ and $\omega J/\psi$ final states with comparable branching fractions [Olive *et al.* (2014)], thus violating isospin symmetry. Hence, the $X(3872)$ particle may not be a simple $c\bar{c}$ state and it has been speculated to be a $D\overline{D}^*$ molecule, a tetraquark state or a charmonium-molecule mixture.

LHCb performed an analysis of the angular correlations in $B \to X(3872)K$, $X(3872) \to \pi^+\pi^- J/\psi$, $J/\psi \to \mu^+\mu$ decays without any assumption about the orbital angular momentum in the $X(3872)$ decay. These correlations carry significant information about the $X(3872)$ quantum numbers. The results confirm that the eigenvalues of total angular momentum, parity and charge-conjugation of the $X(3872)$ state are 1^{++} as displayed in Fig. 5 [LHCb [18] (2015)]. The quantum numbers are consistent with those predicted by the molecular or tetraquark models and with the $\chi_{c1}(2^3P_1)$ charmonium state. Other charmonium states are excluded. No significant D-wave fraction is found, with an upper limit of 4% at 95% CL. This S-wave dominance is expected in charmonium or tetraquark models, in which the $X(3872)$ state has a compact size.

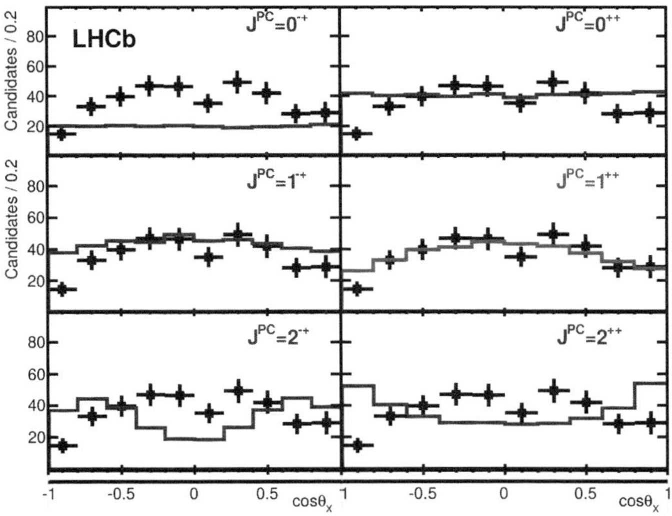

Fig. 5. Distribution of the helicity angle in X for the $X(3872)$ candidates for data (points with error bars) compared to the expected distributions for various $X(3872)$ J^{PC} assignments (solid histograms) (Fig. 4 from [LHCb [18] (2015)]).

First evidence for the $Z(4430)^-$ was presented by the Belle collaboration in 2008. The LHCb measurement analysed resonant structures in $B^0 \to \psi'\pi^+K^-$ decays [LHCb [11] (2014)]. It confirms the existence of a signal with a significance of 13.9 standard deviations and establishes its resonant nature via a phase-shift analysis. The quantum numbers of the state are determined to be $J^P = 1^+$. The distribution of $m^2_{\psi'\pi^-}$ is shown in Fig. 6 together with a fit with one 1^+ resonance. The minimal quark content of this state is $c\bar{c}d\bar{u}$, making this the first unambiguous observation of an exotic particle that is neither a baryon nor a meson.

2.8. *Observation of $J/\psi p$ resonances consistent with pentaquark states*

In the quark model baryons consist of three quarks and mesons are formed of quark–antiquark pairs but the model also allows for the existence of other quark composite states, such as pentaquarks composed of four quarks and an antiquark. Past claims of observations of pentaquark states have been shown to be non-conclusive [Hicks (2012)].

LHCb searched for pentaquark states in $\Lambda_0 \to J/\psi K p$ decays [LHCb [19] (2015)]. This decay is expected to be dominated by $\Lambda^* \to Kp$

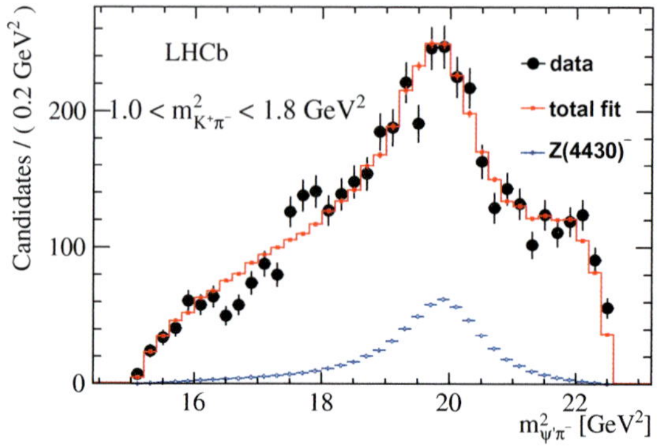

Fig. 6. Distribution of $m^2_{\psi'\pi^-}$ in $B^0 \to \psi'\pi^+K^-$ decays, compared with the fit with one $Z(4430)^-$ resonance with $J^P = 1^+$ (from [LHCb [11] (2014)]).

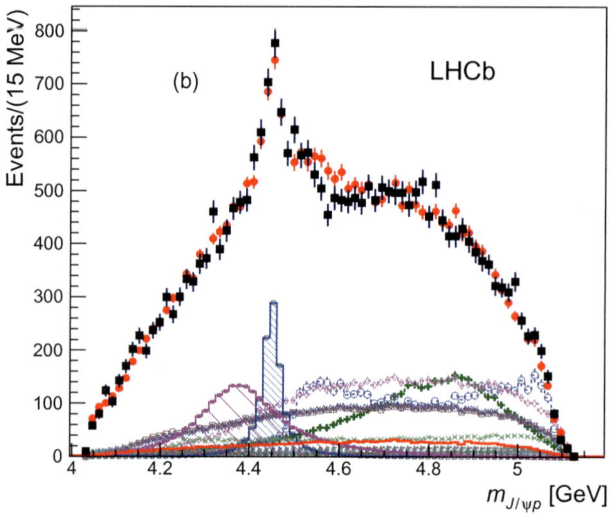

Fig. 7. Fit to $m_{J/\psi p}$ in $\Lambda_0 \to J/\psi K p$ decays for the model with two P_c^+ states. Data are shown as black squares and the fit result as red circles, blue open and purple solid squares show the two P_c^+ states. The Λ^* components is also shown (Fig. 2 from [LHCb [18] (2015)]).

resonances but the invariant mass distribution revealed additional exotic contributions (Fig. 7). Such resonances must have a minimal quark content of *ccuud*, and are thus attributed to charmonium-pentaquark states.

In order to ensure that the structures in the mass distribution originate from resonances and are not due to reflections generated by the Λ^* states, a full amplitude analysis is performed allowing for interference effects between both decay sequences. The amplitude analysis of the three-body final-state reproduces the two-body mass and angular distributions. To obtain a satisfactory fit of the structures seen in the $J/\psi p$ mass spectrum (Fig. 7), it is necessary to include two resonant states. These have been named $P_c(4450)^+$ and $P_c(4380)^+$, the former being clearly visible as a peak in the data with a significance of 12 standard deviations. The lower mass peak with a significance of 9 is required to describe the data fully. $P_c(4380)^+$ has a mass of $4380 \pm 8 \pm 29$ MeV and a width of $205 \pm 18 \pm 86$ MeV, while $P_c(4450)^+$ is narrower, with a mass of $4449.8 \pm 1.7 \pm 2.5$ MeV and a width of $39 \pm 5 \pm 19$ MeV. The parities of the two states are opposite; one state has spin 3/2 and the other 5/2.

2.9. *Measurements with electroweak bosons*

Measurements of W and Z boson production in the forward region constitute a test of QCD at LHC energies and provide valuable input to the knowledge of the parton density functions (PDF) of the proton in a kinematic region uniquely accessible by LHCb [LHCb [20] (2015); LHCb [21] (2015); LHCb [15] (2014); LHCb [4] (2013)]. Measurements at LHCb are sensitive to Bjorken-x values as low as 8×10^{-6} where x is the fractional momentum carried by the struck quark. The study of the W charge asymmetry $A_l = (\sigma(W^+) - \sigma(W^-))/(\sigma(W^+) + \sigma(W^-))$ or the ratio $\sigma(W^+)/\sigma(W^-)$ allows a precise determination of the up/down valence quark ratio and sea quark densities in the proton, as experimental and theoretical uncertainties partially cancel. Figure 8 shows ratios of W and Z cross sections in comparison with predictions at next-to-next-to leading order in QCD using different PDF sets. The experimental uncertainties are better than the uncertainty in the theory calculations.

Recently LHCb has also reported measurements of the associated W boson production with b- or c-jets [LHCb [22] (2015)].

The measurement of the forward-backward charge asymmetry for the process $q\bar{q} \rightarrow Z/\gamma^* \rightarrow \mu^+\mu^-$ as a function of the invariant mass of the di-muon system is used to extract the effective electroweak mixing angle θ_W^{eff}. In this measurement LHCb can take advantage of its forward geometry since the direction of the Z boson is likely to be the direction of the quark due to the asymmetric momentum distribution of the quark and antiquark. The measured forward-backward asymmetry as a function of the di-muon invariant mass is shown in Fig. 9 together with

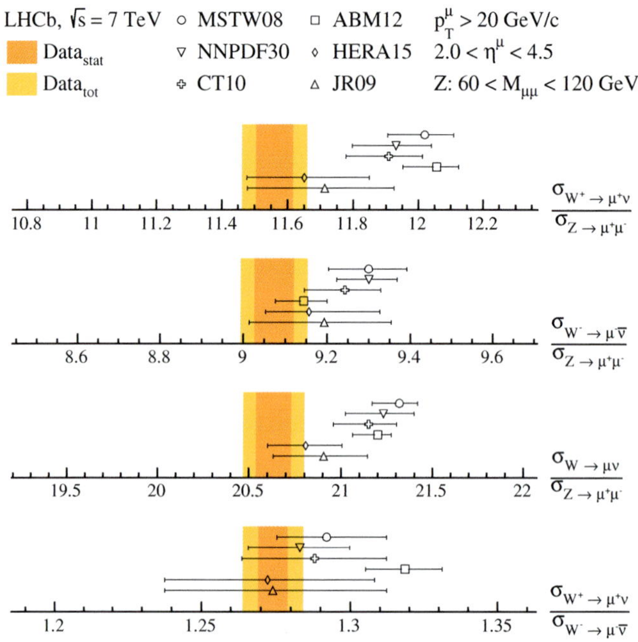

Fig. 8. Ratios of electroweak boson production (Fig. 9 from [LHCb [20] (2015)]) compared to predictions at next-to-next-to-leading in perturbative QCD with six different sets for the parton density functions of the proton. The shaded bands indicate the statistical and total uncertainties on the measurements.

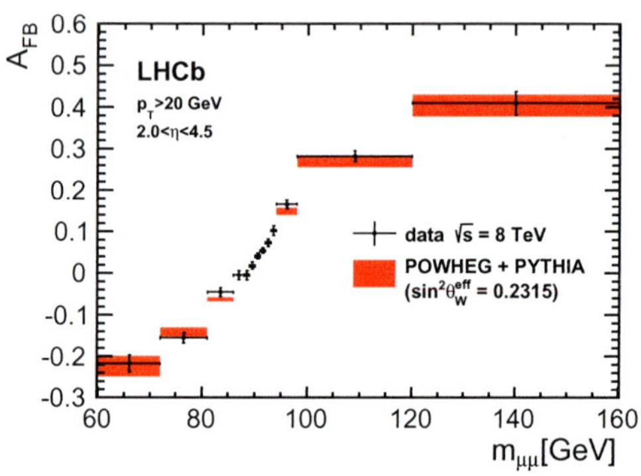

Fig. 9. Measurements of the forward-backward charge asymmetry as a function of the di-muon invariant mass compared to SM predictions (Fig. 2 from [LHCb [23] (2015)]).

the SM prediction. The measurement constrains $\sin^2\theta_W^{\text{eff}}$ to be $\sin^2\theta_W^{\text{eff}} = 0.23142 \pm 0.00073(\text{stat}) \pm 0.00052(\text{syst}) \pm 0.00056(\text{theory})$ [LHCb [23] (2015)]. This result is in agreement with the current world average, and is one of the most precise determinations at hadron colliders to date.

2.10. *First observation of top in the forward region*

The production of top quarks in the forward region is of considerable experimental and theoretical interest. In the SM, four processes make significant contributions to top quark production: $t\bar{t}$ pair production; t- or s-channel production of single top; and single top produced in association with a W boson. Top quarks decay almost entirely via $t \to Wb$.

Events are selected with an isolated muon from the $W \to \mu\nu$ decay and a well separated b-tagged jet [LHCb [24] (2015)]. The jets are tagged as originating from the hadronization of a b-quark by the presence of a secondary vertex (SV) in the jet and the SV direction of flight, defined by the vector from the pp interaction point to the SV position. The dominant background comes from associated production of a W and a b-quark. The top contribution is estimated by a simultaneous fit to the distributions of the charge asymmetry and the sum of the p_T of the muon and the b-jet, shown in Fig. 10. The data cannot be described by the expected direct $W + b$ contribution alone.

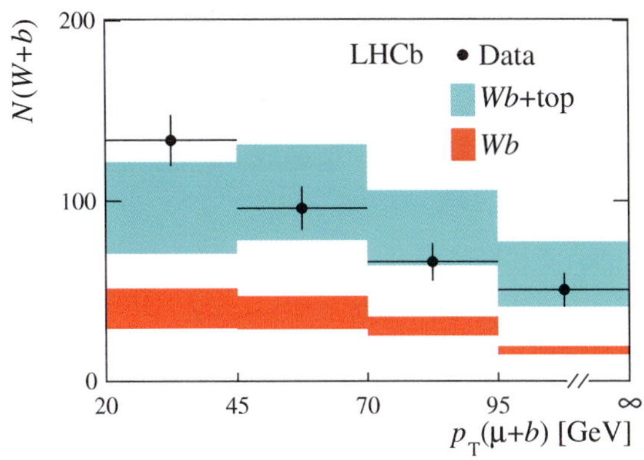

Fig. 10. Results for the $W + b$ yield versus $p_T(\mu + b\text{−jet})$ compared to SM predictions (Fig. 5 from [LHCb [24] (2015)]).

The resulting inclusive top production cross-sections in the fiducial region of the measurement,[1] are $\sigma = 239 \pm 53(\text{stat}) \pm 38(\text{syst})$ fb for 7 TeV, and $\sigma = 289 \pm 43(\text{stat}) \pm 46(\text{syst})$ fb for 8 TeV, in agreement with SM predictions.

2.11. *Central exclusive production*

Central exclusive production (CEP), $pp \to pXp$, in which the protons remain intact and the system X is produced with a rapidity gap on either side, proceeds via the exchange of colourless, neutral particles, either photons or combinations of gluons, for example pomerons which play a critical role in the description of diffraction and soft processes. Experimentally, this leads to a unique signature with a small number of particles in the detector, either produced directly or as decay products, and two rapidity gaps that extend to the outgoing protons which escape through the beam-pipe. CEP allows searches for exotic states in a low-background experimental environment, the study of QCD and the pomeron and to probe the gluon distribution of the proton.

LHCb has published measurements of CEP of J/ψ, $\psi(2S)$ [LHCb [5] (2013)] and Υ [LHCb [25] (2015)] as well as of charmonium pairs [LHCb [12] (2014)]. These measurements probe x down to 5×10^{-6} and are also sensitive to saturation effects in the proton. Figure 11 shows the differential cross section for J/ψ in comparison with predictions. The next-to-leading order prediction gives a better description of the measured shape.

LHCb is particularly suited for CEP measurements as pileup is low and a special trigger with high efficiency for low multiplicity events was implemented. Searches for exclusive production of $X(3872)$, open charm, and di-mesons, using particle identification to detect hadrons in the final states, are also ongoing. Perspectives are excellent for the next running period due to the implementation of a new detection sub-system which can veto activity within $5 < |\eta| < 8$.

2.12. *Results from proton–ion collisions*

In ultra-relativistic heavy-ion collisions, the production of quarkonia is expected to be suppressed with respect to pp collisions, if a quark–gluon plasma, QGP, is created. In proton–nucleus (pA) collisions, where a QGP is

[1] Defined by $p_{\rm T}(\mu) > 25$ GeV/c, $2.0 < \eta(\mu) < 4.5$, $50 < p_{\rm T}(b) < 100$ GeV/c, $2.2 < \eta(b) < 4.2$, $\Delta R(\mu, b) > 0.5$, and $p_{\rm T}(\mu + b) > 20$ GeV/c.

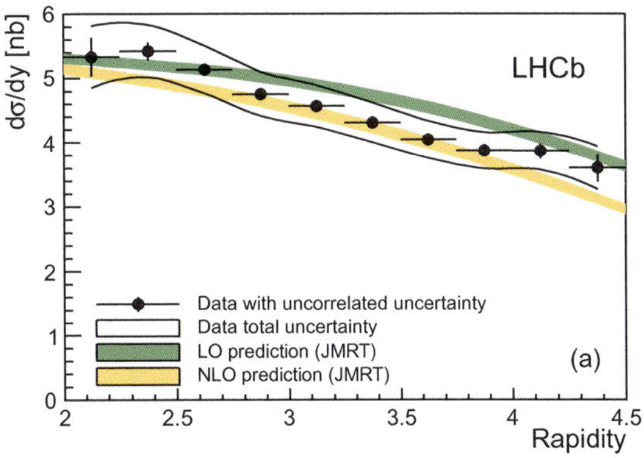

Fig. 11. Cross-section for central exclusive production of J/ψ as a function of rapidity (Fig. 5 from [LHCb [5] (2013)]) compared to leading (LO) and next-to-leading (NLO) order predictions. The band indicates the total uncertainty, most of which is correlated between bins.

not expected to be created suppression can occur due to cold nuclear matter effects, such as nuclear absorption, parton shadowing and parton energy loss in initial and final states. The study of pA collisions therefore provides important input to disentangle QGP from cold nuclear effects, probe nuclear parton distribution functions — which are poorly constrained — and provide a reference sample for nucleus-nucleus collisions.

Results on J/ψ [LHCb [6] (2013)], Υ [LHCb [13] (2014)] and Z [LHCb [14] (2014)] production in proton–lead collisions have been published by LHCb. All three decays are reconstructed in the di-muon channel and the pseudo-proper time is used to separate prompt J/ψ from the J/ψ from b hadron decays.

Nuclear effects can be characterised by the nuclear modification factor $R_{pPb} = \sigma_{pPb}/(A \cdot \sigma_{pp})$ which depends on the production cross-section in pA and pp collisions at the same centre-of-mass energy as well as the atomic number A. It is one if the cross-section in proton–lead collisions is simply a superposition of pp collisions. Figure 12 shows R_{pPb} for prompt J/ψ mesons, J/ψ from b and Υ as function of rapidity. A clear suppression of about 40% at large rapidity is observed for prompt J/ψ while the suppression of J/ψ from b is about 20%. This is the first indication of the suppression of b hadron production in proton-lead collisions. In the case of Υ the data are consistent with a suppression at large rapidity similar to J/ψ from b and a possible enhancement in the backward region. The measurements are in reasonable agreement with predictions.

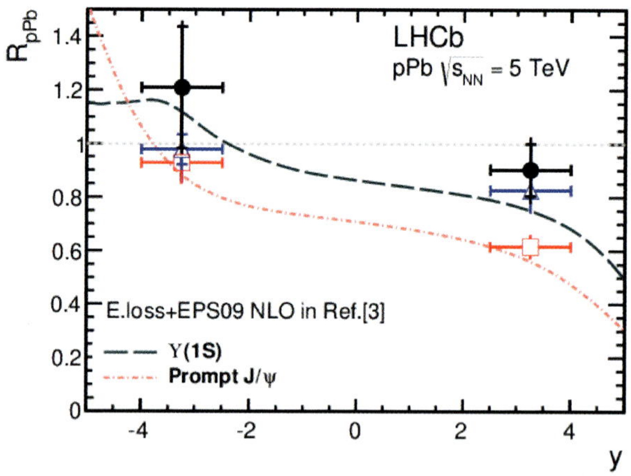

Fig. 12. Nuclear modification factor $R_{pPb} = \sigma_{pPb}/(A \cdot \sigma_{pp})$ for Υ (circles), prompt J/ψ (squares) and J/ψ from b (triangles) as functions of rapidity (Fig. 2 from [LHCb [13] (2014)]).

2.13. *Summary*

The excellent performance of the LHCb detector has allowed the LHCb collaboration to publish a wide range of physics results, demonstrating LHCb's unique role, both as a heavy flavour experiment and as a general purpose detector in the forward region. Despite most of the measurements are in agreement with the SM, there are intriguing discrepancies; many of these measurements are statistically limited.

The LHCb detector will be upgraded in the next long shutdown scheduled to take place 2018–2019 such that all sub-detectors are read out at the bunch crossing frequency of 40 MHz [LHCb [2] (2012)]. The trigger will then be fully based on software with the full event information available for all events. The goal of the upgrade is to operate the detector at a five times higher instantaneous luminosity. Together with a higher trigger efficiency this will increase the yield by a factor 10 for muonic and a factor 20 for hadronic channels.

References

Amhis, Y. *et al.* Heavy Flavor Averaging Group (HFAG) Collaboration (2014). arXiv:1412.7515.

Atwood, D., Gronau, M. and Soni, A., *Phys. Rev. Lett.* **79** (2), 185 (1997).

Bobeth, C., Gorbahn, M., Hermann, T., Misiak, M., Stamou, E. *et al. Phys. Rev. Lett.* **112**, 101801 (2013).

Bozek, A. *et al.* Belle Collaboration (2010). *Phys. Rev.* **D82**, 072005 (2010).
Choi, S. K. *et al.* Belle Colaboration (2003). *Phys. Rev. Lett.* **91**, 262001 (2003).
CMS & LHCb Collaboration (2014). *Nature* **522**, 68 (2015).
Descotes-Genon, S., Hofer, L., Matias, J. and Virto, J. *J. Phys. Conf. Ser.* **631** (1), 012027 (2015).
Descotes-Genon, S., Hurth, T., Matias, J. and Virto, J. *JHEP* **1305**, 137 (2013).
Descotes-Genon, S., Matias, J., Ramon, M. and Virto, J., *JHEP* **01**, 048 (2012).
Gauld, R., Goertz, F. and Haisch, U., *Phys. Rev.* **D89**, 015005 (2013).
Hicks, K. H., *Eur. Phys. J.* **H37**, 1 (2012).
Lees, J. P. *et al. Phys. Rev. Lett.* **109**, 101802 (2012).
LHCb Collaboration (2008). *JINST* **3**, S08005 (2008).
LHCb Collaboration [1] (2012). *Phys. Rev. Lett.* **110**, 021801 (2013).
LHCb Collaboration [2] (2012). *CERN-LHCC-2012-007.*
LHCb Collaboration [3] (2013). *Phys. Rev. Lett.* **111**, 191801 (2013).
LHCb Collaboration [4] (2013). *JHEP* **04**, 091 (2014).
LHCb Collaboration [5] (2013). *J. Phys.* **G41**, 055002 (2013).
LHCb Collaboration [6] (2014). *JHEP* **02**, 072 (2014).
LHCb Collaboration [7] (2014). *Phys. Rev. Lett.* **113**, 151601 (2014).
LHCb Collaboration [8] (2014). *Phys. Rev. Lett.* **112**, 161801 (2014).
LHCb Collaboration [9] (2014). *JHEP* **02**, 121 (2015).
LHCb Collaboration [10] (2014). https://cds.cern.ch/record/1755256.
LHCb Collaboration [11] (2014). *Phys. Rev. Lett.* **112**, 222002 (2014).
LHCb Collaboration [12] (2014). *J. Phys.* **G41**, 115002 (2014).
LHCb Collaboration [13] (2014). *JHEP* **07**, 094 (2014).
LHCb Collaboration [14] (2014). *JHEP* **09**, 030 (2014).
LHCb Collaboration [15] (2014). *JHEP* **12**, 079 (2014).
LHCb Collaboration [16] (2015). https://cds.cern.ch/record/2002772.
LHCb Collaboration [17] (2015). *Phys. Rev. Lett.* **115**, 111803 (2015).
LHCb Collaboration [18] (2015). *Phys. Rev.* **D92**, 011102(R) (2015).
LHCb Collaboration [19] (2015). *Phys. Rev. Lett.* **115**, 07201 (2015).
LHCb Collaboration [20] (2015). *JHEP* **08**, 039 (2015).
LHCb Collaboration [21] (2015). *JHEP* **05**, 109 (2015).
LHCb Collaboration [22] (2015). *Phys. Rev.* **D92**, 052012 (2015).
LHCb Collaboration [23] (2015). *JHEP* **11**, 190 (2015).
LHCb Collaboration [24] (2015). *Phys. Rev. Lett.* **115**, 112001 (2015).
LHCb Collaboration [25] (2015). *JHEP* **09**, 084 (2015).
Lyon, J. and Zwicky, R. arXiv:1406.0566 (2014).
Olive, K. A. *et al.* Particle Data Group Collaboration (2014). *Chin. Phys.* **C38**, 090001 (2014).

Chapter 10

TeV Astrophysics: Probing the Relativistic Universe

Ulisses Barres de Almeida

CBPF — Brazilian Center for Research in Physics
R. Dr. Xavier Sigaud, 150 - Urca,
Rio de Janeiro - RJ, 22290-180, Brazil
ulisses@cbpf.br

We focus our contribution to this volume on the relativistic Universe and on present and future experimental efforts in *Teraelectronvolt Astronomy*, i.e., the observation of cosmic gamma-ray photons above 100 GeV, employing ground-based imaging atmospheric Cherenkov telescopes. The main topics of our contribution are directed to the discussion of the importance of TeV observations for a better understanding of the cosmic-ray content of the galaxy and of the Universe's most energetic astrophysical objects. In particular, we are concerned with cosmic ray accelerators: supernova remnants, pulsars and their environments, active galaxies and black holes in general. In perspective, we will briefly discuss the role of the Cherenkov Telescope Array (CTA) in the current revolution that astroparticle physics is undergoing, emphasising the key position occupied by South America in this context. The 21st Century promises to be a golden age for astroparticle physics, and it is expected that much of the frontier research in relativistic astrophysics in next decades will be associated to advances in this growing field of observational astronomy.

Contents

1. A Window into the Extreme Universe

TeV gamma-rays are the latest astronomical window into the skies. Far from being the last, as we enter the threshold of non-electromagnetic astrophysics,[1] it remains nevertheless the most privileged viewpoint available to date of the extreme Universe. The current generation of satellite and ground-based gamma-ray detectors have achieved a level of sensitivity in the past decade which allowed for the first deep surveys of the Galaxy and the nearby Cosmos to be performed at energies up to 1 TeV and above, providing insight on essentially every kind of known relativistic astrophysical source.[2] Despite some essential information on these extreme engines and processes being dependent on direct access to cosmic particles other than photons, and although much progress has been recently achieved in the detection of energetic cosmic rays and astrophysical neutrinos, a proper particle astronomy is still far off in the future, and a detailed scrutiny of specific sources relies on electromagnetic carriers.

2. Gamma-Ray Eyes

The Jesuit philosopher and scientist Pierre Teilhard de Chardin, an almost exact contemporary of Einstein, once defined the itinerary of natural sciences as "the development of ever more perfect eyes in a world where there is always something more to see."[Bersanelli & Gargantini (2009)] From

[1]Of which the most relevant example is the recent detection of gravitational waves by the LIGO collaboration [Abbott *et al.* (2016)].
[2]For a complete, up-to-date catalogue of TeV sources, please refer to the following repository maintained by the University of Chicago: http://tevcat.uchicago.edu.

the early days of stellar spectroscopy, to the first steps towards the future Cherenkov Telescope Array[3] (CTA), even a superficial gaze is enough to testify that indeed such is the history of astrophysics in the 20th Century.

The first motivated proposal for a gamma-ray astronomy was advanced by Philip Morrison in 1958 [Morrison (1958)], following theoretical expectations on radiative processes and extensive evidence from cosmic ray physics which suggested that a number of high energy processes should take place in astrophysical sources, leading to detectable gamma-ray emission.

The context of Morrison's suggestion was marked, from one side, by the growing development of cosmic-ray physics, which at the time was inaugurating its first particle arrays for studying extensive air showers (EAS), following the discovery of the π-meson by Lattes, Occhialini, and Powell [Lattes *et al.* (1947)]. On the other hand, the ideas of Fermi, describing a potential universal mechanism for the acceleration of cosmic-rays [Fermi (1949)] had been recently advanced, whilst theories of compact objects [Schönberg & Chandrasekhar (1942); Oppenheimer & Snyder (1939); Oppenheimer & Volkoff (1939)] and supernovae [Gamow & Schoenberg (1941)] provided the necessary background where to stage the first acts of a newly-born relativistic astrophysics [Baade & Zwicky (1934)], led by names such as Gamow, Schoenberg, Chandrasekhar and Oppenheimer.

Although the first concrete observational proposal for detecting celestial gamma-rays came from Cocconi in 1959 [Cocconi (1959)], based on the ground-based air-shower technique, the first effective measurements of celestial gamma-radiation were made from satellites. Early space borne gamma-ray astronomy was a success story which culminated with the famous discovery of gamma-ray bursts in 1967 (declassified only in 1973 [Klebesadel *et al.* (1973)]) by the military Vela programme — a result which motivated and guided much of the subsequent development in the field, up until the current days of the Swift and Fermi missions.

Space-based gamma-ray astronomy technology, employing scintillator (Compton telescopes) or solid-state detectors (pair-conversion telescopes), are a direct application of laboratory techniques. In principle, such techniques can be employed in the detection of gamma-rays of very high energy, but the steeply falling fluxes of cosmic gammas above a few ten GeVs rapidly become prohibitive for the $\sim$m^2 detector effective areas available in space.

[3]For a complete view of the project, please refer to the Cherenkov Telescope Array Consortium homepage at https://web.cta-observatory.org.

2.1. *Ground-based gamma-ray astronomy*

Above the threshold of a few GeV, gamma-rays impinging the atmosphere carry enough energy to pair produce in the presence of the atoms of air and give out $e^{\pm}$ pairs which can initiate a large particle cascade via bremsstrahlung and further pair-production, leading to the formation of an EAS detectable from the ground.

Ground-based gamma-ray astronomy is an indirect approach which images the EAS to derive the direction and energy of the primary gamma-ray photon. The experimental techniques are basically two, and involve either the detection of the atmospheric Cherenkov yield of the secondary particles in the shower, or the direct detection of the secondary particles themselves, applying another kind of converter such as scintillators or water tanks. This use of indirect methods to measure the shower's energy content is what allowed gamma-ray astronomers to counter the problem of the low GeV-TeV fluxes, given that the enormous footprints of the EAS make up for effective detector areas of 10^5 m^2, circa.

Even though the potential of the technique was realised quite early on, very-high energy (VHE) astronomers stayed a long time grounded, as development was slow.[4] Its breakthrough was a combination of a joint evolution in hardware technology and imaging analysis techniques.

From the point of view of detector technology, the main challenge was related to the development of fast PMT cameras and electronics, able to integrate the few-ns pulses of the Cherenkov light from the EAS, singling them out from the night sky background. Although this was achieved in the late 60's by Whipple — the first dedicated imagining atmospheric air-Cherenkov telescope (IACT), with a pixelated PMT camera coupled to fast electronics — the first TeV gamma-ray source came only in 1989, with detection of the Crab nebula [Weekes (1989)]. The feat was a direct consequence of the imaging analysis method developed by Michael Hillas in 1985 [Hillas (1985)], which allowed to distinguish the gamma-initiated showers from the thousand times more numerous hadronic shower background.

Nowadays, the current generation of ground-based IACT observatories, H.E.S.S., MAGIC and VERITAS, responsible for leading the field into maturity, use multiple telescopes to image the showers stereoscopically, thus improving sensitivity down to 1% of the Crab nebula flux and pushing

[4]For an up-to-date historical account by one of its pioneers, see [Hillas (2013)].

angular resolution to the arcmin-level, allowing morphological studies of extended sources.

Concerning the particle ground array detectors, the current leading facility in the field, HAWC, uses water tanks to measure the Cherenkov light from the secondaries and adopts techniques of gamma-hadron separation based on measuring the shower shape and timing of the shower front. Sometimes, detectors of this kind are aided by employment of hybrid techniques, such as a muon detector array. Since lepton-initiated showers have very low muon counts, in contrast to hadronic showers where pion production and decay lead to a high muon content, this measurement can be effectively used as background discriminator.

Even if IACTs are much more sensitive nowadays, the two techniques are complementary. Particle ground arrays can work continuously, being therefore ideal instruments for monitoring transient sources and triggering on rare events, whereas atmospheric Cherenkov detectors can only operate during dark nights, leading to an annual duty cycle of circa 1000 hours. Also, the currently larger effective areas of particle ground arrays allow for the detection of higher energy photons than the IACTs. To achieve the low energy thresholds of IACTs, below 100 GeV, particle arrays have to be placed at extreme altitudes of 5 km above sea level to detect the secondaries of the lower energy showers. There are ongoing discussions for projects aiming at this, having in mind a South-American complement to CTA and other active facilities in Latin America [Sidelnik (2014); Matthiae (2015)].

3. Probes of Relativistic Astrophysics

From compact stars and black holes to cosmology and the dark sector of the Universe, experimental astroparticle physics provides some of the best observational insights into the relativistic Universe. This review will concentrate on the compact objects and their role as a source of energetic cosmic particles.

To observe the skies in gamma-rays is to map the relativistic particle populations, the gamma-ray production rate reflecting the density of relativistic particles and targets available for interaction, tracking their way as they diffuse in the magnetised medium. A sky map such as in Fig. 1 shows that efficient cosmic particle acceleration is ubiquitous.

From the point of view of their radiative output, relativistic sources are predominantly non-thermal. Their relevance to the grand picture is readily

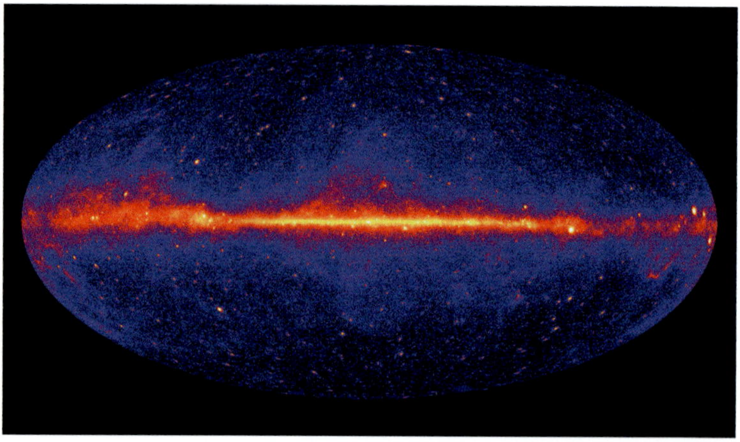

Fig. 1. Gamma-ray all-sky image at GeV energies. Credits: NASA/GSFC and the Fermi-LAT Team.

noticed when realising that, from radio to gamma-rays, relativistic particles make up for a total radiative output which is roughly equivalent to the entire thermal optical energy content of the Extragalactic Background Light (EBL) spectrum [Ressell & Turner (1989)]. For some individual sources, in fact, such as blazars, most of the photon luminosity νF_ν comes from relativistic particles and is seen in the X- to gamma-ray bands.

The downside of having only indirect access to these cosmic-ray (CR) particles via gamma-ray observations is greatly compensated by the fact that we are directly imaging the acceleration sites, bypassing the loss of directional information that so much hinders the study of their origin. Furthermore, the absence of thermal radiation contamination at such high energies and the relatively low intensity of the very-high-energy (VHE) gamma-ray diffuse background, all contribute to provide TeV astronomy with a clear view into the CR production sites.

The non-thermal spectral shape of relativistic sources is the result of interactions of accelerated charged particles — protons, nuclei or electrons — with ambient matter or magnetic and radiation fields. Both electrons and protons (or heavier nuclei) are expected to be accelerated at shock fronts within these powerful engines [Reynolds (2008)], but electrons dominate the radiative output from radio to X-rays, due to their efficient cooling via synchrotron emission. Synchrotron cooling imposes, conversely, a strong limit to the maximum energy attainable by electrons. In purely leptonic models, the gamma-ray emission must be the result of inverse-Compton (IC) scattering of soft photons by the energetic particles. The result is a well correlated double hump structure observed from practically all relativistic sources (see Fig. 2), where the relation between the soft and IC scattered photon frequencies is $\nu_{\rm IC} \propto \gamma^2 \nu_{\rm soft}$, with γ being the electron's Lorentz factor.

Proton emission is expected to contribute to the gamma-ray SEDs of sources where cosmic ray acceleration is taking place, but because heavier particles radiate less by synchrotron [Aharonian (2002)], the dominant channel for gamma-ray generation from protons is photo-pion production or hadronic collisions and subsequent cascading. Both processes lead to secondary mesons which then decay into gammas via the $\pi^0 \rightarrow \gamma\gamma$ channel.

Shock acceleration is efficient for hadrons, the maximum expected energies being $E_{\rm max} \sim \beta_{\rm s} ZBL$, where Z is the particle's atomic number and $\beta_{\rm s}$ the shock speed. For typical galactic source sizes $L \sim$ pc, and magnetic field intensities B of few μG, protons can achieve energies comparable to what is observed for the knee of the cosmic-ray spectrum. One of the ways in which to identify the presence of hadrons contributing to the gamma-ray emission is to look for a direct spectral signature produced from pion decay, which generates a characteristic bump in the SED, showing a steep flux rise in the MeV range not easily replicable by leptonic models.

Sources capable of accelerating particles up to energies of 10^{15}–10^{16} eV are referred to as PeVatrons, and their identification is of crucial importance to understand the origin of the galactic cosmic-rays and its production mechanism. In hadronic models, a gamma ray of energy E_γ is roughly mapping protons in the energy range of $\sim 10–100 E_\gamma$, implying that the direct detection of these PeVatrons requires telescope sensitivities up to ~10 TeV or above. In any case, the most surprising feature of these relativistic sources is that energy conversion to particle acceleration and photons seem to happen at extreme efficiencies, far above anything experimentally achieved at Earth.

3.1. *Opacity and horizon in TeV astrophysics*

The view of the cosmos in VHE gamma-rays is not a clear sight. The low-energy (soft) photon background is a source of opacity via electron–positron pair-production ($\gamma_{\rm TeV} + \gamma_{\rm soft} \rightarrow e^- + e^+$). The effect is significant both in the attenuation by the EBL of the gamma-ray flux during propagation, as well as for internal absorption at the source, preventing the escape of

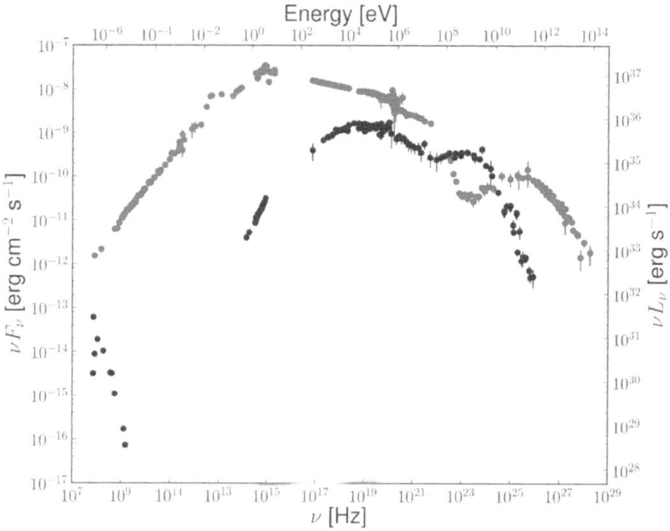

Fig. 2. The overall spectral energy distribution (SED) of the Crab Nebula (grey) and the Crab pulsar (black). Figure reproduced from [Buehler & Blandford (2014)].

gamma-ray photons from their compact production regions. The process is kinematically allowed above an energy threshold $E_\gamma E_{\rm bkg} = m^2c^4$, and peaks at gamma energies of $E_{\rm TeV} \sim 0.9/E_{\rm eV}$, where $E_{\rm eV}$ is the target soft photon energy, measured in eV.

From tens of GeV to hundreds of TeV, different bands contribute to propagation opacity leading to a differential effect on the gamma-ray spectrum. As we move from GeV to TeV energies the target soft photons change from optical to infrared and the mean free path dramatically decreases, As we reach 100 TeV, the gamma-ray horizon shrinks down to intra-Galaxy distances, due to the rising contribution of the CMB (see Fig. 3).

In any case, even for energies as low as 100 GeV, the horizon is still quite limited, and the most distant VHE source detected to date is just below $z \sim 1$: a blazar seen as the first gravitationally lensed object detected in gamma-rays. [Sitarek *et al.* (2015)]. To achieve truly cosmological distances, such as desirable for gamma-ray burst searches, for example, it is necessary to approach the theoretical limits of the atmospheric Cherenkov technique, with energy thresholds of only few tens of GeV.

The use of pair-production opacity during propagation from distant extragalactic sources is also an effective probe of the EBL density within

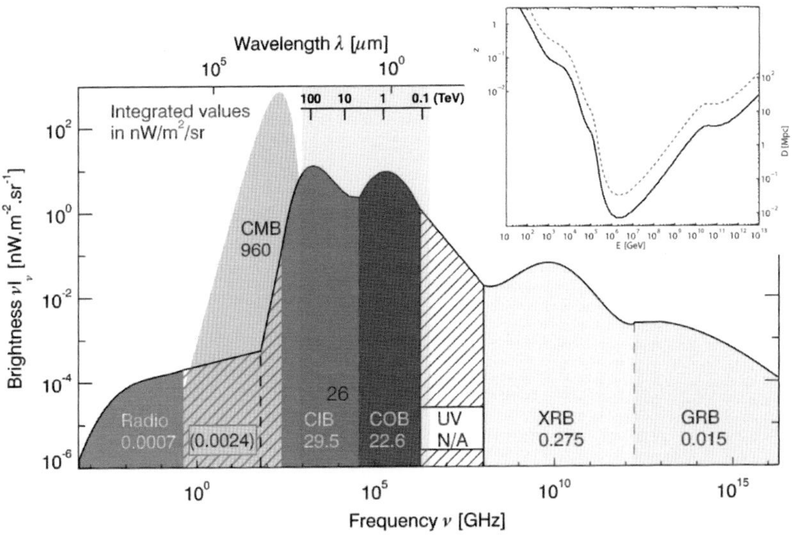

Fig. 3. Schematic representation of the extragalactic background light (EBL) brightness from radio to gamma-rays (adapted from [Dole (2013)]). The blue band marks the range of soft photon targets relevant for pair-creation opacity, with the scale indicating the corresponding peak gamma-ray energy absorbed. The overlaid plot is the resulting gamma-ray horizon [de Angelis *et al.* (2013)].

the optical-IR range. Thanks to measurements by ground-based gamma-ray telescopes, our knowledge of the EBL has much improved in recent years. A more detailed presentation of the status of this topic will be given towards the end of this review (Sec. 7).

As advanced earlier, absorption is also relevant within the sources where gamma-rays are produced, since these are usually compact, radiation-intensive environments. A compactness parameter, l, can be defined as a measure of the region's radiation energy density, $l = L/Rm_ec^3$, where L is the luminosity in the relevant target photon band and R the linear source size; the derived optical depth is therefore $\tau_{\gamma\gamma} \propto \sigma_T l$, where σ_T is the Thomson cross-section.

The phenomenon is particularly interesting in the case of Active Galactic Nuclei (AGN). The gamma-ray production sites in AGNs are known to be compact from the fast variability timescales observed in the light curves of these sources, with linear size $R \approx ct_{\rm var}\Gamma$. In such systems, opacity from the low-energy synchrotron photons within the very sites of gamma-ray production works as a potent absorber of the VHE radiation, implying strong constraints to the plasma's bulk Lorentz Factor Γ, should gamma radiation escape these regions, with consequence to the estimates of the beaming in relativistic outflows.

Depending on the dominant source of radiation, internal source absorption can be highly differential across the gamma-ray band, leading to significant modifications from the original source spectrum. Reprocessing can also happen within the sources, and cascades induced by the e^+/e^- pairs can systematically shift energy from high to lower energy gammas, further adding to the spectral changes [Coppi (1994)] and leading to the formation of (still undetected) pair haloes around sources.

3.2. *Overview of the gamma-ray sky*

The VHE sky map, numbering today little less than 200 objects, is dominated by discrete sources. Particularly rich is the list of gamma-ray emitters which fill our quadrant of the galaxy,[5] a population which has been systematically collected over the past decade by the current generation of IACTs, and specially H.E.S.S., the only such observatory located in the Southern Hemisphere (see Fig. 4 and references therein). Among them are a plethora of compact objects found in all possible systems arising at the endpoints of massive stellar evolution. The most populous members are supernova remnants (SNR) and pulsar wind nebula (PWN), but within the last several years a number of binary systems as well as a handful of pulsars have also been identified as direct gamma-ray emitters.[6]

A rapid inspection of the morphology of these sources as presented in Fig. 4 shows that they are mostly extended, with angular sizes as large as 1°, in contrast with the $\sim 5'$ resolution of the instruments (e.g., [Aharonian *et al.* (2007a)]). This extended character results naturally from the gamma-ray production mechanisms, as particles (generally electrons) diffuse away from the acceleration sites and radiate in interaction with the ambient magnetic and radiation fields. Extended emission has also been detected in association with hadronic particles. In the W 28 complex [Aharonian

[5]Because of the limited sensitivity of current instruments, about 80% of the galactic sources with known distance are less than 8 kpc away, half of the Milky Way remaining virtually non-charted territory in the VHE band.

[6]Our view of the galaxy in gamma-rays is also marked by a number (circa 1/4) of unidentified sources. This is rather due to the low angular resolution of instruments and the crowdedness of fields, which hinders unambiguous identification of counterparts.

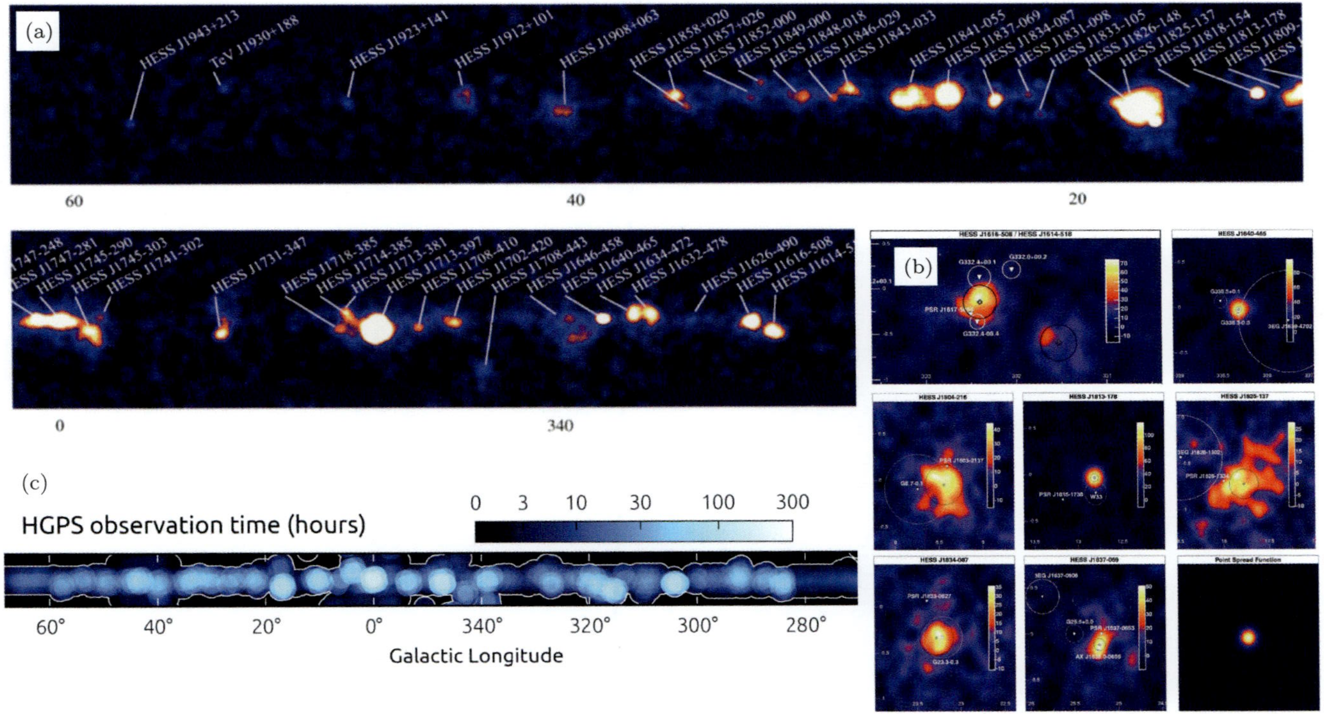

Fig. 4. The H.E.S.S. galactic plane scan (HGPS) to a limit sensitivity of 1% Crab flux above 1 TeV (most sources shown are below 0.5 Crab): (a) A view of the inner galactic plane as of 2013 (from [Carrigan *et al.* (2013)]); (b) Close up map of the morphology of a few sources. White circles mark the location and extension of the associated SNR, and white triangles indicate the position of nearby pulsars (from [Aharonian *et al.* (2005a)]); (c) Chart showing the integration time for the confection of the HGPS between 2005 and 2015 (adapted from [Sitarek *et al.* (2015)]).

et al. (2008)], for example, emission has been detected in positional coincidence with molecular clouds nearby a SNR, suggesting cosmic-ray particles from the SNR collide with nuclear target material in the clouds, a few kpc away from their production sites, giving off gammas via neutral-pion decay.

Finally, the first measurements of the diffuse TeV gamma-ray emission component from the galactic plane at large scale have only recently been presented, by the MILAGRO water Cherenkov experiment at 15 TeV [Abdo *et al.* (2008)], and in 2014 by H.E.S.S., with a lower energy threshold of 1 TeV [Abramowski *et al.* (2014)]. Very prominent at other wavebands, and even the dominant component at lower gamma-ray energies, where bremsstrahlung is the main production mechanism, the galactic diffusive emission at TeVs is a clear signature of hadronic, π^0-decay, which H.E.S.S. estimated to contribute to about a third of the total flux observed [Abramowski *et al.* (2014)], plus some additional contribution from inverse-Compton emission and unresolved sources. Both measurements by H.E.S.S. and MILAGRO are compatible with the expected density of cosmic rays and targets in the galaxy and the conventional propagation models, putting, for the first time, meaningful constraints to the background flux from unresolved sources.

As for the extragalactic sky, almost the totality of the VHE sources are active galactic nuclei (AGN), the luminous nucleus of galaxies where the presence of a central supermassive black hole (SMBH), of masses between $10^6 - 10^{10}$ $M_\odot$, is signalled by activity resulting from the accretion of matter onto the massive object. Although all massive galaxies are expected to harbour such a SMBH at their center, only 1% of galaxies are active, with typical luminosities ranging from 10^{44} to 10^{47} ergs.s^{-1}, or up to 10^4 times the stellar luminosity of a typical galaxy.

4. The Cosmic Ray Content of the Galaxy

4.1. *Supernova remnants and CR accelerators*

The study of SNRs is extremely important in astroparticle physics as they are believed to be the main source of high-energy galactic cosmic-rays [Zwicky (1939)], possibly up to the knee of the spectrum, around 10^{15-17} eV. The basic argument for the suspicion is based on energetic considerations, comparing the estimated energy flux of CRs in the galaxy $\sim 5 \times 10^{40}$ erg/s with the typical kinetic energy output of a SN explosion ($\sim 10^{51}$ erg) plus an expected rate of 2–3 explosions per 100 yr, and a suitable mechanism to convert the macroscopic kinetic energy from the explosion into particle energy. These numbers imply that an efficiency of 10% conversion of kinetic energy into accelerated particles would suffice to account for the observed CR energy flux. First-order Fermi diffusive shock acceleration [Lemoine & Pelletier (2015)] is widely viewed as providing the theoretical background for CR production at SNRs [Bell (1987)], although there remain fundamental questions regarding particle injection and confinement, some of which are currently being probed through observation of young SNRs, such as Tycho [Acciari *et al.* (2011)], Cas A [Acciari *et al.* (2010)] and SN 1987A [Abramowski *et al.* (2015)].

Despite these expectations, hadronic acceleration in SNRs has only recently been unambiguously identified. Low energy MeV data from Fermi has revealed a characteristic spectral signature from pion-decay in two systems which show SNR-cloud interactions, IC 443 and W 44, providing the first direct evidence that cosmic-ray protons of TeV energies are produced in these systems [Ackermann *et al.* (2013)], but leaving open the quest for the highest energy Galactic CRs of PeV energies. Population estimates suggest that no more than a dozen SNR PeV-accelerators could be present in the Galaxy [Gabici & Aharonian (2007)]. This estimate is based on the consideration that PeV particles would remain confined for a relatively short period of ~ 1 kyr, thus within the early stages of the remnant's evolution, before escaping to the interstellar medium.

The highest energy emission from a SNR detected to date was seem from RX J1713.7-3946, a bright, young, nearby source, with an SED extending beyond 10 TeV, but which cannot be unambiguously attributed to protons [Gabici & Aharonian (2015)]. As was the case with the two SNRs observed by Fermi, one fundamental difficulty associated with the direct identification of proton CR accelerators is the fact that it likely depends on the presence of nearby clouds to serve as targets for the hadrons.

In addition to SNR, some larger systems could provide an even more natural environment for PeV CR production. In the past couple of years, the ARGO and H.E.S.S. collaborations have detected, for the first time, the TeV emission component from two "super-bubbles" — blown-up cavities in the interstellar medium as the result of combined activity of multiple SN and stellar winds. These sources, the Cygnus Coccon [Bartoli (2014)]and 30 Dor C [Abramowski *et al.* (2015)], located at the Large Magellanic Cloud (LMC), are true natural CR laboratories. Although in neither of them it is possible to rule out leptonic emission as the dominant source of gammas, observations suggest that extreme conditions must be in place within the super-bubbles, owing to their combination of large sizes (of order 100 pc) and intense, turbulent magnetic field, where successive shocks and particle acceleration episodes may develop. Modelling of the TeV emission as neutron-pion decay reveals that super-bubbles could potentially provide the right conditions for proton acceleration to extreme, 10^{15} eV energies, supporting views that they could be responsible for the highest-energy CRs in the Galaxy [Ferrand & Marcowith (2010)].

Neutrino detection is another avenue to discriminate between hadronic and leptonic scenarios in SNRs. PeV neutrino detection is already within the reach of current observatories, such as IceCube, which saw particles with energies reaching beyond 2 PeV [Aartsen *et al.* (2013a)]. Given the necessary extraterrestrial origin of this high-energy population, its source has been extensively investigated by the IceCube collaboration. Although a cosmogenic (GZK/propagation) origin for the flux has been disfavoured, still a precise astrophysical origin of the PeV events remain unknown [Aartsen *et al.* (2013b)].

4.2. *Galactic Center: The Milky Way's PeVatron source*

The Galactic Center (GC) — the central few hundred pc of the Milky Way — is a complex, active region of the Galaxy, with multiple signatures of particle acceleration taking place within different GeV-TeV sources. It was first imaged in detail in VHE gamma-rays by the H.E.S.S. Collaboration as part of its Galactic Plane Survey, over 10 years ago, and has since them been extensively studied. As can be seen from Figure 5 it presents a complex morphology, likely the combined effect from stellar winds and shocks, and the potential signature of activity from Sgr A* (a radio source positionally coincident with the central SMBH). It is also one of the prime locations to search for Dark Matter gamma-ray signatures in the Cosmos [van Eldik (2015)].

The observations by H.E.S.S. revealed faint, diffuse emission from an extended region reaching over 1° away from the GC itself, well correlated with molecular clouds (MC) in the so-called Galactic Center Ridge, about 200 pc in extension. Since MCs serve as target material for hadronic interactions, these are tracers of CRs diffusing in the GC, leading to estimates for a particle population above TeV energies which is a factor of a few larger than the density seem from the Galactic disk and our local portion of the

Galaxy. This population has been considered as possibly coming from an excess SNR population in the region [Aharonian *et al.* (2007a)]. Broadband radio to gamma-ray leptonic models have also been evaluated.

In addition to the extended signal, the GC observations by H.E.S.S. had enough angular power and sensitivity to resolve some of the sources in the vicinity of the central SMBH. Careful analysis left only Sgr A* itself and a nearby PWN G359.95 as potential counterparts to the gamma-ray GC point source 3EG J1746-2851 [Aharonian *et al.* (2006)]. As early as 2011, additional combined analysis with Fermi-LAT data suggested the presence of a hadronic component in the emission, weakening the case for the PWN emission and leaving Sgr A* itself as the most likely candidate source for the central accelerator [Chernyakova *et al.* (2011)].

Such was the status of these studies until very recently, when the H.E.S.S. collaboration published a breakthrough result, pointing to the presence of a population of PeV protons as responsible for the VHE gamma-ray emission from the GC, finally shedding some light on the long-sought Galactic PeV accelerator [Abramowski *et al.* (2016)]. The arc-min resolution of these new observations were able to locate the PeV particle signature within the central 10 pc from the GC, which points the SMBH Sgr A* as the Milky Way's sole-detected PeVatron source.

Connected to this topic is the existence of a giant, gamma-ray bright structure some 60 kpc across, which was revealed by Fermi/LAT, in the form of bubbles growing out for 50° in angular size below and above the GC [Ackermann *et al.* (2014)](see Figure 1). These so-called "Fermi bubbles," detected between 1–50 GeV, whose edges shine bright also in ROSAT X-ray images, are large cosmic ray reservoirs, with an estimated energy content of about 10^{55} ergs. In comparison to the total luminous output of the Galaxy, of 10^{44} erg/s, they could indicate a past state of activity of the GC, such as an earlier accretion event onto the central SMBH lasting up to few Myr, or nuclear starburst activity from sometime within the past 10 Myr. Their present-date gamma-ray luminosity is $\sim 4 \times 10^{37}$ erg/s.

Spectral modelling of the bubble favours the presence of high-energy electrons emitting via synchrotron and inverse Compton, but a significant component of hadrons cannot be excluded [Ackermann *et al.* (2014)]. In fact, the presence of the "Fermi bubbles" suggest that Sgr A* would have gone through a past period of enhanced activity and outburst [Su *et al.* (2010)], as supported also by X-ray observations [Clavel *et al.* (2013)]. Although Sgr A*'s present emission rate is seen as insufficient to provide a significant contribution to the Galactic cosmic-ray content, such past activity some 1–10 Myr ago could well be able to account for the PeVatron cosmic rays observed today at Earth, thus providing a convincing explanation for the Galactic CR spectrum, which does not rely exclusively on SNR sources.

4.3. *A sidenote: Starburst galaxies*

Due to the limited sensitivity of current instruments, ground-based gamma-ray telescopes have no direct access to individual SNRs outside the Milky Way. In addition to the super-bubbles detected at the LMC, collective activity of extragalactic supernovae has been observed in VHEs from two nearby starburst galaxies: NGC 253, detected by H.E.S.S. [Acero *et al.* (2009)] and M82, detected by VERITAS [Acciari *et al.* (2009)], both in 2009. More recently, the LAT instrument onboard Fermi observed a large sample of infrared luminous galaxies, detecting two additional starburst galaxies at GeVs, NGC 1068 and NGC 4945 [Ackermann *et al.* (2012)].

Starbursts are galaxies exhibiting an increased rate of supernovae in their central regions due to enhanced star formation, and distinguishable by their high IR luminosity. Their observation is a privileged sight for studying cosmic-rays and their production in galactic accelerators. Their extended star-forming regions work as giant CR calorimeters that can be used to estimate the efficiency of energy channelling into CR particles. Estimates from the four sources detected in gamma-rays suggest a calorimetric efficiency of 30–50%, if emission is dominated by pion-decay following hadronic interactions with the ambient medium. This efficiency, along with estimates from undetected sources, lead to a prediction that starburst galaxies could account for up to 20% of the isotropic diffuse gamma-ray background (IGRB) [Ackermann *et al.* (2012)].

Finally, the average estimated density of CR particles in these systems (with energies above ~TeV) is around three orders of magnitude larger than the present galactic cosmic-ray density measured at the Solar System. This hints on the potential importance of CRs for the energy budget during phases of high-star formation in galaxies, as they transport energy away from acceleration sites by diffusion, and populate the inter and intra-galactic space with nucleosynthesis products, affecting galactic evolution.

5. Gamma-Ray Emission from Compact Sources

5.1. *Pulsar and their environments*

Some supernovae give origin to rapidly spinning neutron stars. Pulsars dissipate their rotational energy giving off a relativistic wind of electrons and positrons which can strongly influence the morphology and evolution of the remnant, as the expanding pulsar wind nebula (PWN) may interact with the reverse shock coming from the SNR. PWN are among the most numerous and best studied VHE gamma-ray sources in the sky, and although shocks within PWN are known to accelerate electrons to extreme energies (e.g., [Aleksic (2015)]), they are unlikely hadronic accelerators.

The best studied pulsar is the Crab, along with its associated nebula. The latter is in fact the first TeV source ever detected, by the pioneering experiment Whipple in 1989 [Weekes (1989)], and being the strongest stable VHE source in the sky, it is also used as a reference calibrator source. Perhaps because it is so well known, the Crab is also a very complex object, and inasmuch as the topic is interesting, we cannot detain ourselves too much to it and refer the reader to an excellent recent review by Buehler & Blandford, which includes some extended discussion on the Crab nebula flares systematically detected since 2008 by Fermi/LAT in the MeV–GeV band [Buehler & Blandford (2014)].

The Crab is the brightest of three pulsars from which pulsed VHE gamma-ray emission (at or above a few tens of GeV) has been detected [Aliu (2008)] — the list includes also the Vela [Leung *et al.* (2014)] and Geminga [Abdo *et al.* (2010)] pulsars, seem only by the LAT instrument onboard Fermi. Although the gamma-ray pulsed emission is most likely leptonic and must originate within the pulsar magnetosphere, the emission mechanism is still debatable. For the Crab pulsar, MAGIC recently detected pulsed photons up to TeV energies [Ahnen *et al.* (2016)]. The VHE photons require a population of electrons with extreme Lorentz factors and demand a revision of current pulsar magnetospheric models, capable of narrow-pulsed, coherent, smooth power-law emission over 4 orders of magnitude in energy at gamma-rays.

The stringent constraints of this impressive measurement rules out synchro-curvature radiation as a probable source for the emission, leaving the inverse-Compton mechanism — either within the outer magnetospheric gap or by the pulsar wind — as the most likely source for the gamma-rays.

5.2. *Other stellar end-products*

Up to now we presented an overview of isolated neutron stars and their nebular products. Nevertheless, massive stars are commonly found in associations and many compact sources at the end of stellar evolution are in binary systems. Given the extreme properties of such systems (with strong radiation fields, dense stellar winds, and intense magnetic field) and their compact character (up to a few tens AU, with orbital periods as short as a few years or less, depending on the nature of the compact object and its companion — see Table 1 in [Dubus (2015)] for a summary), binaries undergo changes in relatively short timescales which allow for a detailed probing of an evolving relativistic stellar system.

As is the case with pulsars and SNRs, hadronic emission from binaries is expected to be efficient only where target matter is available for hadronic interactions with protons escaping the system, such as in nearby clouds. Additionally, because electrons cool fast in the presence of intense radiation and magnetic fields, the gamma-ray signature from leptons is expected to modulate strongly with the system's orbital period, leading to an excellent probe of internal source's properties and structures.

Although Fermi/LAT has detected a much larger number of binary gamma-ray emitters in the MeV–GeV range, there is only a handful of so-called high-mass gamma-ray binaries which produce VHE emission. In the vast majority of Fermi binaries, the compact object is usually identified with a pulsar[7] meaning that gamma-ray observations are a privileged way of identifying and studying these binary massive star systems which survived the supernovae explosion.

The VHE-emitting binary systems all have unconfirmed compact companions to the massive O–B star. With the only exception of PSR B1259-63 [Aharonian *et al.* (2005b)], the nature of all other systems is still unidentified. In systems where a pulsar is the compact object, gamma-ray emission is likely the result of interaction between the pulsar and the stellar wind, which shock somewhere along the orbit, accelerating particles that lead to

[7] Refer to the webpage maintained by SLAC https://confluence.slac.stanford.edu/display/GLAMCOG/Public+List+of+LAT-Detected+Gamma-Ray+Pulsars for a complete, up-to-date list of Fermi detected pulsars. Notice that of the circa 160 objects listed as of today, about 1/3 are found in binary systems. Conversely, in radio, the proportion of pulsars identified in binary systems is of only 1% [Lorimer (2008)].

gamma-ray emission. Such configuration leads to strong modulation of the VHE emission along the orbital period, which is a common characteristic to the other 4 VHE emitting binaries [Dubus (2013)]. Despite that, the non-detection of direct pulsed emission from the system does not allow for a conclusive stand on the nature of the compact object.

Accretion onto a compact object (neutron star or black hole) is another potential energy source for particle acceleration in binary systems, and an alternative to the binary pulsar scenario (see Sec. 5.2). Microquasars, which were discovered by Mirabel & Rodriguez in 1994 [Mirabel & Rodriguez (1994)] are the galactic analogs of Active Galactic Nuclei (AGN), and have been favoured by some authors as a promising scenario to explain the gamma-ray observations of objects such as LS I +61°303 [Massi *et al.* (2012)] and LS 5039 [Paredes *et al.* (2000)] thanks to putative, yet unconfirmed, morphological evidence of radio outflows akin to jets being present.

5.3. *Looking for galactic black holes with gamma-rays*

Given the uncertainty on the nature of the compact objects in the aforementioned systems, the only two firm gamma-ray bright microquasars in the galaxy remain Cyg X-1 [Malyshev *et al.* (2013)] and Cyg X-3 [Abdo *et al.* (2009)]. Both systems are composed of a black hole candidate orbited by a giant star that gives off mass via strong winds which are then accreted to originate out-flowing relativistic pc-scale jets. Scaled-down version of the giant AGNs, microquasars are extremely interesting systems from the point of view of relativistic astrophysics, where the jet activity and its relation to the accretion state of the source seen in X-rays can be closely studied as it evolves in timescales of hours to days [Fender (2002)].

MeV–GeV gamma-rays are detected only when radio emission is present, directly associating it to the relativistic jet. Orbital modulation in the gamma-ray emission of Cyg X-3, arising from changes in the density of soft photons from the massive star that are up-scattered via inverse-Compton, indicate that the HE emission originates within the jet, at distances comparable to the orbital separation between the stars, away from the black hole engine. Such a scenario suggests particle acceleration occurring at recolimation shocks downstream in the outflow [Zdziarski *et al.* (2012)], akin to what is seen from AGN knots, and establishing an important parallel between the dynamics of both systems — leaving unsupported, for specific sources, some proposals for black hole magnetospheric emission of gamma-rays [Sitarek *et al.* (2015)].

Among the two objects, only Cyg X-1 presents some evidence of VHE emission, marginally detected by MAGIC in 2007, during a short outburst [Albert *et al.* (2007)], and never re-observed. A complicating factor might be the fact that both systems are very tight, and emission within or close to the orbital radius of the massive companion would suffer from strong γ–γ opacity, not being detectable at VHE.

Bypassing the problem of local absorption, a singular case of VHE emission associated to a micro-quasar has been recently reported from SS 433, where the persistent HE photon emission likely originate not from within the system but in hadronic interaction with the surrounding media [Bordas *et al.* (2015)]. Here, the hadrons would originate from a baryon-loaded jet, a characteristic shared by not many known galactic jets — and not expected for luminous AGN jets which are most likely leptonic or Poynting flux dominated at launch — so that it is difficult to gauge the potential contribution of such systems to the total CR content of the galaxy.

6. Active Galaxies and Supermassive Black Holes

Second only to GRBs — still undetected at VHEs — AGNs are among the brightest astronomical phenomena in the cosmos, and have provided us with some of the most luminous events ever observed at any wavelength. Their lasting periods of activity — the AGN phase of a galaxy has an expected duration of at least a few ten million years, e.g. [Shankar *et al.* (2009)] — means they are fundamental players in the evolution of galaxies and galaxy clusters, dramatically affecting their dynamics by funnelling gravitational potential energy from the small, 100 AU-scale regions of the accretion disc's innermost orbits [Doeleman (2012)], away to distances of several kpcs, at the endpoints of the extragalactic jets [Fabian (2012); King & Pounds (2015)].

AGNs are extremely complex objects, and their radiative output is markedly anisotropic, which implies in observational characteristics that greatly depend on the observer's viewpoint. The zoo of active galaxies is interpreted within the framework of a so-called unified model [Antonucci (1993); Urry & Padovani (1995)], a grand division existing between radio-loud (circa 10% of the population) and radio-quiet objects, characterised by the presence or not of a relevant contribution from a relativistic jet and associated radio lobe.

VHE-emitting AGN are mostly radio-loud objects of the blazar class, that is, those for which the jet axis is nearly aligned to the observer's

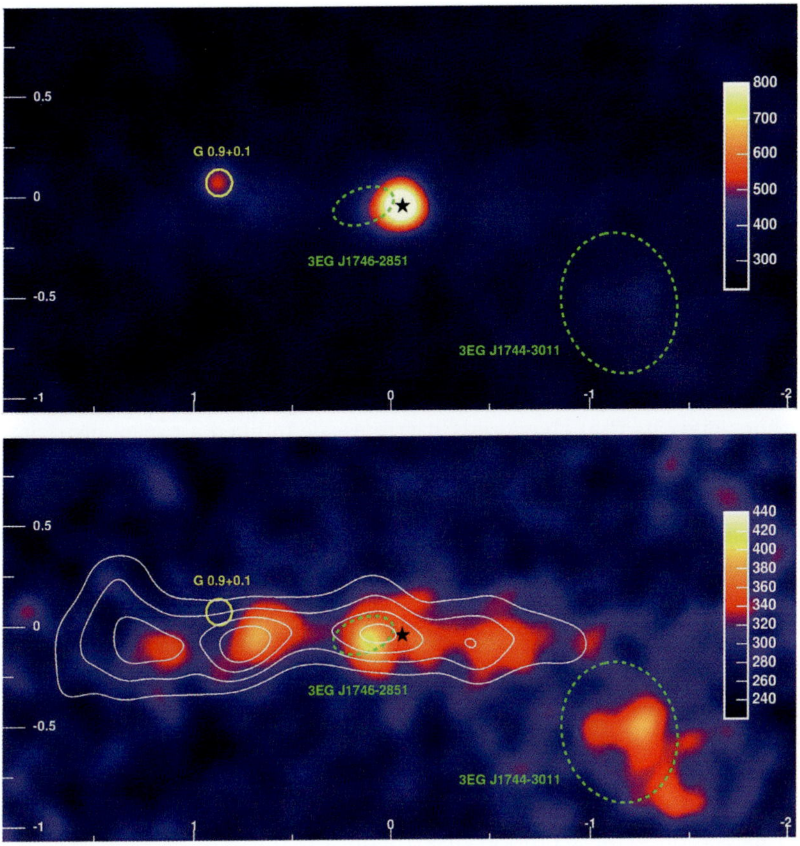

Fig. 5. VHE images of the GC region by the H.E.S.S. telescopes [Aharonian *et al.* (2006)], showing (top) gamma-ray count map, and (bottom) the same map after subtraction of the two dominant point sources, showing an extended band of gamma-ray emission. The white contours indicate CS emission, which traces molecular gas density. The position of Sgr A* is indicated by a black star.

line-of-sight (l.o.s.). Although the exact sites of gamma-ray emission and radiation mechanisms are not completely defined, it is broadly understood that emission comes from the jet or its base, and is produced by a population of relativistic electrons (with some potential contribution from hadrons) accelerated *in situ* at internal shocks. Because of the relativistic character of the outflowing plasma, Doppler boosting — $D = 1/\Gamma(1 - \beta \cos\theta)$, where Γ is the bulk Lorentz factor of the flow and θ the jet angle to the l.o.s. — will play an important role, and both the gamma-ray luminosity (power is enhanced by D^4) and source variability (enhanced by a factor D) will be greatly increased.

In fact, the most dramatic outburst of gamma-rays ever, observed by H.E.S.S. from the blazar PKS 2155-304 in 2006 [Aharonian *et al.* (2007b)], presented a 15-times increase in luminosity, within variability episodes of roughly one minute. By causality arguments (the size of the variable emission region must be inferior to $R \approx ct_{var}\delta$), Doppler factors approaching 100 were enforced for the flow, close to what is inferred for GRBs, and in stark contrast to what is observed in radio VLBI, where the fastest individual jet components do not exceed $\Gamma \sim 50$ [Lister *et al.* (2009)] — a situation referred to as the "Doppler Crisis" of TeV blazars (see e.g., [Lyutikov & Lister (2010)]). An alternative to the problem which was advanced by some authors suggests that the variability timescale implied a causal dissociation from direct changes in the central engine,[8] being rather related to events downstream in the jet [Begelman *et al.* (2008)]. Both hypothesis demanded profound revisions of jet models, and the VHE gamma-ray observations of blazars prompted a recent revival of the field, probing deep into the dynamics of relativistic jets and internal shock particle acceleration.

The onset of accretion and AGN activity around massive black holes can be studied via the observation of so-called tidal disruption events (TDE). Firstly theorised by Martin Rees in 1988 [Rees (1988)] and investigated in X-rays for over a decade (e.g., [Halpern *et al.* (2004)]), the phenomenon first caught attention of the gamma-ray community by the detection of a swift transient GRB candidate event in 2011, later classified as the onset of accretion of a tidally disrupted star, followed by the launch of a newly formed relativistic jet, due to its persistent emission and characteristic light curve profile [Bloom (2011); Aleksic (2013)]. Although rare, and despite the fact that no other tidal disruption has been so clearly detected and studied in gamma rays since then, TDEs remain a prime avenue for searching and studying intermediate-mass black holes. A small tidal-disruption event, from the gas cloud G2, with the potential for triggering accretion episodes from Sgr A* has been followed up in 2012–13 within our own Galactic Center, but finally revealed no signal [Gillessen *et al.* (2012)]. Nevertheless, monitoring of the GC should continue, given

[8]The Schwarzschild radius of a $10^9 M_\odot$ SMBH is ~ 10 AU or circa 100 light-minutes.

the potentially unique insights that such a nearby event could give on AGN activity from SMBHs.

6.1. *Blazars and the gamma-ray emission from jets*

To observe blazars is to look down the barrel of a relativistic jet, no other external emission usually being relevant, at least in gamma-rays. Blazar models assume the emission to arise from a compact blob of high-energy particles [Tavecchio *et al.* (1998)] moving downstream and varying in luminosity as new particles are injected in the jet or re-accelerated in situ via internal shocks [Rieger & Mannheim (2002)]. As with other non-thermal sources, the SED follows the same general characteristics described in Sec. 3.

Phenomenological studies have revealed the existence of a continuous spectral trend within the blazar population, the so-called blazar sequence [Fossati *et al.* (1998)], which has served as a systematic basis with which to approach the study of these objects, although its physical relevance is not yet fully understood [Ghisellini & Tavecchio (2008)]. Compton dominance, the ratio of the inverse-Compton to the synchrotron peak luminosities is an intrinsic feature of this classification, and indicates that in some VHE-emitting blazars the bulk of the radiative luminosity is emitted in the gamma-rays, making these observations fundamental to the understanding of jet energetics [Finke (2013)].

Blazar emission is predominantly modelled as leptonic in origin, with a still undetermined contribution from hadrons [Boettcher *et al.* (2013)]. This is due to uncertainties in the jet composition and degeneracies in the SED modelling, although extreme, short timescale variability and multi-band correlations currently seem to disfavour hadronic models. Conversely, hadronic emission could generate episodes of so-called "orphan flares", where variability is observed at high-energies without a low-energy counterpart, a feature unaccounted for in traditional leptonic models (e.g., [Boettcher (2005); Tavani *et al.* (2015)]).

The general characteristics of blazars were derived applying single-zone models, meaning that an individual blob is assumed responsible for the gross emission and variability observed [Tavecchio *et al.* (1998); Katarzynski (2001)]. More recently, novel complexity revealed in the variability and its correlations in large multi-band campaigns suggest that multi-zone emission has to be considered (see e.g. [Marscher (2014)]), and might influence our view of different source states. The author of this review has proposed that a careful analysis of the polarisation variability in optical and radio might help distinguishing when single or multi-zone models are required, and provides additional physical basis for the modelling [Barres de Almeida *et al.* (2014)]. In that work, analysis of the optical polarisation data revealed a hidden correlation between the X-ray and optical variability in PKS 2155-304, which was argued could have implications to the interpretation of multi-band variability correlations and in particular "orphan flares".

A fundamental question in the physics of AGN jets refers to the region of energy dissipation and radiation emission, that is, the sites of particle acceleration and gamma-ray production — the so-called "blazar zone" [Pacciani *et al.* (2014)]. Due to very extreme variability timescales being observed, simply associating the inferred size of the emitting region from causality arguments with the jet cross-section — which for typical opening angles $\theta \sim 1/\Gamma$ would translate into distances from the jet base of $d \approx \Gamma R$, R being the size of the emitting region — is unsatisfactory, and in fact, location and size of the emitting region are to be regarded as un-associated, in principle. More promising is to look for multi-wavelength correlations and spectral features, for example, which can tie the emission to some region of the AGN where an external photon field required by the modelling is present [Pacciani *et al.* (2014)]. Very promising is also the programme proposed by [Svetlana *et al.* (2013)] to correlate multi-wavelength SED features with optical and radio polarimetric data to associate the event of gamma-ray outbursts with specific VLBI features resolved in the jets.

This combination of simultaneous multi-wavelength observations with optical polarisation and radio VLBI images seems indeed to be the most promising approach for a phenomenological determination of the sites of energy dissipation and high-energy emission in jets. In 2008, a first successful result was obtained by Marscher *et al.* for the prototypical BL Lac [Marscher *et al.* (2008)]. The authors concluded that the gamma-ray outburst of BL Lac originated from the inner jet, within $\sim pc$ from the central engine, when the flow reaches the standing shock at the "radio core". Subsequent works applied to other objects have also associated the gamma-ray emission from blazars with VLBI features, where flow collimation and consequent particle acceleration takes place in the pc-scale jet (e.g., [Marscher *et al.* (2010); Abdo *et al.* (2010b)]).

The physics of extragalactic jets is more detailedly reviewed here: [Bykov *et al.* (2012); Rieger *et al.* (2013)]. We see fit nevertheless to add a brief note about M 87, the best natural laboratory of jet physics and the most studied radio galaxy and extragalactic gamma-ray source in the sky, as it was the target of multiple multi-wavelength campaigns over the past decade [Abramowski *et al.* (2010)].

Despite much detailed study and the unique linear resolution images of its jet, the origin and location of gamma-ray emission in M87 is still elusive. Radio/TeV contemporaneous outburst events have been observed in two different epochs to correlate respectively with (a) activity at the unresolved jet base [Abramowski *et al.* (2010)] within 0.1 pc in linear projected distance from the central black hole (or $<100\ R_{\mathrm{Sch}}$ as revealed by the Event Horizon Telescope [Akiyama *et al.* (2015)]), and (b) the peculiar knot HST-1, a variable, brightened portion of plasma some 120 pc from the nucleus [Acciari *et al.* (2009b)], whose precise origin or nature is still unknown.

Although the detailed mechanism remains unclear, the correlation with VLBI radio structures and activity suggest that gamma-ray emission is directly associated to the creation of blobs of plasma which are subsequently accelerated to high Lorentz factors in the inner jet and propagate downstream. Future monitoring and multi-wavelength observations of flaring activity with improved sensitivity and angular resolution should help elucidating the circumstances and sites in which intense energy dissipation in jets may happen, leading to gamma-ray outbursts, and thus constraining better the energetics and origins of these kpc-scale relativistic structures.

6.2. *Gamma-rays from blazars as cosmological probes*

The census of AGN detected by current VHE instruments numbers little more than 50 sources, mostly blazars, spread in redshift up to $z = 0.944$. The most distant source detected, the blazar S3 0218+357, was observed by Fermi/LAT and MAGIC [Ahnen *et al.* (2016)] as the first gamma-ray lensed source ever seen, in a result which opens new perspectives for extragalactic astrophysics at VHE.

The downside of a relatively short horizon of VHE gamma-ray observations, which currently limits our capacity to probe the TeV Universe beyond $z \sim 1$ due to extragalactic background light (EBL) absorption is greatly compensated by the power that the effect conveys as a cosmological probe. As discussed in Sec. 3.1, the differential absorption of gamma-ray photons by the IR-optical background allows ground-based instruments to provide direct limits for the density of the EBL in this spectral range.

In this range, the EBL is the result of the sum of starlight (direct and reprocessed by dust) emitted by galaxies throughout cosmic history, containing therefore important information regarding the evolution of baryonic matter in the Universe and structure formation. EBL density estimations using $\gamma - \gamma$ pair-production absorption provide upper limits which are complementary to evidence based on direct galaxy counts, meaning that the combination of both concurs for a precise detection of the EBL.

Since the technique is based on extrapolation of the unabsorbed blazar spectrum at source, which is unknown, some uncertainty is present which can only be circumvented by improvements on blazar models. Alternatively, the internal source spectrum can be constrained by combined population studies, or extending simultaneous spectral measurements to lower energies, below few tens of GeV, where propagation absorption is negligible. The current status of measurements has been comprehensively summarised by a recent statistical analysis of the full spectral sample available [Biteau & Williams (2015)], between 0.26 and 105 μm, from mid-UV to far-IR, showing good agreement to direct galaxy counts lower-limits. Particularly relevant is their analysis for wavelengths greater than 1.2 μm, where a significant contribution from Pop III stars during re-ionisation would be expected.

6.3. *The source of extragalactic cosmic rays*

Although the new, large-scale observatories such as Pierre Auger and Telescope Array have provided us with crucial data on UHECR particles, the origin of cosmic rays with energies beyond 10^{18} eV is unknown, and remain a fundamental question of cosmic ray physics. For energies beyond the EeV, the extragalactic origin can be firmly argued from energetic considerations which reflect on size scale of confinement during production and acceleration. However, no attempts to find a correlation of even the highest energy particles, whose magnetic deflection during propagation should be minimal, has been conclusive. For energies above 57 EeV, only marginal correlation is observed between the arrival direction of particles and nearby AGN which serve as proxies of matter distribution within the GZK horizon of ~75 Mpc [Abreu *et al.* (2010)].

AGNs, as well as galaxy clusters, the latter by virtue of their sheer sizes, are in fact prime locations in which to look for astrophysical signatures of UHECR acceleration with gamma-rays, but for the moment no clue is at hand. Due to the very limited gamma-ray horizon above TeVs, direct search is drastically limited to only the nearest sources. CTA, with its improved sensitivity at VHE and greater angular resolution, should be able to add something to the topic, specially by scrutinising important, nearby extragalactic sources with resolved morphology in gamma-rays, such as Fermi/LAT was able to do for the radiogalaxy Cen A [Abdo *et al.* (2010c,d)]. Although GRBs could be likely sites of UHECR production [Wick *et al.* (2004)], direct observational evidence is much challenging due to the short lifetime of the sources compared to the particle propagation time and flux suppression from their cosmological distances.

Neutrinos provide us with a direct signature of hadronic processes taking place at astrophysical sources, and could be an ancillary probe into the origin of extragalactic CRs. Two recent works by Padovani *et al.* [Padovani *et al.* (2015, 2016)] have taken on IceCube's discovery of high-energetic neutrinos to investigate the possible role of blazars as neutrino sources, considering a hadronic mechanism to model the observed SED of the sources. The authors concluded that blazars could in principle explain the PeV neutrino background inferred from IceCube. Although it is still early days, combined multi-messenger approaches such as this one might prove efficacious in looking for the source of extragalactic cosmic rays.

7. In Perspective: CTA and the Future of Astroparticle Physics in South America

The next few decades will see unprecedented progress in astroparticle physics. In space, NASA's Fermi satellite should continue working towards 2020, leaving behind the legacy of an unveiled gamma-ray sky. On the ground, the current facilities H.E.S.S., MAGIC and VERITAS, which revolutionised the field, will open space for the new Cherenkov Telescope Array (CTA), the definitive ground-based atmospheric Cherenkov observatory, and the first to operate with open time to the entire astronomical community [Acharya *et al.* (2013)].

CTA, which is currently in its prototyping phase is expected to install the first instruments on site, for commissioning and testing, by 2017. The two sites which will host the worldwide facility have been chosen in 2015, and will be spread over the two Hemispheres for full sky coverage: the principal site, with access to the Galaxy and the Galactic Center, will be located in the European Southern Observatory (ESO) at Cerro Paranal, in Chile, at an altitude of 2.8 km, whereas the northern site, more dedicated to extragalactic science, will be located in the Canary Island of La Palma, at

Fig. 6. Artist's impression of the final constructed Cherenkov Telescope Array. Image credits: G. Perez, SMM, IAC.

the *Observatorio Roque de Los Muchachos*, of IAC, the *Instituto Astrofísico de Canárias*, 2.2 km above sea level.

CTA will have unprecedented sensitivity over an energy range spanning 5 orders of magnitude, from 10s of GeV to 100s of TeV, a large portion of which is uncharted territory. In order to achieve such broad goals, layout optimisation studies suggest a hybrid solution, with the combination of three different telescope types (see Fig. 6), totalling circa 100 instruments distributed over the two sites. The total effective area of the observatory will be of 10 km^2 in the South and circa 3 km^2 in the Northern Hemisphere. A large field of view of 7° will be adopted in the middle energy range around 1 TeV for surveys. As a result, CTA will have at least an order of magnitude increase in sensitivity relative to current instruments over all the VHE spectrum, accessing truly cosmological distances at its low-energy threshold and probing also the PeVatron range, up to about 300 TeV. CTA should be fully operational within the first years of the following decade.

Expected to be the leading astroparticle physics observatory of its time, CTA will not be alone, and its activities will be complemented by a number of facilities. Among them, already-existing observatories, such as the water-Cherenkov instrument HAWC, recently inaugurated in Mexico in 2015, will operate full time in wide-sky coverage to sample the variable VHE sky between 100 GeV and 100 TeV. From the point of view of multi-messenger astronomy, astroparticle physicists will also count with the continued operation of the cosmic ray observatories Pierre Auger, in Argentina, and Telescope Array, in the US, with operation guaranteed beyond 2020, as well as the IceCube Antarctic Neutrino facility.

Last, but not least, there exist proposals to install a wide-field, high-altitude particle sampler for low-energy threshold gamma-ray observations in South America, aiming to expand the activities of facilities such as HAWC to the Southern Hemisphere, functioning as a more direct complement to CTA. The LATTES project [Assis *et al.* (2016)], a collaboration between researchers from Brazil, Italy and Portugal is the most advanced among such initiatives, currently under early-stage hardware development. It is being considered within an international context to take advantage of the high altitude desert plateaus found in the Andes to develop a network of Astroparticle Physics facilities in the Continent.

Acknowledgements

I would like to thank my friends and colleagues from the MAGIC Collaboration and the CTA Consortium, in whose company I have been carrying out research on TeV Astrophysics for the past 5 years. Specially, I would like to thank Profs. Alessandro de Angelis and Masahiro Teshima, and Dr. David Paneque for their comments, which helped improve this manuscript. I acknowledge the support of the Brazilian government through the CNPq Fellowship "Bolsa de Produtividade", ref. 309606/2013-6, and the support of the government of Rio de Janeiro through a FAPERJ "Young Scientist Fellowship", ref. 203.312/2016. My research on gamma-ray astronomy is partially supported by a FAPERJ Research Grant APQ1 111.581/2013, a FAPERJ Thematic Grant, ref. 110.148/2013, and a CNPq Universal Grant, ref. 475010/2013-2.

References

Aartsen, M. G. *et al.* (IceCube Collab.) (2013a). *Science*, **342**, 1242856.
Aartsen, M. G. *et al.* (IceCube Collab.) (2013). *Phys. Rev. D*, **88**, 112008.
Abbott, P. M. *et al.* (LIGO Collab.) (2016). *Phys. Rev. Lett.*, **116**, 061102.
Abdo, A. A. *et al.* (MILAGRO Collab.) (2008). *Astrophys. J.*, **688**, 1078–1083.
Abdo, A. A. *et al.* (Fermi-LAT Collab.) (20089). *Science*, **326**, 1512.
Abdo, A. A. *et al.* (Fermi-LAT Collab.) (2010). *Astrophys. J.*, **720**, 272–283.
Abdo, A. A. *et al.* (Fermi-LAT Collab.) (2010b). *Nature*, **463**, 919–923.
Abdo, A. A. *et al.* (Fermi-LAT Collab.) (2010c). *Science*, **328**, 725.
Abdo, A. A. *et al.* (Fermi-LAT Collab.) (2010d). *Astrophys. J.*, **719**, 1433.
Abramowski, A. *et al.* (2010). *Astrophys. J.*, **746**, 151, 18.
Abramowski, A. *et al.* (H.E.S.S. Collab.) (2014). *Phys. Rev. D*, **90**, 122007.
Abramowski, A. *et al.* (H.E.S.S. Collab.) (2015). *Science*, **347**, 406.

Abramowski, A. *et al.* (H.E.S.S. Collab.) (2016). *Nature*, **531**, 476–479.
Abreu, A. *et al.* (Pierre Auger Collab.) (2010). *Astrop. Phys.*, **34**, 314–326.
Acciari, V. A. *et al.* (VERITAS Collab.) (2009). *Nature*, **462**, 770–772.
Acciari, V. A. *et al.* (2009b). *Nature*, **462**, 770–772.
Acero, F. *et al.* (H.E.S.S. Collab.) (2009). *Science*, **326**, 1080.
Acero, F. *et al.* (CTA Consortium) (2013). *Astrop. Phys.*, **43**, 276–286.
Acharya, B. S. *et al.* (CTA Consortium) (2013). *Astrop. Phys.*, **43**, 3–18.
Ackermann, M. *et al.* (Fermi-LAT Collab.) (2012). *Astrophys. J.*, **755**, 164, 23.
Ackermann, M. *et al.* (Fermi-LAT Collab.) (2013). *Science*, **339**, 807–811
Ackermann, M. *et al.* (Fermi-LAT Collab.) (2014). *Astrophys. J.*, **739**, 64, 34.
Acciari, V. A. *et al.* (VERITAS Collab.) (2011). *Astrophys. J.*, **714**, 163.
Acciari, V. A. *et al.* (VERITAS Collab.) (2011). *Astrophys. J.*, **730**, L20.
Aharonian, F. A. (2002). *MNRAS* **332**, 215–230.
Aharonian, F. *et al.* (H.E.S.S. Collab.) (2005a). *Science*, **307**, 1938–1942.
Aharonian, F. *et al.* (H.E.S.S. Collab.) (2005b). *Astron. & Astrophys.*, **442**, 1–10.
Aharonian, F. *et al.* (H.E.S.S. Collab.) (2006). *Nature*, **439**, 695–698.
Aharonian, F. *et al.* (H.E.S.S. Collab.) (2007a).*Astrophys. J.*, **661**, 236–249.
Aharonian, F. *et al.* (H.E.S.S. Collab.) (2007b). *Astrophys. J. Lett.*, **664**, L71–L74.
Aharonian, F. *et al.* (H.E.S.S. Collab.) (2007c). *Nature*, **439**, 695.
Aharonian, F. *et al.* (H.E.S.S. Collab.) (2008). *Astron. & Astrophys.*, **481**, 401–410.
Ahnen, M. L. *et al.* (MAGIC Collab.) (2016). *Astron. & Astrophys.*, **595**, A98.
Akiyama, K. *et al.* (2015). *Astrophys. J.*, **807**, 150, 11.
Albert, J. *et al.* (MAGIC Collab.) (2007). *Astrophys. J.*, **665**, L51–L54.
Aliu, E. *et al.* (MAGIC Collab.) (2008). *Science*, **322**, 1221.
Aleksic, J. *et al.* (MAGIC Collab.) (2013). *Astron. & Astrophys.*, **552**, A112, 6.
Aleksic, J. *et al.* (MAGIC Collab.) (2015). *J. High Energy Phys.*, **5**, 30–38.
Antonucci, R. (1993). *Ann. Rev. Astron. & Astrophys.*, **31**, 473–521.
Assis, P. *et al.* (2016). *APh*, submitted. arXiv:1607.03051.
Baade, W. and Zwicky, F. (1934). *Proc. Natl. Acad. Sci.* **5**, 254–259.
Baade, W. and Zwicky, F. (1934). *Proc. Natl. Acad. Sci.* **5**, 259–263.
Bartoli, B. *et al.* (ARGO-YBJ Collab.) (2014). *Astrophys. J.*, **790**, 152–157.
Barres de Almeida, U. *et al.* (2014). *MNRAS*, **441**, 2885–2890.
Begelman, M. C. *et al.* (2008). *MNRAS*, **384**, L19-L23.
Bell, A. R. (1987). *MNRAS*, **225**, 615.
Bersanelli, M. and Gargantini, M. (2009). *From Galileo to Gell-Mann*, Bowden, J. Trns. (Templeton Press).
Biteau, J. and Williams, D. A. (2015). *Astrophys. J.*, **812**, 60, 26.
Bloom, J. S. (2011). *Science*, **333**, 203.
Boettcher, M. (2005). *Astrophys. J.*, **621**, 176–180.
Boettcher, M. *et al.*(2013). *Astrophys. J.*, **768**, 54, 14.
Bordas, P. *et al.* (2015). *Astrophys. J. Lett.*, **807**, L8, 5.
Buehler, R. and Blandford, R. (2014). *Rep. Prog. Phys.*, **77**, 066901.
Bykov. *et al.* (2012). *Space Sci. Rev.*, **173**, 309–339.
Carrigan, S. *et al.* (2013). *Proc. 48th Rencontres de Moriond (La Thuile).*

Chernyakova, M. *et al.* (2011). *Astrophys. J.*, **726**, 60, 8.
Clavel, M. *et al.* (2013). *Astron. & Astrophys.*, **558**, 32.
Cocconi, G. (1959). *Proceedings of the 7th ICRC (Moscow)* **2**, 309.
Coppi, P. S. (1993). *Towards a Major Atmospheric Cherenkov Detector III*, T. Kifune (eds.) (Universal Academy Press: Tokio), p. 139.
de Angelis, A., Galanti, G. and Roncadelli, M. (2013). *MNRAS*, **432**, 3245–3249.
Doeleman, J. S. (2012). *Science*, **338**, 355–358.
Dole, H. (2013). *Review of the Cosmic Infrared Background*, www.ias.u-psud.fr/dole/astrophysics/Dole_EBL_20130919_CASPAR_DESY_EBLpdf.
Dubus, G. (2013). *Astron. & Astrophys.*, **21**, 64.
Dubus, G. (2015). *Comp. Rend. Phys.*, **16**, 661–673.
Fabian, A. C. (2012). *Ann. Rev. Astron. & Astrophys.*, **50**, 455–489.
Fender, R. (2002). *Lecture Notes in Physics*, (Springer Verlag), **589**, 101.
Fermi, E. (1949). *Phys. Rev.*, **75**, 1169.
Ferrand, G. and Marcowith, A. (2010). *Astron. & Astrophys.*, **510**, A101.
Finke, J. D. (2013). *Astrophys. J.*, **763**, 134, 11.
Fossati, G. *et al.* (1998). *MNRAS*, **299**, 433–448.
Gabici, S. and Aharonian, F. A. (2007). *Astrophys. J.*, **665**, L131-L134.
Gabici, S. and Aharonian, F. A. (2015). *RICAP 2014.*, arXiv:1502.00644.
Gamow, G. and Schoenberg, M. (1941). *Phys. Rev.*, **59**, 539.
Gillessen, S. *et al.* (2012). *Nature*, **481**, 51–54.
Ghisellini, G. and Tavecchio, F. (2008). *MNRAS*, **387**, 1669–1680.
Halpern, J. P., Gezari, S. and Komossa, S. (2004). *Astrophys. J.*, **604**, 572–578.
Hillas, A. M. (1985). *Proc. 19th ICRC.* (La Jolla), **3**, 445.
Hillas, A. M. (2013). *Astrop. Phys.*, **43**, 19–43.
Hinton, J. A. and Hofmann, W. (2009). *Ann. Rev. A&A*, **47**, 523–565.
Katarzynski, K., Sol, H. and Kus, A. (2001). *Astron. & Astrophys.*, **367**, 809-825.
King, A. and Pounds, K. (2015). *Ann. Rev. Astron. & Astrophys.*, **53**, 115–154.
Klebesadel, R. W. *et al.* (1973). *Astrophys. J. Lett.*, **182**, L85.
Lattes, C. *et al.* (1947). *Nature*, **160**, 7, 453–456, 486–492.
Lemoine, M. and Pelletier, G. (2015). *Comp. Rend. Phys.*, **16**, 6, 587–704.
Leung, G. C. K. *et al.* (2014). *Astrophys. J. Lett.*, **797**, L13.
Lister, M. L. *et al.* (2009). *Astrophys. J. Lett.*, **138**, 1874–1892.
Lorimer, D. R. (2008).*Liv. Rev. Relat.*, **11**, 8.
Lyutikov, M. and Lister, M. (2010). *Astrophys. J.*, **722**, 197–203.
Malyshev, D. *et al.* (2013). *MNRAS*, **434**, 2380–2389.
Marscher, A. P. *et al.* (2008). *Nature*, **452**, 966–969.
Marscher, A. P. *et al.* (2010). *Astrophys. J.*, **710**, L126–L131.
Marscher, A. P. (2014). *Astrophys. J.*, **780**, 87, 10.
Matthiae, G. (2015). *Proc. of 2CL Meeting* (Rio de Janeiro), in press.
Massi, M. *et al.* (2012). *Astron. & Astrophys.*, **540**, A142.
Mirabel, I. F. and Rodriguez, L. F. (1994). *Nature*, **371**, 46.
Morrison, P. (1958). *Il Nuovo Cimento*, **7**, 858–865.
Oppenheimer, J. R. and Snyder, H. (1939). *Phys. Rev*, **56**, 455.

Oppenheimer, J. R. and Volkoff, G. M. (1939). *Phys. Rev*, **55**, 374–381.
Pacciani, L. *et al.* (2014). *Astrophys. J.*, **790**, 45, 17.
Padovani, P. *et al.* (2015). *MNRAS*, **452**, 1877–1887.
Padovani, P. *et al.* (2016). *MNRAS*, **457**, 3582–3592.
Paredes, J. M. *et al.*(2000). *Science*, **288**, 2340–2342.
Ressell, M. T. and Turner, M. S. (1989). http://purl.fdlp.gov/GPO/gpo48234.
Rees, M. J. (1988). *Nature*, **333**, 523–528.
Reynolds, S. P. (2008). *Ann. Rev. Astron. & Astrophys.*, **46**, 89–126.
Rieger, F. M. and Mannheim, K. (2002). *Astron. & Astrophys*, **396**, 833–846.
Rieger, F. M. *et al.* (2013). *Front. Phys.* **8**, 714–747.
Schönberg, M. and Chandrasekhar, S. (1942). *Astrophys. J.*, **96**, 161–172.
Shankar, F., Bernardi, M. and Haiman, Z. (2009). *Astrophys. Phys.*, **694**, 867–878.
Sidelnik, I. (2014). arXiv:1407.8117.
Sitarek, J. *et al.* (2015). *Proc. 34th ICRC* (The Hague), arxiv:1508.04580.
Sol, H. *et al.* (2013). *Astrop. Phys.*, **43**, 215–240.
Su, M., Slatyer, T. R. and Finkbeiner, D. P. (2010). *Astrophys. J.*, **724**, 1044–1082.
Svetlana, S. *et al.* (2013). *Astrophys. J.*, **773**, 147, 27.
Tavani, M. Vittorini, V. and Cavaliere, A. (2015). *Astrophys. J.*, **814**, 51, 13.
Tavecchio, F., Maraschi, L. and Ghisellini, G. (1998). *Astrophys. J.*, **509**, 608–619.
Urry, M. and Padovani, P. (1995). *Publ. Astron. Soc. Pac.*, **107**, 803–845.
van Eldik, C. (2015). *Astrop. Phys.*, **171**, 45–70.
Weekes, T. C. (1989).*Astrophys. J.*, **342**, 379–395
Wick, S. D. *et al.* (2004). *Astrop. Phys.*, **21**, 125–148.
Zdziarski, A. *et al.*(2012). *MNRAS*, **421**, 2956–2968.
Zwicky, F. (1939). *Proc. Nat. Acad. Sci.*, **25**, 338-344.

Chapter 11

Observation of Gravitational Waves from a Binary Black Hole Merger

B. P. Abbott *et al.*

LIGO Scientific Collaboration and Virgo Collaboration
*Full author list given at the end of this contribution**

Albert Einstein's general theory of relativity, first published a century ago, was described by physicist Max Born as "the greatest feat of human thinking about nature." We report on two major scientific breakthroughs involving key predictions of Einstein's theory: the first direct detection of gravitational waves and the first observation of the collision and merger of a pair of black holes. This cataclysmic event, producing the gravitational-wave signal GW150914, took place in a distant galaxy more than one billion light years from the Earth. It was observed on September 14, 2015 by the two detectors of the Laser Interferometer Gravitational-Wave Observatory (LIGO), arguably the most sensitive scientific instruments ever constructed. LIGO estimated that the peak gravitational-wave power radiated during the final moments of the black hole merger was more than ten times greater than the combined light power from all the stars and galaxies in the observable Universe. This remarkable discovery marks the beginning of an exciting new era of astronomy as we open an entirely new, gravitational-wave window on the Universe.[1]

Contents

*Contribution submitted by Spokesperson, LIGO Scientific Collaboration (lsc-spokesperson@ligo.org).

[1]See more at http://ligo.org/science/Publication-GW150914/index.php and at http://public.virgo-gw.eu/the-virgo-collaboration/.

1. Introduction and Background

Gravitational waves are 'ripples' in space-time produced by some of the most violent events in the cosmos, such as the collisions and mergers of massive compact stars. Their existence was predicted by Einstein in 1916, when he showed that accelerating massive objects would shake space-time so much that waves of distorted space would radiate from the source. These ripples travel at the speed of light through the Universe, carrying with them information about their cataclysmic origins, as well as invaluable clues to the nature of gravity itself.

Over the past few decades astronomers have amassed strong supporting evidence that gravitational waves exist, chiefly by studying their effect on the motions of tightly orbiting pairs of stars in our galaxy. The results of these indirect studies agree extremely well with Einstein's theory — with their orbits shrinking, exactly as predicted, due to the emission of gravitational wave energy. Nevertheless the *direct* detection of gravitational waves as they reach the Earth has been hugely anticipated by the scientific community as this breakthrough would provide new and more stringent ways to test general relativity under the most extreme conditions and open up an entirely novel way to explore the Universe.

In the same year that Einstein predicted gravitational waves, the physicist Karl Schwarzschild showed that Einstein's work permitted the existence of black holes: bizarre objects which are so dense and so compact that not even light can escape their gravitational field. Although by definition we cannot directly 'see' light from a black hole, astronomers have gathered a great deal of circumstantial evidence for their existence by studying the effects of black hole candidates on their immediate surroundings. For example, it is thought that most galaxies in the Universe, including the Milky Way, contain a supermassive black hole at their center — with masses millions or even billions of times that of the Sun. There is also evidence of many black hole candidates with much lower masses (ranging from a few to a few dozen times the Sun's mass), believed to be the remnants of dead stars that have undergone a cataclysmic explosion known as a core-collapse supernova.

Alongside this substantial progress in the indirect *observation* of black holes, there have been dramatic improvements in our *theoretical*

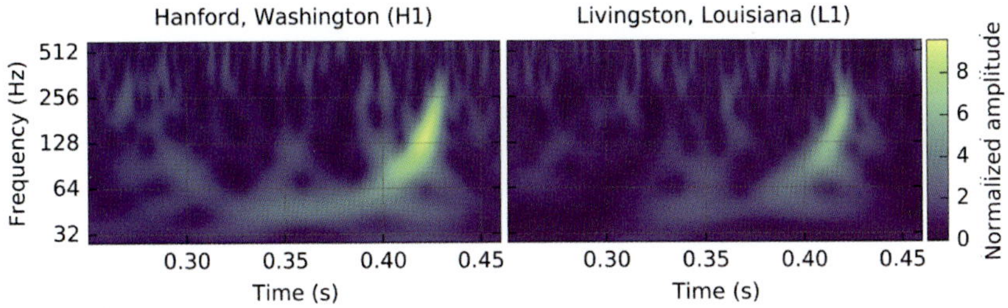

Fig. 1. (Adapted from Fig. 1 of our publication [Abbott *et al.* (2016)]). The gravitational wave event GW150914 observed by the LIGO Hanford (H1, left panel) and LIGO Livingston (L1, right panel) detectors. The two plots show how the gravitational wave strain (see below) produced by the event in each LIGO detector varied as a function of time (in seconds) and frequency (in hertz, or number of wave cycles per second). Both plots show the frequency of GW150914 sweeping sharply upwards, from 35 Hz to about 150 Hz over two tenths of a second. GW150914 arrived first at L1 and then at H1 about seven thousandths of a second later — consistent with the time taken for light, or gravitational waves, to travel between the two detectors.

understanding of these bizarre objects — including, over the past decade, some remarkable advances in modeling a pair of black holes (referred to as a binary) through several close orbits before they finally merge. These computer models have allowed us to construct precise gravitational waveforms — i.e. the pattern of gravitational waves emitted by the black holes as they approach ever closer and finally merge into a single, larger black hole — in accordance with the predictions of general relativity. The direct observation of a binary black hole merger would therefore provide a powerful cosmic laboratory for testing Einstein's theory.

2. The LIGO Detectors

LIGO is the world's largest gravitational wave observatory and one of the world's most sophisticated physics experiments. Composed of two giant laser interferometers located thousands of kilometers apart, one in Livingston, Louisiana and the other in Hanford,[2] Washington, LIGO uses the physical properties of light and of space itself to detect gravitational waves — a concept first proposed in the early 1960's and the 1970's.

[2]Note of the Editor: Fig. 1 shows the gravitational wave event GW150914 recently observed by the LIGO Hanford (H1, left panel) and LIGO Livingston (L1, right panel) detectors.

A set of initial interferometers was completed by the early 2000's, including TAMA300 in Japan, GEO600 in Germany, LIGO in the United States and Virgo in Italy. Combinations of these detectors made joint observations between 2002 and 2011, but did not detect any gravitational wave sources. After undergoing major upgrades, in 2015 the LIGO detectors began operation as Advanced LIGO: the first of a significantly more sensitive global network of advanced detectors.

An interferometer like LIGO consists of two "arms" (each one 4 km long) at right angles to each other, along which a laser beam is shone and reflected by mirrors (suspended as test masses) at each end. When a gravitational wave passes by, the stretching and squashing of space causes the arms of the interferometer alternately to lengthen and shrink, one getting longer while the other gets shorter and then vice versa. As the interferometers' arms change lengths, the laser beams take a different time to travel through the arms — which means that the two beams are no longer "in step" (or in phase) and what we call an interference pattern is produced. This is why we refer to the LIGO detectors as "interferometers".

The difference between the two arm lengths is proportional to the strength of the passing gravitational wave, referred to as the gravitational-wave strain, and this number is mind-bogglingly small. For a gravitational wave typical of what we can detect, we expect the strain to be about 1/10,000th the width of a proton! However LIGO's interferometers are so sensitive that they can measure even such tiny amounts.

Figure 2 shows a simplified diagram of an Advanced LIGO detector.

To successfully detect a gravitational wave event like GW150914 the LIGO detectors need to combine astounding sensitivity with an ability to isolate *real* signals from sources of instrument noise: tiny disturbances, due to e.g. environmental effects or the behavior of the instruments themselves, that could mimic — or indeed simply overwhelm — the signature strain pattern we are looking for. This is a key reason why there are two Advanced LIGO detectors, as it allows us to distinguish gravitational waves from local instrumental or environmental effects: only a real gravitational wave signal would appear in *both* detectors — albeit separated by a few thousandths of a second, to account for the time taken for light (or a gravitational wave) to travel between the two detector sites.

Inset (b) in Fig. 2 shows how the instrument noise in the LIGO detectors depends on frequency. We can see that the instrument noise is lowest in the 'sweet spot' around a few hundred hertz, but increases sharply at both low and high frequencies. There are also a number of narrow spikes

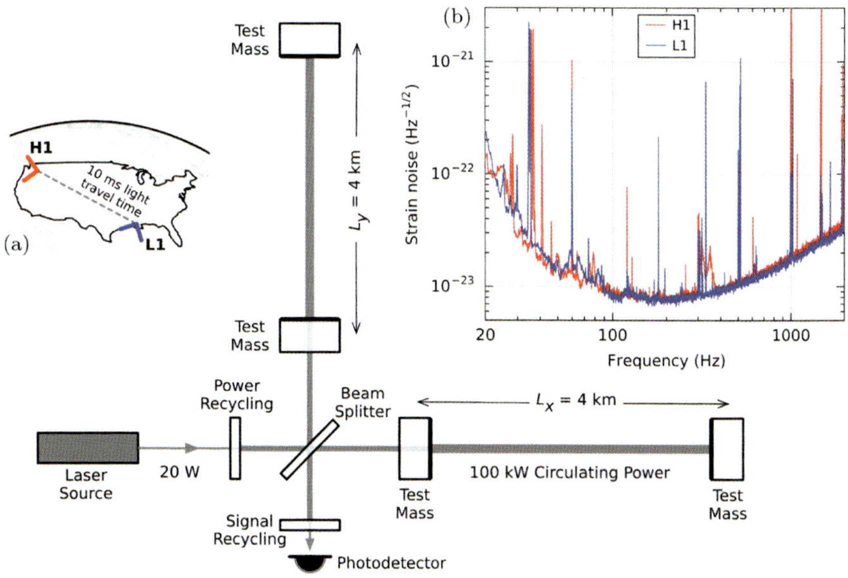

Fig. 2. Simplified diagram of an Advanced LIGO detector (not to scale), including several of the key enhancements to the basic design: an optical cavity that reflects the laser light back and forth many times in each arm, multiplying the effect of the gravitational wave on the phase of the laser light; a power recycling mirror that increases the power of the laser in the interferometer as a whole; a signal recycling mirror that further optimizes the signal extracted at the photodetector. These enhancements boost the power of the laser in the optical cavity by a factor of 5000, and increase the total amount of time that the signal spends circulating in the interferometer. Inset (a), on the left, shows the locations and orientations of the two LIGO observatories, and indicates the light travel time between them. Inset (b) shows how the instrument strain noise varied with frequency in each detector near to the time of the event. The lower the instrument noise, the higher the detectors' sensitivity. The tall spikes indicate narrow frequency ranges where the instrument noise is particularly large.

where the instrument noise is particularly large, due to e.g. vibration of the fibers that suspend the mirrors and test masses in each interferometer.

Reaching the much greater sensitivity of Advanced LIGO required the upgrading of almost every aspect of the initial LIGO design. These upgrades included:

- Significantly increasing the laser power, to reduce the main source of high frequency noise;
- Redesigning the recyclying cavities to better contain the spatial distribution of the laser light;

- Using larger, heavier fused silica test masses, to reduce the random motions of the mirrors;
- Suspending the test masses using fused silica fibers, to reduce their thermal noise;
- Suspending the test masses with a four-stage pendulum, improving their seismic isolation;
- Using an active "measure and cancel" strategy for reducing the impact of ground motions.

Operating a network of two or more detectors also lets us 'triangulate' the direction in the sky from which a gravitational wave arrives, by studying the difference in arrival time at each detector. The more detectors in one's network, the better the sky position of a gravitational wave source can be localised. In 2016 the Advanced Virgo detector,[3] in Italy, will join the global network, and other advanced interferometers are planned for the future. For more details see e.g. our publication [Abbott *et al.* (2016)].

3. Our LIGO Observations and What They Mean

On September 14, 2015 at 09:50:45 Greenwich Mean Time the LIGO Hanford and Livingston Observatories both detected a signal from GW150914 [Abbott *et al.* (2016)]. The signal was identified first by what we call *low-latency* search methods that are designed to analyse the detector data very promptly, looking for evidence of a gravitational-wave-like pattern but without modeling the precise details of the waveform. These prompt searches reported the candidate event within only three minutes of the signals arriving at the detectors. The gravitational-wave strain data acquired by the LIGO interferometers was then compared with an extensive bank of theoretically predicted waveforms — a process known as matched filtering — with the goal of finding the waveform that best matched the data.

Figure 3 presents key results of these detailed analyses — all of which firmly point to GW150914 being produced by the coalescence of two black holes [Abbott *et al.* (2016)]. The middle part of the figure shows our reconstruction of the gravitational-wave strain, as seen by the Hanford detector. Note, in particular, the impressive agreement between this pattern (shown in grey) and (shown in red) a waveform for two coalescing black holes consistent with our data, computed using general relativity.

[3]Note from the Editor: You can read more about the remarkable technology that underpins Advanced LIGO at http://tinyurl.com/ALIGO-upgrades-pdf.

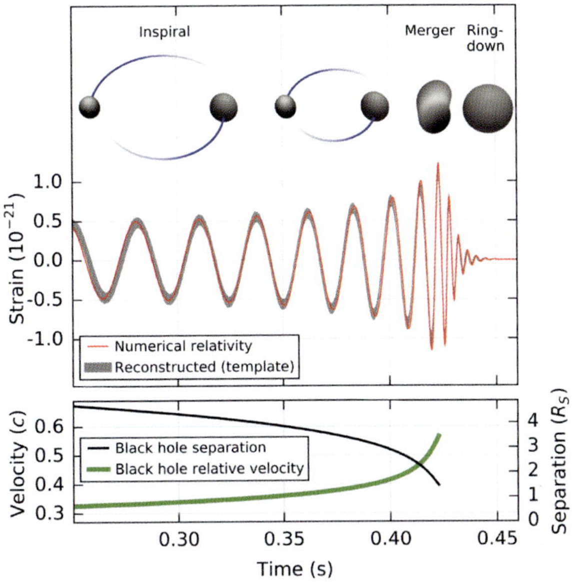

Fig. 3. Some key results of our analysis of GW150914, comparing the reconstructed gravitational-wave strain (as seen by H1 at Hanford) with the predictions of the best-matching waveform computed from general relativity, over the three stages of the event: inspiral, merger and ringdown. Also shown are the separation and velocity of the black holes, and how they change as the merger event unfolds [Abbott *et al.* (2016)].

Images of the black hole horizons at various stages of this computation are shown at the top of Fig. 3: the *inspiral*, as the two black holes approach each other; the *merger* as the black holes join together and the subsequent *ringdown*, as the single black hole that has newly formed briefly oscillates before settling down.

Comparing the strain data with theoretical predictions allows us to test whether general relativity is able to fully describe the event. It passes this test with flying colors: all of our observations are consistent with the predictions of general relativity.

We can also use the data to estimate the specific physical characteristics of the system that produced GW150914, including the masses of its two black holes before the merger, the mass of the single post-merger black hole, and the distance of the event.

Our results indicate that GW150914 was produced by the merger of two black holes with masses of about 36 times and 29 times the mass of the Sun respectively, and that the post-merger black hole had a mass of about 62 times the Sun's mass. Moreover, we infer that the final black hole is *spinning* — such rotating black holes were first predicted theoretically in 1963 by mathematician Roy Kerr. Finally, our results indicate that the GW150914 occurred at a distance of more than one billion light years. So the LIGO detectors have observed a truly remarkable event that happened a long time ago in a galaxy far, far away!

If we compare the masses of the pre- and post-merger black holes, we see that the coalescence converted about three times the mass of the Sun (or nearly six million trillion trillion kilograms) into gravitational-wave energy, most of it emitted in a fraction of a second. By contrast the Sun converts a mere *two billionths of one trillionth* of its mass into electromagnetic radiation every second. In fact, the gravitational-wave power radiated by GW150914 was more than ten times greater than the combined *luminosity* (i.e. the light power) of every star and galaxy in the observable Universe.

4. How Do We Know GW150914 was a Black Hole Merger?

Our estimated pre-merger masses of the two components in GW150914 make a very strong argument that they are *both* black holes — particularly when we also consider the enormous velocity and tiny separation of the two components, as shown in the lower part of Fig. 3 [Abbott *et al.* (2016)]. In this figure indicative velocities of the two components are seen to be significant fractions of the speed of light. Similarly their approximate separation is shown to be just a few times the characteristic size of a black hole, known as its Schwarzschild radius.

These graphs imply that the two components were only a few hundred kilometers apart just before they merged, i.e. when the gravitational-wave frequency was about 150 Hz. Black holes are the only known objects compact enough to get this close together *without* merging. Based on our estimated total mass for the two components, a pair of neutron stars would not be massive enough, and a black hole–neutron star pair would have already merged at a *lower* frequency than 150 Hz.

5. Are We Sure that GW150914 was a Real Astrophysical Event?

The short answer is "yes", but of course this is a crucial question and the LIGO Scientific Collaboration and Virgo Collaboration have together made a huge effort to address it, carrying out a variety of independent and thorough checks — all of which contribute to strengthening the detection case for GW150914.

Firstly, as we noted already, the time delay between the observations made at each LIGO detector was consistent with the light travel time between the two sites. Also, as seen in Fig. 1, the Hanford and Livingston signals showed a similar pattern, as would be expected given the near alignment of the two interferometers, and were strong enough to 'stand out' against the background noise around the time of the event — like a burst of laughter heard above the background chatter of a crowded room.

Understanding this background noise is an essential part of our analysis and involves monitoring a vast array of environmental data recorded at both sites: ground motions, temperature variations and power grid fluctuations to name just a few. In parallel, many data channels monitor in real time the status of the interferometers — checking, for example, that the various laser beams are properly centred. If any of these environmental or instrumental channels indicated a problem, then the detector data would be discarded. However, despite exhaustive studies, no such data quality problems were found at the time of the event.

But perhaps GW150914 was a rare noise fluctuation, which happened to occur simply by chance with similar characteristics at both sites? To reject this possibility we need to work out just how rare such a fluctuation would be: the less often it could occur by chance, the more confidently we can rule out this scenario in favour of the alternative — that GW150914 was indeed a real gravitational wave event [Abbott *et al.* (2016)].

To carry out this statistical analysis we used 16 days' worth of stable, high quality detector strain data from the month following the event. GW150914 was indeed by far the strongest signal observed in either detector during that period. We then introduced a series of artificial time shifts between the H1 and L1 data, effectively creating a much longer data set in which we could search for apparent signals that were as strong (or stronger) than GW150914. By using only time shifts greater than 10 milliseconds (the light travel time between the detectors) we ensured that these artificial data sets contained no real signals, but only coincidences in noise. We can then see, in the very long artificial data set, how often a coincidence mimicking GW150914 would appear. This analysis gives us the false alarm rate: how often we could expect to measure such a seemingly loud event that was really just a noise fluctuation (i.e. a 'false alarm').

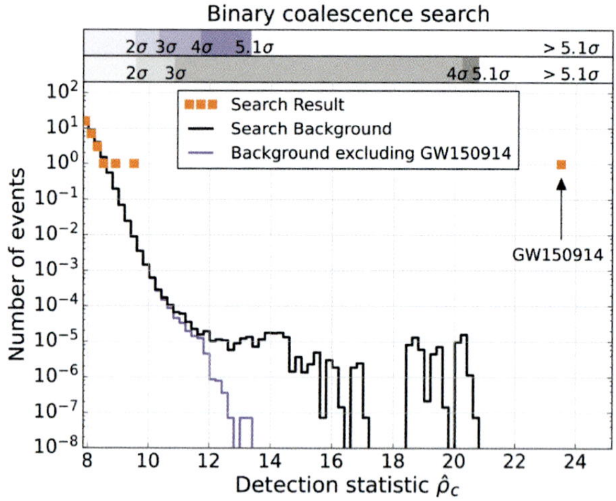

Fig. 4. (Adapted from Fig. 4 of our publication [Abbott *et al.* (2016)]). Results from our binary coalescence search quantifying how rare GW150914 was compared with false 'events' resulting from noise fluctuations. This search concluded that a noise event mimicking GW150914 would be extremely rare — less than one occurrence in about 200,000 years of data like this — a value that corresponds to a detection significance of more than 5 'sigma'.

Figure 4 (adapted from Fig. 4 of our publication [Abbott *et al.* (2016)]) shows the result of this statistical analysis, for one of the searches carried out on our detector data. The solid black and purple curves represent the 'background': the number of coincidental noise 'events' that we estimate would be produced (under slightly different assumptions) for different strengths of signal. The orange boxes represent what we actually saw, without the artificial time shifts. The key message of this figure is how far away the observed event GW150914 is from the background noise. This means that a noise event mimicking GW150914 would be *exceedingly rare* — indeed we expect an event as strong as GW150914 to appear by chance only once in about 200,000 years of such data! This false alarm rate can be translated into a number of 'sigma' (denoted by σ), which is commonly used in statistical analysis to measure the significance of a detection claim. This search identifies GW150914 as a real event, with a significance of more than 5 sigma.

6. Conclusions and Outlook

The first direct detection of gravitational waves and the first observation of a binary black hole merger are remarkable achievements [Abbott *et al.*

(2016)], but they represent only the first page of an exciting new chapter in astronomy.

The next decade will see further improvements to the Advanced LIGO detectors and extension of the global detector network to include Advanced Virgo in Italy, KAGRA in Japan, and a possible third LIGO detector in India.

This enhanced global network will significantly improve our ability to locate the positions of gravitational-wave sources in the sky and estimate more accurately their physical properties. The nascent field of gravitational-wave astronomy has a very bright future!

Editor's note (taken from https://www.ligo.caltech.edu/system/media_files/binaries/301/original/detection-science-summary.pdf):
FURTHER INFORMATION. LIGO Scientific Collaboration homepage (includes link to our main publication, published in Physical Review Letters): http://www.ligo.org. Advanced Virgo homepage: http://public.virgo-gw.eu/language/en/. Some of the companion papers to our main publication:

- Observing gravitational-wave transient GW150914 with minimal assumptions: https://dcc.ligo.org/P1500229/;
- GW150914: First results from the search for binary black hole coalescence with Advanced LIGO: https://dcc.ligo.org/P1500269/;
- Astrophysical implications of the binary black hole merger GW150914: https://dcc.ligo.org/P1500262/;
- Localization and broadband follow-up of the gravitational-wave candidate G184098: https://dcc.ligo.org/P1500227/;
- GW150914: a black-hole binary coalescence as predicted by general relativity: https://dcc.ligo.org/P1500213/;
- The rate of binary black hole mergers inferred from Advanced LIGO observations surrounding GW150914: https://dcc.ligo.org/P1500217/;
- Properties of the binary black hole merger GW150914: https://dcc.ligo.org/P1500218/. LIGO Open Science Center (with access to GW150914 data): https://losc.ligo.org/about/.

Full author list

B.P. Abbott[1], R. Abbott[1], T.D. Abbott[2], M. R. Abernathy[1], F. Acernese[3,4], K. Ackley[5], C. Adams[6], T. Adams[7], P. Addesso[3],

R.X. Adhikari[1], V.B. Adya[8], C. Affeldt[8], M. Agathos[9], K. Agatsuma[9], N. Aggarwal[10], O.D. Aguiar[11], L. Aiello[12,13], A. Ain[14], P. Ajith[15], B. Allen[8,16,17], A. Allocca[18,19], P.A. Altin[20], S.B. Anderson[1], W.G. Anderson[16], K. Arai[1], M.A. Arain[5], M.C. Araya[1], C.C. Arceneaux[21], J.S. Areeda[22], N. Arnaud[23], K.G. Arun[24], S. Ascenzi[25,13], G. Ashton[26], M. Ast[27], S.M. Aston[6], P. Astone[28], P. Aufmuth[8], C. Aulbert[8], S. Babak[29], P. Bacon[30], M.K.M. Bader[9], P.T. Baker[31], F. Baldaccini[32,33], G. Ballardin[34], S.W. Ballmer[35], J.C. Barayoga[1], S. E. Barclay[36], B.C. Barish[1], D. Barker[37], F. Barone[3,4], B. Barr[36], L. Barsotti[10], M. Barsuglia[30], D. Barta[38], J. Bartlett[37], M.A. Barton[37], I. Bartos[39], R. Bassiri[40], A. Basti[18,19], J.C. Batch[37], C. Baune[8], V. Bavigadda[34], M. Bazzan[41,42], B. Behnke[29], M. Bejger[43], C. Belczynski[44], A.S. Bell[36], C.J. Bell[36], B.K. Berger[1], J. Bergman[37], G. Bergmann[8], C.P.L. Berry[45], D. Bersanetti[46,47], A. Bertolini[9], J. Betzwieser[6], S. Bhagwat[35], R. Bhandare[48], I.A. Bilenko[49], G. Billingsley[1], J. Birch[6], R. Birney[50], O. Birnholtz[8], S. Biscans[10], A. Bisht[8,17], M. Bitossi[34], C. Biwer[35], M.A. Bizouard[23], J.K. Blackburn[1], C.D. Blair[51], D.G. Blair[51], R.M. Blair[37], S. Bloemen[52], O. Bock[8], T.P. Bodiya[10], M. Boer[53], G. Bogaert[53], C. Bogan[8], A. Bohe[29], P. Bojtos[54], C. Bond[45], F. Bondu[55], R. Bonnand[7], B.A. Boom[9], R. Bork[1], V. Boschi[18,19], S. Bose[56,14], Y. Bouffanais[30], A. Bozzi[34], C. Bradaschia[19], P.R. Brady[16], V.B. Braginsky[49], M. Branchesi[57,58], J.E. Brau[59], T. Briant[60], A. Brillet[53], M. Brinkmann[8], V. Brisson[23], P. Brockill[16], A.F. Brooks[1], D.A. Brown[35], D.D. Brown[45], N.M. Brown[10], C.C. Buchanan[2], A. Buikema[10], T. Bulik[44], H.J. Bulten[61,9], A. Buonanno[29,62], D. Buskulic[7], C. Buy[30], R.L. Byer[40], M. Cabero[8], L. Cadonati[63], G. Cagnoli[64,65], C. Cahillane[1], J. Calderón Bustillo[66,63], T. Callister[1], E. Calloni[67,4], J.B. Camp[68], K.C. Cannon[69], J. Cao[70], C.D. Capano[8], E. Capocasa[30], F. Carbognani[34], S. Caride[71], J. Casanueva Diaz[23], C. Casentini[25,13], S. Caudill[16], M. Cavaglià[21], F. Cavalier[23], R. Cavalieri[34], G. Cella[19], C.B. Cepeda[1], L. Cerboni Baiardi[57,58], G. Cerretani[18,19], E. Cesarini[25,13], R. Chakraborty[1], T. Chalermsongsak[1], S.J. Chamberlin[72], M. Chan[36], S. Chao[73], P. Charlton[74], E. Chassande-Mottin[30], H.Y. Chen[75], Y. Chen[76], C. Cheng[73], A. Chincarini[47], A. Chiummo[34], H.S. Cho[77], M. Cho[62], J.H. Chow[20], N. Christensen[78], Q. Chu[51], S. Chua[60], S. Chung[51], G. Ciani[5], F. Clara[37], J.A. Clark[63], F. Cleva[53], E.

Coccia[25,12,13], P.-F. Cohadon[60], A. Colla[79,28], C.G. Collette[80], L. Cominsky[81], M. Constancio Jr.[11], A. Conte[79,28], L. Conti[42], D. Cook[37], T.R. Corbitt[2], N. Cornish[31], A. Corsi[71], S. Cortese[34], C.A. Costa[11], M.W. Coughlin[78], S.B. Coughlin[82], J.-P. Coulon[53], S.T. Countryman[39], P. Couvares[1], E.E. Cowan[63], D.M. Coward[51], M.J. Cowart[6], D.C. Coyne[1], R. Coyne[71], K. Craig[36], J.D.E. Creighton[16], T.D. Creighton[83], J. Cripe[2], S.G. Crowder[84], A.M. Cruise[45], A. Cumming[36], L. Cunningham[36], E. Cuoco[34], T. Dal Canton[8], S.L. Danilishin[36], S. D'Antonio[13], K. Danzmann[17,8], N.S. Darman[85], C.F. Da Silva Costa[5], V. Dattilo[34], I. Dave[48], H.P. Daveloza[83], M. Davier[23], G.S. Davies[36], E.J. Daw[86], R. Day[34], S. De[35], D. DeBra[40], G. Debreczeni[38], J. Degallaix[65], M. De Laurentis[67,4], S. Deléglise[60], W. Del Pozzo[45], T. Denker[8,17], T. Dent[8], H. Dereli[53], V. Dergachev[1], R.T. DeRosa[6], R. De Rosa[67,4], R. DeSalvo[87], S. Dhurandhar[14], M.C. Díaz[83], L. Di Fiore[4], M. Di Giovanni[79,28], A. Di Lieto[18,19], S. Di Pace[79,28], I. Di Palma[29,8], A. Di Virgilio[19], G. Dojcinoski[88], V. Dolique[65], F. Donovan[10], K.L. Dooley[21], S. Doravari[6,8], R. Douglas[36], T.P. Downes[16], M. Drago[8,89,90], R.W.P. Drever[1], J.C. Driggers[37], Z. Du[70], M. Ducrot[7], S.E. Dwyer[37], T.B. Edo[86], M.C. Edwards[78], A. Effler[6], H.-B. Eggenstein[8], P. Ehrens[1], J. Eichholz[5], S.S. Eikenberry[5], W. Engels[76], R.C. Essick[10], T. Etzel[1], M. Evans[10], T.M. Evans[6], R. Everett[72], M. Factourovich[39], V. Fafone[25,13,12], H. Fair[35], S. Fairhurs[91], X. Fan[70], Q. Fang[51], S. Farinon[47], B. Farr[75], W.M. Farr[45], M. Favata[88], M. Fays[91], H. Fehrmann[8], M.M. Fejer[40], D. Feldbaum[5], I. Ferrante[18,19], E.C. Ferreira[11], F. Ferrini[34], F. Fidecaro[18,19], L.S. Finn[72], I. Fiori[34], D. Fiorucci[30], R.P. Fisher[35], R. Flaminio[65,92], M. Fletcher[36], H. Fong[69], J.-D. Fournier[53], S. Franco[23], S. Frasca[79,28], F. Frasconi[19], M. Frede[8], Z. Frei[54], A. Freise[45], R. Frey[59], V. Frey[23], T.T. Fricke[8], P. Fritschel[10], V.V. Frolov[6], P. Fulda[5], M. Fyffe[6], H.A.G. Gabbard[21], J.R. Gair[93], L. Gammaitoni[32,33], S.G. Gaonkar[14], F. Garufi[67,4], A. Gatto[30], G. Gaur[94,95], N. Gehrels[68], G. Gemme[47], B. Gendre[53], E. Genin[34], A. Gennai[19], J. George[48], L. Gergely[96], V. Germain[7], Abhirup Ghosh[15], Archisman Ghosh[15], S. Ghosh[52,9], J.A. Giaime[2,6], K.D. Giardina[6], A. Giazotto[19], K. Gill[97], A. Glaefke[36], J.R. Gleason[5], E. Goetz[98], R. Goetz[5], L. Gondan[54], G. González[2], J.M. Gonzalez Castro[18,19], A. Gopakumar[99], N.A. Gordon[36], M.L. Gorodetsky[49], S.E. Gossan[1], M.

Gosselin[34], R. Gouaty[7], C. Graef[36], P.B. Graff[62], M. Granata[65], A. Grant[36], S. Gras[10], C. Gray[37], G. Greco[57,58], A.C. Green[45], R.J.S. Greenhalgh[100], P. Groot[52], H. Grote[8], S. Grunewald[29], G.M. Guidi[57,58], X. Guo[70], A. Gupta[14], M.K. Gupta[95], K.E. Gushwa[1], E.K. Gustafson[1], R. Gustafson[98], J.J. Hacker[22], B.R. Hall[56], E.D. Hall[1], G. Hammond[36], M. Haney[99], M.M. Hanke[8], J. Hanks[37], C. Hanna[72], M.D. Hannam[91], J. Hanson[6], T. Hardwick[2], J. Harms[57,58], G.M. Harry[101], I.W. Harry[29], M.J. Hart[36], M.T. Hartman[5], C.-J. Haster[45], K. Haughian[36], J. Healy[102], J. Heefner[1,a], A. Heidmann[60], M.C. Heintze[5,6], G. Heinzel[8], H. Heitmann[53], P. Hello[23], G. Hemming[34], M. Hendry[36], I.S. Heng[36], J. Hennig[36], A.W. Heptonstall[1], M. Heurs[8,17], S. Hild[36], D. Hoak[103], K.A. Hodge[1], D. Hofman[65], S.E. Hollitt[104], K. Holt[6], D.E. Holz[75], P. Hopkins[91], D.J. Hosken[104], J. Hough[36], E.A. Houston[36], E.J. Howell[51], Y.M. Hu[36], S. Huang[73], E.A. Huerta[105,82], D. Huet[23], B. Hughey[97], S. Husa[66], S.H. Huttner[36], T. Huynh-Dinh[6], A. Idrisy[72], N. Indik[8], D.R. Ingram[37], R. Inta[71], H.N. Isa[36], J.-M. Isac[60], M. Isi[1], G. Islas[22], T. Isogai[10], B.R. Iyer[15], K. Izumi[37], M.B. Jacobson[1], T. Jacqmin[60], H. Jang[77], K. Jani[63], P. Jaranowski[106], S. Jawahar[107], F. Jiménez-Forteza[66], W.W. Johnson[2], N.K. Johnson-McDaniel[15], D.I. Jones[26], R. Jones[36], R.J.G. Jonker[9], L. Ju[51], K. Haris[108], C.V. Kalaghatgi[24,91], V. Kalogera[82], S. Kandhasamy[21], G. Kang[77], J.B. Kanner[1], S. Karki[59], M. Kasprzack[2,23,34], E. Katsavounidis[10], W. Katzman[6], S. Kaufer[17], T. Kaur[51], K. Kawabe[37], F. Kawazoe[8,17], F. Kéfélian[53], M.S. Kehl[69], D. Keitel[8,66], D.B. Kelley[35], W. Kells[1], R. Kennedy[86], D.G. Keppel[8], J.S. Key[83], A. Khalaidovski[8], F.Y. Khalili[49], I. Khan[12], S. Khan[91], Z. Khan[95], E.A. Khazanov[109], N. Kijbunchoo[37], C. Kim[77], J. Kim[110], K. Kim[111], Nam-Gyu Kim[77], Namjun Kim[40], Y.-M. Kim[110], E.J. King[104], P.J. King[37], D.L. Kinzel[6], J.S. Kissel[37], L. Kleybolte[27], S. Klimenko[5], S.M. Koehlenbeck[8], K. Kokeyama[2], S. Koley[9], V. Kondrashov[1], A. Kontos[10], S. Koranda[16], M. Korobko[27], W.Z. Korth[1], I. Kowalska[44], D.B. Kozak[1], V. Kringel[8], B. Krishnan[8], A. Królak[112,113], C. Krueger[17], G. Kuehn[8], P. Kumar[69], R. Kumar[36], L. Kuo[73], A. Kutynia[112], P. Kwee[8], B.D. Lackey[35], M. Landry[37], J. Lange[102], B. Lantz[40], P.D. Lasky[114], A. Lazzarini[1], C. Lazzaro[63,42], P. Leaci[29,79,28], S. Leavey[36], E.O. Lebigot[30,70], C.H. Lee[110], H.K. Lee[111], H.M. Lee[115], K. Lee[36], A. Lenon[35], M.

Leonardi[89,90], J.R. Leong[8], N. Leroy[23], N. Letendre[7], Y. Levin[114], B.M. Levine[37], T.G.F. Li[1], A. Libson[10], T.B. Littenberg[116], N.A. Lockerbie[107], J. Logue[36], A.L. Lombardi[103], L.T. London[91], J.E. Lord[35], M. Lorenzini[12,13], V. Loriette[117], M. Lormand[6], G. Losurdo[58], J.D. Lough[8,17], C.O. Lousto[102], G. Lovelace[22], H. Lück[17,8], A.P. Lundgren[8], J. Luo[78], R. Lynch[10], Y. Ma[51], T. MacDonald[40], B. Machenschalk[8], M. MacInnis[10], D.M. Macleod[2], F. Magaña-Sandoval[35], R.M. Magee[56], M. Mageswaran[1], E. Majorana[28], I. Maksimovic[117], V. Malvezzi[25,13], N. Man[53], I. Mandel[45], V. Mandic[84], V. Mangano[36], G.L. Mansell[20], M. Manske[16], M. Mantovani[34], F. Marchesoni[118,33], F. Marion[7], S. Márka[39], Z. Márka[39], A.S. Markosyan[40], E. Maros[1], F. Martelli[57,58], L. Martellini[53], I.W. Martin[36], R.M. Martin[5], D.V. Martynov[1], J.N. Marx[1], K. Mason[10], A. Masserot[7], T.J. Massinger[35], M. Masso-Reid[36], F. Matichard[10], L. Matone[39], N. Mavalvala[10], N. Mazumder[56], G. Mazzolo[8], R. McCarthy[37], D.E. McClelland[20], S. McCormick[6], S.C. McGuire[119], G. McIntyre[1], J. McIver[1], D.J. McManus[20], S.T. McWilliams[105], D. Meacher[72], G.D. Meadors[29,8], J. Meidam[9], A. Melatos[85], G. Mendell[37], D. Mendoza-Gandara[8], R.A. Mercer[16], E. Merilh[37], M. Merzougui[53], S. Meshkov[1], C. Messenger[36], C. Messick[72], P.M. Meyers[84], F. Mezzani[28,79], H. Miao[45], C. Michel[65], H. Middleton[45], E.E. Mikhailov[120], L. Milano[67,4], J. Miller[10], M. Millhouse[31], Y. Minenkov[13], J. Ming[29,8], S. Mirshekari[121], C. Mishra[15], S. Mitra[14], V.P. Mitrofanov[49], G. Mitselmakher[5], R. Mittleman[10], A. Moggi[19], M. Mohan[34], S.R.P. Mohapatra[10], M. Montani[57,58], B.C. Moore[88], C.J. Moore[122], D. Moraru[37], G. Moreno[37], S.R. Morriss[83], K. Mossavi[8], B. Mours[7], C.M. Mow-Lowry[45], C.L. Mueller[5], G. Mueller[5], A.W. Muir[91], Arunava Mukherjee[15], D. Mukherjee[16], S. Mukherjee[83], N. Mukund[14], A. Mullavey[6], J. Munch[104], D.J. Murphy[39], P.G. Murray[36], A. Mytidis[5], I. Nardecchia[25,13], L. Naticchioni[79,28], R.K. Nayak[123], V. Necula[5], K. Nedkova[103], G. Nelemans[52,9], M. Neri[46,47], A. Neunzert[98], G. Newton[36], T.T. Nguyen[20], A.B. Nielsen[8], S. Nissanke[52,9], A. Nitz[8], F. Nocera[34], D. Nolting[6], M.E.N. Normandin[83], L.K. Nuttall[35], J. Oberling[37], E. Ochsner[16], J. O'Dell[100], E. Oelker[10], G.H. Ogin[124], J.J. Oh[125], S.H. Oh[125], F. Ohme[91], M. Oliver[66], P. Oppermann[8], Richard J. Oram[6], B. O'Reilly[6], R. O'Shaughnessy[102], C.D. Ott[76], D.J. Ottaway[104], R.S. Ottens[5], H. Overmier[6], B.J. Owen[71], A. Pai[108],

S.A. Pai[48], J.R. Palamos[59], O. Palashov[109], C. Palomba[28], A. Pal-Singh[27], H. Pan[73], Y. Pan[62], C. Pankow[82], F. Pannarale[91], B.C. Pant[48], F. Paoletti[34,19], A. Paoli[34], M.A. Papa[29,16,8], H.R. Paris[40], W. Parker[6], D. Pascucci[36], A. Pasqualetti[34], R. Passaquieti[18,19], D. Passuello[19], B. Patricelli[18,19], Z. Patrick[40], B.L. Pearlstone[36], M. Pedraza[1], R. Pedurand[65], L. Pekowsky[35], A. Pele[6], S. Penn[126], A. Perreca[1], H.P. Pfeiffer[69,29], M. Phelps[36], O. Piccinni[79,28], M. Pichot[53], M. Pickenpack[8], F. Piergiovanni[57,58], V. Pierro[87], G. Pillant[34], L. Pinard[65], I.M. Pinto[87], M. Pitkin[36], J.H. Poeld[8], R. Poggiani[18,19], P. Popolizio[34], A. Post[8], J. Powell[36], J. Prasad[14], V. Predoi[91], S.S. Premachandra[114], T. Prestegard[84], L.R. Price[1], M. Prijatelj[34], M. Principe[87], S. Privitera[29], R. Prix[8], G.A. Prodi[89,90], L. Prokhorov[49], O. Puncken[8], M. Punturo[33], P. Puppo[28], M. Pürrer[29], H. Qi[16], J. Qin[51], V. Quetschke[83], E.A. Quintero[1], R. Quitzow-James[59], F.J. Raab[37], D.S. Rabeling[20], H. Radkins[37], P. Raffai[54], S. Raja[48], M. Rakhmanov[83], C. R. Ramet[6], P. Rapagnani[79,28], V. Raymond[29], M. Razzano[18,19], V. Re[25], J. Read[22], C.M. Reed[37], T. Regimbau[53], L. Rei[47], S. Reid[50], D.H. Reitze[1,5], H. Rew[120], S.D. Reyes[35], F. Ricci[79,28], K. Riles[98], N.A. Robertson[1,36], R. Robie[36], F. Robinet[23], A. Rocchi[13], L. Rolland[7], J.G. Rollins[1], V.J. Roma[59], J.D. Romano[83], R. Romano[3,4], G. Romanov[120], J.H. Romie[6], D. Rosińska[127,43], S. Rowan[36], A. Rüdiger[8], P. Ruggi[34], K. Ryan[37], S. Sachdev[1], T. Sadecki[37], L. Sadeghian[16], L. Salconi[34], M. Saleem[108], F. Salemi[8], A. Samajdar[123], L. Sammut[85,114], L.M. Sampson[82], E.J. Sanchez[1], V. Sandberg[37], B. Sandeen[82], G.H. Sanders[1], J.R. Sanders[98,35], B. Sassolas[65], B.S. Sathyaprakash[91], P.R. Saulson[35], O. Sauter[98], R.L. Savage[37], A. Sawadsky[17], P. Schale[59], R. Schilling[8,b], J. Schmidt[8], P. Schmidt[1,76], R. Schnabel[27], R.M.S. Schofield[59], A. Schönbeck[27], E. Schreiber[8], D. Schuette[8,17], B.F. Schutz[91,29], J. Scott[36], S.M. Scott[20], D. Sellers[6], A.S. Sengupta[94], D. Sentenac[34], V. Sequino[25,13], A. Sergeev[109], G. Serna[22], Y. Setyawati[52,9], A. Sevigny[37], D.A. Shaddock[20], T. Shaffer[37], S. Shah[52,9], M.S. Shahriar[82], M. Shaltev[8], Z. Shao[1], B. Shapiro[40], P. Shawhan[62], A. Sheperd[16], D.H. Shoemaker[10], D.M. Shoemaker[63], K. Siellez[53,63], X. Siemens[16], D. Sigg[37], A.D. Silva[11], D. Simakov[8], A. Singer[1], L.P. Singer[68], A. Singh[29,8], R. Singh[2], A. Singhal[12], A.M. Sintes[66], B.J.J. Slagmolen[20], J.R. Smith[22], M.R. Smith[1], N.D. Smith[1],

R.J.E. Smith[1], E.J. Son[125], B. Sorazu[36], F. Sorrentino[47], T. Souradeep[14], A.K. Srivastava[95], A. Staley[39], M. Steinke[8], J. Steinlechner[36], S. Steinlechner[36], D. Steinmeyer[8,17], B.C. Stephens[16], S.P. Stevenson[45], R. Stone[83], K.A. Strain[36], N. Straniero[65], G. Stratta[57,58], N.A. Strauss[78], S. Strigin[49], R. Sturani[121], A.L. Stuver[6], T.Z. Summerscales[128], L. Sun[85], P.J. Sutton[91], B.L. Swinkels[34], M.J. Szczepańczyk[97], M. Tacca[30], D. Talukder[59], D.B. Tanner[5], M. Tápai[96], S.P. Tarabrin[8], A. Taracchini[29], R. Taylor[1], T. Theeg[8], M.P. Thirugnanasambandam[1], E.G. Thomas[45], M. Thomas[6], P. Thomas[37], K.A. Thorne[6], K.S. Thorne[76], E. Thrane[114], S. Tiwari[12], V. Tiwari[91], K.V. Tokmakov[107], C. Tomlinson[86], M. Tonelli[18,19], C.V. Torres[83,c], C.I. Torrie[1], D. Töyrä[45], F. Travasso[32,33], G. Traylor[6], D. Trifirò[21], M.C. Tringali[89,90], L. Trozzo[129,19], M. Tse[10], M. Turconi[53], D. Tuyenbayev[83], D. Ugolini[130], C.S. Unnikrishnan[99], A.L. Urban[16], S.A. Usman[35], H. Vahlbruch[17], G. Vajente[1], G. Valdes[83], M. Vallisneri[76], N. van Bakel[9], M. van Beuzekom[9], J.F.J. van den Brand[61,9], C. Van Den Broeck[9], D.C. Vander-Hyde[35,22], L. van der Schaaf[9], J.V. van Heijningen[9], A.A. van Veggel[36], M. Vardaro[41,42], S. Vass[1], M. Vasúth[38], R. Vaulin[10], A. Vecchio[45], G. Vedovato[42], J. Veitch[45], P.J. Veitch[104], K. Venkateswara[131], D. Verkindt[7], F. Vetrano[57,58], A. Viceré[57,58], S. Vinciguerra[45], D.J. Vine[50], J.-Y. Vinet[53], S. Vitale[10], T. Vo[35], H. Vocca[32,33], C. Vorvick[37], D. Voss[5], W.D. Vousden[45], S.P. Vyatchanin[49], A.R. Wade[20], L.E. Wade[132], M. Wade[132], S.J. Waldman[10], M. Walker[2], L. Wallace[1], S. Walsh[16,8,29], G. Wang[12], H. Wang[45], M. Wang[45], X. Wang[70], Y. Wang[51], H. Ward[36], R.L. Ward[20], J. Warner[37], M. Was[7], B. Weaver[37], L.-W. Wei[53], M. Weinert[8], A.J. Weinstein[1], R. Weiss[10], T. Welborn[6], L. Wen[51], P. Weßels[8], T. Westphal[8], K. Wette[8], J.T. Whelan[102,8], S.E. Whitcomb[1], D.J. White[86], B.F. Whiting[5], K. Wiesner[8], C. Wilkinson[37], P.A. Willems[1], L. Williams[5], R.D. Williams[1], A.R. Williamson[91], J.L. Willis[133], B. Willke[17,8], M.H. Wimmer[8,17], L. Winkelmann[8], W. Winkler[8], C.C. Wipf[1], A.G. Wiseman[16], H. Wittel[8,17], G. Woan[36], J. Worden[37], J.L. Wright[36], G. Wu[6], J. Yablon[82], I. Yakushin[6], W. Yam[10], H. Yamamoto[1], C.C. Yancey[62], M.J. Yap[20], H. Yu[10], M. Yvert[7], A. Zadrożny[112], L. Zangrando[42], M. Zanolin[97], J.-P. Zendri[42], M. Zevin[82], F. Zhang[10], L. Zhang[1], M. Zhang[120], Y. Zhang[102], C. Zhao[51], M. Zhou[82], Z. Zhou[82], X.J. Zhu[51], M.E.

Zucker[1,10], S.E. Zuraw[103], and J. Zweizig[1] (LIGO Scientific Collaboration and Virgo Collaboration).

Institutions

[1] LIGO, California Institute of Technology, Pasadena, California 91125, USA; [2] Louisiana State University, Baton Rouge, Louisiana 70803, USA; [3] Università di Salerno, Fisciano, I-84084 Salerno, Italy; [4] INFN, Sezione di Napoli, Complesso Universitario di Monte S. Angelo, I-80126 Napoli, Italy; [5] University of Florida, Gainesville, Florida 32611, USA; [6] LIGO Livingston Observatory, Livingston, Louisiana 70754, USA; [7] Laboratoire d'Annecy-le-Vieux de Physique des Particules (LAPP), Université Savoie Mont Blanc, CNRS/IN2P3, F-74941 Annecy-le-Vieux, France; [8] Albert-Einstein-Institut, Max-Planck-Institut für Gravitationsphysik, D-30167 Hannover, Germany; [9] Nikhef, Science Park, 1098 XG Amsterdam, Netherlands; [10] LIGO, Massachusetts Institute of Technology, Cambridge, Massachusetts 02139, USA; [11] Instituto Nacional de Pesquisas Espaciais, 12227-010 São José dos Campos, São Paulo, Brazil; [12] INFN, Gran Sasso Science Institute, I-67100 L'Aquila, Italy; [13] INFN, Sezione di Roma Tor Vergata, I-00133 Roma, Italy; [14] Inter-University Centre for Astronomy and Astrophysics, Pune 411007, India; [15] International Centre for Theoretical Sciences, Tata Institute of Fundamental Research, Bangalore 560012, India; [16] University of Wisconsin-Milwaukee, Milwaukee, Wisconsin 53201, USA; [17] Leibniz Universität Hannover, D-30167 Hannover, Germany; [18] Università di Pisa, I-56127 Pisa, Italy; [19] INFN, Sezione di Pisa, I-56127 Pisa, Italy; [20] Australian National University, Canberra, Australian Capital Territory 0200, Australia; [21] The University of Mississippi, University, Mississippi 38677, USA; [22] California State University Fullerton, Fullerton, California 92831, USA; [23] LAL, Université Paris-Sud, CNRS/IN2P3, Université Paris-Saclay, Orsay, France; [24] Chennai Mathematical Institute, Chennai, India 603103; [25] Università di Roma Tor Vergata, I-00133 Roma, Italy; [26] University of Southampton, Southampton SO17 1BJ, United Kingdom; [27] Universität Hamburg, D-22761 Hamburg, Germany; [28] INFN, Sezione di Roma, I-00185 Roma, Italy; [29] Albert-Einstein-Institut, Max-Planck-Institut für Gravitationsphysik, D-14476 Potsdam-Golm, Germany; [30] APC, AstroParticule et Cosmologie, Universié Paris Diderot, CNRS/IN2P3, CEA/Irfu, Observatoire de Paris, Sorbonne Paris Cité, F-75205 Paris Cedex 13, France; [31] Montana State University, Bozeman, Montana 59717, USA; [32] Università di Perugia,

I-06123 Perugia, Italy; [33] INFN, Sezione di Perugia, I-06123 Perugia, Italy; [34] European Gravitational Observatory (EGO), I-56021 Cascina, Pisa, Italy; [35] Syracuse University, Syracuse, New York 13244, USA; [36] SUPA, University of Glasgow, Glasgow G12 8QQ, United Kingdom; [37] LIGO Hanford Observatory, Richland, Washington 99352, USA; [38] Wigner RCP, RMKI, H-1121 Budapest, Konkoly Thege Miklós út 29-33, Hungary; [39] Columbia University, New York, New York 10027, USA; [40] Stanford University, Stanford, California 94305, USA; [41] Università di Padova, Dipartimento di Fisica e Astronomia, I-35131 Padova, Italy; [42] INFN, Sezione di Padova, I-35131 Padova, Italy; [43] CAMK-PAN, 00-716 Warsaw, Poland; [44] Astronomical Observatory Warsaw University, 00-478 Warsaw, Poland; [45] University of Birmingham, Birmingham B15 2TT, United Kingdom; [46] Università degli Studi di Genova, I-16146 Genova, Italy; [47] INFN, Sezione di Genova, I-16146 Genova, Italy; [48] RRCAT, Indore MP 452013, India; [49] Faculty of Physics, Lomonosov Moscow State University, Moscow 119991, Russia; [50] SUPA, University of the West of Scotland, Paisley PA1 2BE, United Kingdom; [51] University of Western Australia, Crawley, Western Australia 6009, Australia; [52] Department of Astrophysics/IMAPP, Radboud University Nijmegen, P.O. Box 9010, 6500 GL Nijmegen, Netherlands; [53] Artemis, Université Côte d'Azur, CNRS, Observatoire Côte d'Azur, CS 34229, Nice cedex 4, France; [54] MTA Eötvös University, "Lendulet" Astrophysics Research Group, Budapest 1117, Hungary; [55] Institut de Physique de Rennes, CNRS, Université de Rennes 1, F-35042 Rennes, France; [56] Washington State University, Pullman, Washington 99164, USA; [57] Università degli Studi di Urbino "Carlo Bo" I-61029 Urbino, Italy; [58] INFN, Sezione di Firenze, I-50019 Sesto Fiorentino, Firenze, Italy; [59] University of Oregon, Eugene, Oregon 97403, USA; [60] Laboratoire Kastler Brossel, UPMC-Sorbonne Universités, CNRS, ENS-PSL Research University, Collège de France, F-75005 Paris, France; [61] VU University Amsterdam, 1081 HV Amsterdam, Netherlands; [62] University of Maryland, College Park, Maryland 20742, USA; [63] Center for Relativistic Astrophysics and School of Physics, Georgia Institute of Technology, Atlanta, Georgia 30332, USA; [64] Institut Lumière Matière, Université de Lyon, Université Claude Bernard Lyon 1, UMR CNRS 5306, 69622 Villeurbanne, France; [65] Laboratoire des Matériaux Avancés (LMA), IN2P3/CNRS, Université de Lyon, F-69622 Villeurbanne, Lyon, France;

[66] Universitat de les Illes Balears, IAC3–IEEC, E-07122 Palma de Mallorca, Spain; [67] Università di Napoli "Federico II", Complesso Universitario di Monte S. Angelo, I-80126 Napoli, Italy; [68] NASA/Goddard Space Flight Center, Greenbelt, Maryland 20771, USA; [69] Canadian Institute for Theoretical Astrophysics, University of Toronto, Toronto, Ontario M5S 3H8, Canada; [70] Tsinghua University, Beijing 100084, China; [71] Texas Tech University, Lubbock, Texas 79409, USA; [72] The Pennsylvania State University, University Park, Pennsylvania 16802, USA; [73] National Tsing Hua University, Hsinchu City, 30013 Taiwan, Republic of China; [74] Charles Sturt University, Wagga Wagga, New South Wales 2678, Australia; [75] University of Chicago, Chicago, Illinois 60637, USA; [76] Caltech CaRT, Pasadena, California 91125, USA; [77] Korea Institute of Science and Technology Information, Daejeon 305-806, Korea; [78] Carleton College, Northfield, Minnesota 55057, USA; [79] Università di Roma "La Sapienza", I-00185 Roma, Italy; [80] University of Brussels, Brussels 1050, Belgium; [81] Sonoma State University, Rohnert Park, California 94928, USA; [82] Northwestern University, Evanston, Illinois 60208, USA; [83] The University of Texas Rio Grande Valley, Brownsville, Texas 78520, USA; [84] University of Minnesota, Minneapolis, Minnesota 55455, USA; [85] The University of Melbourne, Parkville, Victoria 3010, Australia; [86] The University of Sheffield, Sheffield S10 2TN, United Kingdom; [87] University of Sannio at Benevento, I-82100 Benevento, Italy and INFN, Sezione di Napoli, I-80100 Napoli, Italy; [88] Montclair State University, Montclair, New Jersey 07043, USA; [89] Università di Trento, Dipartimento di Fisica, I-38123 Povo, Trento, Italy; [90] INFN, Trento Institute for Fundamental Physics and Applications, I-38123 Povo, Trento, Italy; [91] Cardiff University, Cardiff CF24 3AA, United Kingdom; [92] National Astronomical Observatory of Japan, 2-21-1 Osawa, Mitaka, Tokyo 181-8588, Japan; [93] School of Mathematics, University of Edinburgh, Edinburgh EH9 3FD, United Kingdom; [94] Indian Institute of Technology, Gandhinagar Ahmedabad Gujarat 382424, India; [95] Institute for Plasma Research, Bhat, Gandhinagar 382428, India; [96] University of Szeged, Dóm tér 9, Szeged 6720, Hungary; [97] Embry-Riddle Aeronautical University, Prescott, Arizona 86301, USA; [98] University of Michigan, Ann Arbor, Michigan 48109, USA; [99] Tata Institute of Fundamental Research, Mumbai 400005, India; [100] Rutherford Appleton Laboratory, HSIC, Chilton, Didcot, Oxon OX11 0QX, United Kingdom; [101]

American University, Washington, D.C. 20016, USA; [102] Rochester Institute of Technology, Rochester, New York 14623, USA; [103] University of Massachusetts-Amherst, Amherst, Massachusetts 01003, USA; [104] University of Adelaide, Adelaide, South Australia 5005, Australia; [105] West Virginia University, Morgantown, West Virginia 26506, USA; [106] University of Bia ystok, 15-424 Bia ystok, Poland; [107] SUPA, University of Strathclyde, Glasgow G1 1XQ, United Kingdom; [108] IISER-TVM, CET Campus, Trivandrum Kerala 695016, India; [109] Institute of Applied Physics, Nizhny Novgorod, 603950, Russia; [110] Pusan National University, Busan 609-735, Korea; [111] Hanyang University, Seoul 133-791, Korea; [112] NCBJ, 05-400 Swierk-Otwock, Poland; [113] IM-PAN, 00-956 Warsaw, Poland; [114] Monash University, Victoria 3800, Australia; [115] Seoul National University, Seoul 151-742, Korea; [116] University of Alabama in Huntsville, Huntsville, Alabama 35899, USA; [117] ESPCI, CNRS, F-75005 Paris, France; [118] Università di Camerino, Dipartimento di Fisica, I-62032 Camerino, Italy; [119] Southern University and A&M College, Baton Rouge, Louisiana 70813, USA; [120] College of William and Mary, Williamsburg, Virginia 23187, USA; [121] Instituto de Física Teórica, University Estadual Paulista/ICTP South American Institute for Fundamental Research, São Paulo SP 01140-070, Brazil; [122] University of Cambridge, Cambridge CB2 1TN, United Kingdom; [123] IISER-Kolkata, Mohanpur, West Bengal 741252, India; [124] Whitman College, 345 Boyer Avenue, Walla Walla, Washington 99362 USA; [125] National Institute for Mathematical Sciences, Daejeon 305-390, Korea; [126] Hobart and William Smith Colleges, Geneva, New York 14456, USA; [127] Janusz Gil Institute of Astronomy, University of Zielona Góra, 65-265 Zielona Góra, Poland; [128] Andrews University, Berrien Springs, Michigan 49104, USA; [129] Università di Siena, I-53100 Siena, Italy; [130] Trinity University, San Antonio, Texas 78212, USA; [131] University of Washington, Seattle, Washington 98195, USA; [132] Kenyon College, Gambier, Ohio 43022, USA; [133] Abilene Christian University, Abilene, Texas 79699, USA.

Reference

B. P. Abbott *et al.* (LIGO Scientific Collaboration and Virgo Collaboration). *Phys. Rev. Lett.* **116**, 061102 (2016).

Index